L.W. Nixon Library
Butler Community College
901 South Haverhill Road
El Dorado, Kansas 67042-3280

Design of Automatic Machinery

MECHANICAL ENGINEERING
A Series of Textbooks and Reference Books

Founding Editor

L. L. Faulkner

*Columbus Division, Battelle Memorial Institute
and Department of Mechanical Engineering
The Ohio State University
Columbus, Ohio*

1. *Spring Designer's Handbook*, Harold Carlson
2. *Computer-Aided Graphics and Design*, Daniel L. Ryan
3. *Lubrication Fundamentals*, J. George Wills
4. *Solar Engineering for Domestic Buildings*, William A. Himmelman
5. *Applied Engineering Mechanics: Statics and Dynamics*, G. Boothroyd and C. Poli
6. *Centrifugal Pump Clinic*, Igor J. Karassik
7. *Computer-Aided Kinetics for Machine Design*, Daniel L. Ryan
8. *Plastics Products Design Handbook, Part A: Materials and Components; Part B: Processes and Design for Processes*, edited by Edward Miller
9. *Turbomachinery: Basic Theory and Applications*, Earl Logan, Jr.
10. *Vibrations of Shells and Plates*, Werner Soedel
11. *Flat and Corrugated Diaphragm Design Handbook*, Mario Di Giovanni
12. *Practical Stress Analysis in Engineering Design*, Alexander Blake
13. *An Introduction to the Design and Behavior of Bolted Joints*, John H. Bickford
14. *Optimal Engineering Design: Principles and Applications*, James N. Siddall
15. *Spring Manufacturing Handbook*, Harold Carlson
16. *Industrial Noise Control: Fundamentals and Applications*, edited by Lewis H. Bell
17. *Gears and Their Vibration: A Basic Approach to Understanding Gear Noise*, J. Derek Smith
18. *Chains for Power Transmission and Material Handling: Design and Applications Handbook*, American Chain Association
19. *Corrosion and Corrosion Protection Handbook*, edited by Philip A. Schweitzer
20. *Gear Drive Systems: Design and Application*, Peter Lynwander
21. *Controlling In-Plant Airborne Contaminants: Systems Design and Calculations*, John D. Constance
22. *CAD/CAM Systems Planning and Implementation*, Charles S. Knox
23. *Probabilistic Engineering Design: Principles and Applications*, James N. Siddall

24. *Traction Drives: Selection and Application*, Frederick W. Heilich III and Eugene E. Shube
25. *Finite Element Methods: An Introduction*, Ronald L. Huston and Chris E. Passerello
26. *Mechanical Fastening of Plastics: An Engineering Handbook*, Brayton Lincoln, Kenneth J. Gomes, and James F. Braden
27. *Lubrication in Practice: Second Edition*, edited by W. S. Robertson
28. *Principles of Automated Drafting*, Daniel L. Ryan
29. *Practical Seal Design*, edited by Leonard J. Martini
30. *Engineering Documentation for CAD/CAM Applications*, Charles S. Knox
31. *Design Dimensioning with Computer Graphics Applications*, Jerome C. Lange
32. *Mechanism Analysis: Simplified Graphical and Analytical Techniques*, Lyndon O. Barton
33. *CAD/CAM Systems: Justification, Implementation, Productivity Measurement*, Edward J. Preston, George W. Crawford, and Mark E. Coticchia
34. *Steam Plant Calculations Manual*, V. Ganapathy
35. *Design Assurance for Engineers and Managers*, John A. Burgess
36. *Heat Transfer Fluids and Systems for Process and Energy Applications*, Jasbir Singh
37. *Potential Flows: Computer Graphic Solutions*, Robert H. Kirchhoff
38. *Computer-Aided Graphics and Design: Second Edition*, Daniel L. Ryan
39. *Electronically Controlled Proportional Valves: Selection and Application*, Michael J. Tonyan, edited by Tobi Goldoftas
40. *Pressure Gauge Handbook*, AMETEK, U.S. Gauge Division, edited by Philip W. Harland
41. *Fabric Filtration for Combustion Sources: Fundamentals and Basic Technology*, R. P. Donovan
42. *Design of Mechanical Joints*, Alexander Blake
43. *CAD/CAM Dictionary*, Edward J. Preston, George W. Crawford, and Mark E. Coticchia
44. *Machinery Adhesives for Locking, Retaining, and Sealing*, Girard S. Haviland
45. *Couplings and Joints: Design, Selection, and Application*, Jon R. Mancuso
46. *Shaft Alignment Handbook*, John Piotrowski
47. *BASIC Programs for Steam Plant Engineers: Boilers, Combustion, Fluid Flow, and Heat Transfer*, V. Ganapathy
48. *Solving Mechanical Design Problems with Computer Graphics*, Jerome C. Lange
49. *Plastics Gearing: Selection and Application*, Clifford E. Adams
50. *Clutches and Brakes: Design and Selection*, William C. Orthwein
51. *Transducers in Mechanical and Electronic Design*, Harry L. Trietley
52. *Metallurgical Applications of Shock-Wave and High-Strain-Rate Phenomena*, edited by Lawrence E. Murr, Karl P. Staudhammer, and Marc A. Meyers
53. *Magnesium Products Design*, Robert S. Busk
54. *How to Integrate CAD/CAM Systems: Management and Technology*, William D. Engelke
55. *CAM Design and Manufacture: Second Edition; with CAM design software for the IBM PC and compatibles*, disk included, Preben W. Jensen
56. *Solid-State AC Motor Controls: Selection and Application*, Sylvester Campbell

57. *Fundamentals of Robotics*, David D. Ardayfio
58. *Belt Selection and Application for Engineers*, edited by Wallace D. Erickson
59. *Developing Three-Dimensional CAD Software with the IBM PC*, C. Stan Wei
60. *Organizing Data for CIM Applications*, Charles S. Knox, with contributions by Thomas C. Boos, Ross S. Culverhouse, and Paul F. Muchnicki
61. *Computer-Aided Simulation in Railway Dynamics*, by Rao V. Dukkipati and Joseph R. Amyot
62. *Fiber-Reinforced Composites: Materials, Manufacturing, and Design*, P. K. Mallick
63. *Photoelectric Sensors and Controls: Selection and Application*, Scott M. Juds
64. *Finite Element Analysis with Personal Computers*, Edward R. Champion, Jr., and J. Michael Ensminger
65. *Ultrasonics: Fundamentals, Technology, Applications: Second Edition, Revised and Expanded*, Dale Ensminger
66. *Applied Finite Element Modeling: Practical Problem Solving for Engineers*, Jeffrey M. Steele
67. *Measurement and Instrumentation in Engineering: Principles and Basic Laboratory Experiments*, Francis S. Tse and Ivan E. Morse
68. *Centrifugal Pump Clinic: Second Edition, Revised and Expanded*, Igor J. Karassik
69. *Practical Stress Analysis in Engineering Design: Second Edition, Revised and Expanded*, Alexander Blake
70. *An Introduction to the Design and Behavior of Bolted Joints: Second Edition, Revised and Expanded*, John H. Bickford
71. *High Vacuum Technology: A Practical Guide*, Marsbed H. Hablanian
72. *Pressure Sensors: Selection and Application*, Duane Tandeske
73. *Zinc Handbook: Properties, Processing, and Use in Design*, Frank Porter
74. *Thermal Fatigue of Metals*, Andrzej Weronski and Tadeusz Hejwowski
75. *Classical and Modern Mechanisms for Engineers and Inventors*, Preben W. Jensen
76. *Handbook of Electronic Package Design*, edited by Michael Pecht
77. *Shock-Wave and High-Strain-Rate Phenomena in Materials*, edited by Marc A. Meyers, Lawrence E. Murr, and Karl P. Staudhammer
78. *Industrial Refrigeration: Principles, Design and Applications*, P. C. Koelet
79. *Applied Combustion*, Eugene L. Keating
80. *Engine Oils and Automotive Lubrication*, edited by Wilfried J. Bartz
81. *Mechanism Analysis: Simplified and Graphical Techniques, Second Edition, Revised and Expanded*, Lyndon O. Barton
82. *Fundamental Fluid Mechanics for the Practicing Engineer*, James W. Murdock
83. *Fiber-Reinforced Composites: Materials, Manufacturing, and Design, Second Edition, Revised and Expanded*, P. K. Mallick
84. *Numerical Methods for Engineering Applications*, Edward R. Champion, Jr.
85. *Turbomachinery: Basic Theory and Applications, Second Edition, Revised and Expanded*, Earl Logan, Jr.
86. *Vibrations of Shells and Plates: Second Edition, Revised and Expanded*, Werner Soedel
87. *Steam Plant Calculations Manual: Second Edition, Revised and Expanded*, V. Ganapathy
88. *Industrial Noise Control: Fundamentals and Applications, Second Edition, Revised and Expanded*, Lewis H. Bell and Douglas H. Bell

89. *Finite Elements: Their Design and Performance*, Richard H. MacNeal
90. *Mechanical Properties of Polymers and Composites: Second Edition, Revised and Expanded*, Lawrence E. Nielsen and Robert F. Landel
91. *Mechanical Wear Prediction and Prevention*, Raymond G. Bayer
92. *Mechanical Power Transmission Components*, edited by David W. South and Jon R. Mancuso
93. *Handbook of Turbomachinery*, edited by Earl Logan, Jr.
94. *Engineering Documentation Control Practices and Procedures*, Ray E. Monahan
95. *Refractory Linings: Thermomechanical Design and Applications*, Charles A. Schacht
96. *Geometric Dimensioning and Tolerancing: Applications and Techniques for Use in Design, Manufacturing, and Inspection*, James D. Meadows
97. *An Introduction to the Design and Behavior of Bolted Joints: Third Edition, Revised and Expanded*, John H. Bickford
98. *Shaft Alignment Handbook: Second Edition, Revised and Expanded*, John Piotrowski
99. *Computer-Aided Design of Polymer-Matrix Composite Structures*, edited by Suong Van Hoa
100. *Friction Science and Technology*, Peter J. Blau
101. *Introduction to Plastics and Composites: Mechanical Properties and Engineering Applications*, Edward Miller
102. *Practical Fracture Mechanics in Design*, Alexander Blake
103. *Pump Characteristics and Applications*, Michael W. Volk
104. *Optical Principles and Technology for Engineers*, James E. Stewart
105. *Optimizing the Shape of Mechanical Elements and Structures*, A. A. Seireg and Jorge Rodriguez
106. *Kinematics and Dynamics of Machinery*, Vladimír Stejskal and Michael Valásek
107. *Shaft Seals for Dynamic Applications*, Les Horve
108. *Reliability-Based Mechanical Design*, edited by Thomas A. Cruse
109. *Mechanical Fastening, Joining, and Assembly*, James A. Speck
110. *Turbomachinery Fluid Dynamics and Heat Transfer*, edited by Chunill Hah
111. *High-Vacuum Technology: A Practical Guide, Second Edition, Revised and Expanded*, Marsbed H. Hablanian
112. *Geometric Dimensioning and Tolerancing: Workbook and Answerbook*, James D. Meadows
113. *Handbook of Materials Selection for Engineering Applications*, edited by G. T. Murray
114. *Handbook of Thermoplastic Piping System Design*, Thomas Sixsmith and Reinhard Hanselka
115. *Practical Guide to Finite Elements: A Solid Mechanics Approach*, Steven M. Lepi
116. *Applied Computational Fluid Dynamics*, edited by Vijay K. Garg
117. *Fluid Sealing Technology*, Heinz K. Muller and Bernard S. Nau
118. *Friction and Lubrication in Mechanical Design*, A. A. Seireg
119. *Influence Functions and Matrices*, Yuri A. Melnikov
120. *Mechanical Analysis of Electronic Packaging Systems*, Stephen A. McKeown
121. *Couplings and Joints: Design, Selection, and Application, Second Edition, Revised and Expanded*, Jon R. Mancuso
122. *Thermodynamics: Processes and Applications*, Earl Logan, Jr.

123. *Gear Noise and Vibration*, J. Derek Smith
124. *Practical Fluid Mechanics for Engineering Applications*, John J. Bloomer
125. *Handbook of Hydraulic Fluid Technology*, edited by George E. Totten
126. *Heat Exchanger Design Handbook*, T. Kuppan
127. *Designing for Product Sound Quality*, Richard H. Lyon
128. *Probability Applications in Mechanical Design*, Franklin E. Fisher and Joy R. Fisher
129. *Nickel Alloys*, edited by Ulrich Heubner
130. *Rotating Machinery Vibration: Problem Analysis and Troubleshooting*, Maurice L. Adams, Jr.
131. *Formulas for Dynamic Analysis*, Ronald L. Huston and C. Q. Liu
132. *Handbook of Machinery Dynamics*, Lynn L. Faulkner and Earl Logan, Jr.
133. *Rapid Prototyping Technology: Selection and Application*, Kenneth G. Cooper
134. *Reciprocating Machinery Dynamics: Design and Analysis*, Abdulla S. Rangwala
135. *Maintenance Excellence: Optimizing Equipment Life-Cycle Decisions*, edited by John D. Campbell and Andrew K. S. Jardine
136. *Practical Guide to Industrial Boiler Systems*, Ralph L. Vandagriff
137. *Lubrication Fundamentals: Second Edition, Revised and Expanded*, D. M. Pirro and A. A. Wessol
138. *Mechanical Life Cycle Handbook: Good Environmental Design and Manufacturing*, edited by Mahendra S. Hundal
139. *Micromachining of Engineering Materials*, edited by Joseph McGeough
140. *Control Strategies for Dynamic Systems: Design and Implementation*, John H. Lumkes, Jr.
141. *Practical Guide to Pressure Vessel Manufacturing*, Sunil Pullarcot
142. *Nondestructive Evaluation: Theory, Techniques, and Applications*, edited by Peter J. Shull
143. *Diesel Engine Engineering: Thermodynamics, Dynamics, Design, and Control*, Andrei Makartchouk
144. *Handbook of Machine Tool Analysis*, Ioan D. Marinescu, Constantin Ispas, and Dan Boboc
145. *Implementing Concurrent Engineering in Small Companies*, Susan Carlson Skalak
146. *Practical Guide to the Packaging of Electronics: Thermal and Mechanical Design and Analysis*, Ali Jamnia
147. *Bearing Design in Machinery: Engineering Tribology and Lubrication*, Avraham Harnoy
148. *Mechanical Reliability Improvement: Probability and Statistics for Experi-mental Testing*, R. E. Little
149. *Industrial Boilers and Heat Recovery Steam Generators: Design, Applications, and Calculations*, V. Ganapathy
150. *The CAD Guidebook: A Basic Manual for Understanding and Improving Computer-Aided Design*, Stephen J. Schoonmaker
151. *Industrial Noise Control and Acoustics*, Randall F. Barron
152. *Mechanical Properties of Engineered Materials*, Wolé Soboyejo
153. *Reliability Verification, Testing, and Analysis in Engineering Design*, Gary S. Wasserman
154. *Fundamental Mechanics of Fluids: Third Edition*, I. G. Currie
155. *Intermediate Heat Transfer*, Kau-Fui Vincent Wong

156. *HVAC Water Chillers and Cooling Towers: Fundamentals, Application, and Operation*, Herbert W. Stanford III
157. *Gear Noise and Vibration: Second Edition, Revised and Expanded*, J. Derek Smith
158. *Handbook of Turbomachinery: Second Edition*, Revised and Expanded, edited by Earl Logan, Jr., and Ramendra Roy
159. *Piping and Pipeline Engineering: Design, Construction, Maintenance, Integrity, and Repair*, George A. Antaki
160. *Turbomachinery: Design and Theory*, Rama S. R. Gorla and Aijaz Ahmed Khan
161. *Target Costing: Market-Driven Product Design*, M. Bradford Clifton, Henry M. B. Bird, Robert E. Albano, and Wesley P. Townsend
162. *Fluidized Bed Combustion*, Simeon N. Oka
163. *Theory of Dimensioning: An Introduction to Parameterizing Geometric Models*, Vijay Srinivasan
164. *Handbook of Mechanical Alloy Design*, edited by George E. Totten, Lin Xie, and Kiyoshi Funatani
165. *Structural Analysis of Polymeric Composite Materials*, Mark E. Tuttle
166. *Modeling and Simulation for Material Selection and Mechanical Design*, edited by George E. Totten, Lin Xie, and Kiyoshi Funatani
167. *Handbook of Pneumatic Conveying Engineering*, David Mills, Mark G. Jones, and Vijay K. Agarwal
168. *Clutches and Brakes: Design and Selection, Second Edition*, William C. Orthwein
169. *Fundamentals of Fluid Film Lubrication: Second Edition*, Bernard J. Hamrock, Steven R. Schmid, and Bo O. Jacobson
170. *Handbook of Lead-Free Solder Technology for Microelectronic Assemblies*, edited by Karl J. Puttlitz and Kathleen A. Stalter
171. *Vehicle Stability*, Dean Karnopp
172. *Mechanical Wear Fundamentals and Testing: Second Edition, Revised and Expanded*, Raymond G. Bayer
173. *Liquid Pipeline Hydraulics*, E. Shashi Menon
174. *Solid Fuels Combustion and Gasification*, Marcio L. de Souza-Santos
175. *Mechanical Tolerance Stackup and Analysis*, Bryan R. Fischer
176. *Engineering Design for Wear,* Raymond G. Bayer
177. *Vibrations of Shells and Plates: Third Edition, Revised and Expanded*, Werner Soedel
178. *Refractories Handbook*, edited by Charles A. Schacht
179. *Practical Engineering Failure Analysis*, Hani M. Tawancy, Anwar Ul-Hamid, and Nureddin M. Abbas
180. *Mechanical Alloying and Milling*, C. Suryanarayana
181. *Mechanical Vibration: Analysis, Uncertainties, and Control, Second Edition, Revised and Expanded*, Haym Benaroya
182. *Design of Automatic Machinery*, Stephen J. Derby
183. *Practical Fracture Mechanics in Design: Second Edition, Revised and Expanded*, Arun Shukla
184. *Practical Guide to Designed Experiments*, Paul D. Funkenbusch

Additional Volumes in Preparation

Mechanical Engineering Software

Spring Design with an IBM PC, Al Dietrich

Mechanical Design Failure Analysis: With Failure Analysis System Software for the IBM PC, David G. Ullman

Design of Automatic Machinery

Stephen J. Derby
Rensselaer Polytechnic Institute
Troy, New York, U.S.A.

MARCEL DEKKER

NEW YORK

670.427 DER 2005

Derby, Stephen J.
Design of automatic
machinery.

Although great care has been taken to provide accurate and current information, neither the author(s) nor the publisher, nor anyone else associated with this publication, shall be liable for any loss, damage, or liability directly or indirectly caused or alleged to be caused by this book. The material contained herein is not intended to provide specific advice or recommendations for any specific situation.

Trademark notice: Product or corporate names may be trademarks or registered trademarks and are used only for identification and explanation without intent to infringe.

Library of Congress Cataloging-in-Publication Data
A catalog record for this book is available from the Library of Congress.

ISBN: 0-8247-5369-0

This book is printed on acid-free paper.

Headquarters
Marcel Dekker, 270 Madison Avenue, New York, NY 10016, U.S.A.
tel: 212-696-9000; fax: 212-685-4540

Distribution and Customer Service
Marcel Dekker, Cimarron Road, Monticello, New York 12701, U.S.A.
tel: 800-228-1160; fax: 845-796-1772

World Wide Web
http://www.dekker.com

The publisher offers discounts on this book when ordered in bulk quantities. For more information, write to Special Sales/Professional Marketing at the headquarters address above.

Copyright © 2005 by Marcel Dekker. All Rights Reserved.

Neither this book nor any part may be reproduced or transmitted in any form or by any means, electronic or mechanical, including photocopying, microfilming, and recording, or by any information storage and retrieval system, without permission in writing from the publisher.

Current printing (last digit):
10 9 8 7 6 5 4 3 2 1

PRINTED IN THE UNITED STATES OF AMERICA

Preface

This text addresses the need for today's industry to develop automation. For most industries—and, in many ways, for entire countries—to compete in the global economic environment, it is essential to become cost-competitive. To do so requires that a person or team have the knowledge to either design a new automatic machine or write specifications for the purchase of a machine to perform an automated process.

The process of designing automation is not a new one, but there has been a void of current texts that are useful tools. Although there is a good supply of books dedicated to machine design, including such topics as bearing supports, dynamic balancing, and modes of vibration, these books do not instruct the reader on options for the actual design of an automated process. Many people in the automation field have learned from experienced co-workers or through the school of hard knocks. Some of today's implemented automation projects are regarded as miraculously brilliant, while others appear to be cursed. This range of unknown performance can be greatly reduced if the practice of designing automation is approached by thinking both methodically and outside of the box.

This book covers the design process for automation, starting with the all-important understanding of the process to be performed by a machine. This step is often the most misunderstood and poorly addressed, and therefore leads to uncertain results. The steps in automation design parallel those of many college capstone design classes, but with a different view. The process, not the product, is the focus.

This book focuses on the many existing automation concepts used in recent history, and shows that there are often several machine configurations that will work satisfactorily. There is no one correct answer to automation problems, but my experiences and case studies, as well as those of my associates, will hopefully allow readers to gain insight beyond their own years. I have included several major case studies of my startup automation companies that have lessons both for operational success and for disappointments when the market for sales goes sour.

At the end of most chapters the book presents problems that are very open-ended. Thus, there is no answer key to selected problems at the back of the book, nor a solutions manual. There are also project assignments for most chapters that allow the reader to systematically work through an automation project of his own, or to apply to the list present in Appendix B, Projects.

The reader is assumed to have a technical background, either as a college senior or graduate student in mechanical or electrical engineering or as an engineering technology student. Or the reader might be a practicing engineer who has worked his way up from the CAD terminal or drawing board. This book does not attempt to restate all the available theory and examples relation to dynamics, calculating rotational moments of inertia, machine elements, and bearings derived in earlier college courses. It assumes that the reader has encountered these topics in other courses or work-related experience and can use these topics when needed in the course of performing the projects in this text or in the creation of a real machine.

The reader is also required to have access to the Web, to investigate the current vendors and component options for the concepts presented. Although the concepts will not change over time, the vendors can be bought or sold in a very short time, and are therefore not listed as an integral part of the discussions.

Many people contributed to the material in this text. These include students and staff at Rensselaer Polytechnic Institute, including the Flexible Manufacturing Center, engineers at American Dixie and Cambridge Valley Machining, and good friends at Distributed Robotics and elsewhere. Some of the contributors who provided material for case studies or technical discussion are:

    Mr. Merritt Bell
    Mr. Bernhard Bringmann
    Mr. David Brown
    Mr. Peter Caratzas
    Mr. Clay Cooper
    Mr. Paul Crilly
    Dr. Ryan Durante
    Mr. Van Judd
    Mr. Ned Kirchner

Preface

Dr. John McFadden
Mr. Raymond Puffer
Mr. Jesse Ruppel
Mr. Donald Schneider, Sr.
Mr. Matt Simon
Ms. Jane Somers

A special contribution was made by my daughter, Melanie Derby, who created the majority of the computer-generated concept drawings, and did so with great joy and earnest.

Many people have influenced me—and thus this text—over the past almost half century. This started with my parents, who let me build wooden robots in the garage, and who always knew that automation would be my calling. And over the last 25 years I have been influenced by my 100 master's and Ph.D. students and countless graduate and undergraduate students, who always seemed to wonder "why" or "why not," and together struggled to search for a better solution.

I thank the late Dr. Joseph Duffy, who demonstrated during my Ph.D. work the world of robot kinematics, and to Mr. Clay Cooper, who showed me the real world of automation during a sabbatical in the late 1990s. Thanks to Mr. Ned Kirchner, the first to step forward to assist me in bringing my inventions to market. Thanks also to Dr. John McFadden, Mr. David Brown, and the other participants of Distributed Robotics, LLC, who shared both the joy of creative genius and the frustrating times.

I would like to thank Mr. John Corrigan of Marcel Dekker who inspired me to write this text. His vision on where the world needed to be in terms of automation was my motivation to start this endeavor.

Finally, thanks must be given to my wife, Sharon, for all her encouragement, daughters Laura and Melanie, and all of my family and friends who assisted me in the writing of this book.

Soli Deo Gloria
Stephen J. Derby

Contents

Preface   *iii*

Chapter 1 Introduction   1

   1.1 Why Automate?   3
   1.2 Book Topics   4
   1.3 Conceptual Design and This Text   5

Chapter 2 Steps to Automation   7

   2.1 What is Automation?   7
      2.1.1 Automation: Webster's Definition   8
      2.1.2 Black Box Approach   8
      2.1.3 The Expert Knows Best   9
      2.1.4 Call It What You Will   9
      2.1.5 Systematic Design of Components   10
      2.1.6 Generalized Automatic Machine   10
   2.2 Automation Design Process   10
      2.2.1 Look at Similar Automation Processes and Machines   12
      2.2.2 Looking at Human Performing an Operation   13
      2.2.3 Try New Things   14
      2.2.4 Apply a Mechatronics Strategy   16
   2.3 Knowing the Process   18

	2.3.1 Wound Capacitor Example	18
2.4	Process Example: Peanut Butter Chocolate Kiss Cookies	21
	2.4.1 Process Possibilities	22
	2.4.2 How to Proceed	25
2.5	Conclusions	25
	Problems	25
	Project Assignment	26
	References	27

Chapter 3 Justifying Automation — 28

3.1	Traditional Project Cost Justification for a Purchase	29
	3.1.1 Fast Food Worker	30
	3.1.2 U.S. Postal Worker	31
	3.1.3 Which Market to Approach?	32
3.2	Traditional Costing Estimating for Building and Selling Automation	32
	3.2.1 Room Refinishing Example	32
	3.2.2 Impact on a Machine Builder	35
	3.2.3 Cost Structure for an Automation Builder	36
3.3	Win–Win Purchasing Philosophy	38
3.4	Maximum Profit Cost Estimating for Building and Selling Automation	39
3.5	Justifying Flexible Automation over Hard Automation	40
3.6	Intellectual Property, Patents, and Trade Secrets	41
	3.6.1 Patents	42
	3.6.2 Trademarks	43
	3.6.3 Copyrights	44
	3.6.4 Trade Secrets	44
	3.6.5 Intellectual Property for Automation	44
3.7	Conclusions	45
	Problems	45
	Project Assignment	46
	Reference	47

Chapter 4 The Automation Design Process — 48

4.1	System Specifications	49
4.2	Brainstorming	50
4.3	Machine Classification by Function	51
	4.3.1 Assembly Machines	52

Contents

	4.3.2 Inspection Machines	52
	4.3.3 Test Machines	54
	4.3.4 Packaging Machines	55
4.4	Machine Classification by Transfer Method	55
	4.4.1 Linear Indexing	55
	4.4.2 Linear Continuous	58
	4.4.3 Linear Asynchronous	61
	4.4.4 Rotary Indexing	62
	4.4.5 Rotary Continuous	65
	4.4.6 Robot Centered	66
	4.4.7 Dedicated Custom Design	68
	4.4.8 Modular/Flexible Design	69
4.5	Machine Configuration Trade-offs	72
4.6	Mechanisms Toolbox	74
4.7	TBBL Automation Project	75
	4.7.1 The Need	76
	4.7.2 The Product	76
	4.7.3 System Goals	76
	4.7.4 Case Study Number 1: Case Opening	81
	4.7.5 Case Study Number 2: Label Insertion and Printing	85
	4.7.6 Case Study Number 3: Crossed Four-Bar BMC Unloader	89
4.8	Conclusions	93
Problems		94
Project Assignment		95
References		95

Chapter 5 Industrial Robots   97

5.1	Handling of Parts with Robotics and Automation	98
	5.1.1 Hold Onto a Part as Long as Possible	100
5.2	Selecting a Robot Arm	101
5.3	Generic Robot Types	103
	5.3.1 Kinematic Solutions	105
5.4	Robot Workspace Analysis	106
5.5	Robot Mechanical Actuators	107
	5.5.1 Robotic Linear Modules	108
5.6	Industrial Robot Applications	110
	5.6.1 Welding: Spot and Arc	110
	5.6.2 Spray Painting	111
	5.6.3 Dispensing	112

	5.6.4	Assembly/Material Handling	112
	5.6.5	Packaging	113
	5.6.6	Food Processing	113
	5.6.7	Drug Discovery	114
	5.6.8	Deburring and Polishing	114
	5.6.9	Machine Loading/Unloading	115
5.7	Case Study Number 1: Machine Loading/Unloading		117
	5.7.1	Where Did Things Go Astray	119
5.8	Case Study Number 2: Pants Pressing Robot		120
	5.8.1	Project Assessment	122
5.9	Conclusions		122
Problems			123
Project Assignment			124
References			124

Chapter 6 Workstations   125

6.1	When is it a Workstation?		127
6.2	Workstation Basics		128
	6.2.1	Structural Members	128
	6.2.2	Bearing Devices	130
6.3	Drive Mechanisms		136
6.4	Case Study Number 1: TBBL Workstation Design		141
	6.4.1	Moving the Case by Translation or Sliding	143
	6.4.2	Use a Scissors Jack Approach	146
	6.4.3	Inflate an Air Bladder	147
	6.4.4	Using Magnetic Power	148
	6.4.5	Using a Four-Bar Linkage	150
	6.4.6	Conclusions	152
6.5	Case Study Number 2: Automated Screwdriver Workstation Design		152
	6.5.1	Automatic Screwdriver Workstations	154
	6.5.2	Conclusions	155
6.6	Machine Design and Safety		155
	6.6.1	Pinch Points	156
	6.6.2	Lockout/Tagout	158
	6.6.3	Warning Labels	158
	6.6.4	Risk Assessment	158
	6.6.5	Safety Responsibility After Delivery	159
	6.6.6	Safety Standards are Just a Starting Point	160
	6.6.7	Real-Life Accidents	160

Contents

6.7 Conclusions	160
Problems	160
Project Assignment	161
References	162

Chapter 7   Feeders and Conveyors · 163

7.1 Feeders	165
7.1.1 Escapement Feeders	166
7.1.2 Vibratory Bowl Feeder	168
7.1.3 Centripetal Feeder	170
7.1.4 Flexible Feeders	171
7.2 Conveyors	172
7.2.1 Segmented Conveyors	172
7.2.2 Other Conveyor Options	174
7.2.3 Timing Screws	176
7.2.4 Star Wheels	177
7.3 Accumulators	178
7.4 Pick and Place Feeders	179
7.5 Case Study Number 1: Dropping Cookies	180
7.5.1 Bull Nose Conveyor Solution	181
7.5.2 Experimental Testing	182
7.5.3 Project Assessment	183
7.6 Case Study Number 2: Feeding of TBBL Cases	184
7.6.1 Conclusions	187
7.7 Case Study Number 3: Donut Loader Machine	187
7.7.1 Design Approaches	188
7.7.2 Dual Cam Track Mechanisms	189
7.7.3 High-Speed Donut Loader Details	190
7.7.4 Conclusions	192
7.8 Conclusions	192
Problems	192
Project Assignment	193
References	193

Chapter 8   Actuators · 194

8.1 Types of Actuators	194
8.2 Application Concerns	199
8.2.1 Amount of Work to be Done: Power = Force × Velocity + Friction	199

8.2.2	Frictional Losses	199
8.2.3	Dispensing Frictional Materials	201
8.2.4	Gearing and Inertia	201
8.2.5	Conveyors	203
8.2.6	Intermittent Motion	204
8.3	Pneumatics	204
8.3.1	Advanced Pneumatics Devices	207
8.3.2	Pneumatics Vacuum Generators	207
8.4	Hydraulics	209
8.4.1	Hydraulics in Automation Today	211
8.5	Electric Motors	211
8.5.1	Electric Servomotor	212
8.5.2	Electric Stepper Motor	213
8.5.3	Motor Sizing	215
8.5.4	Motor Selection	216
8.5.5	Motion Profiles	218
8.6	Amplifiers, Drivers, and Tuning	219
8.7	Case Study Number 1: Stepper Motor Sizing	221
8.8	Case Study Number 2: Servomotor Sizing	225
8.8.1	Proposed Improvements	226
8.8.2	Hardware Requirements	227
8.8.3	Basic Overall Concept	228
8.8.4	Actuator Sizing	231
8.8.5	Sizing the Amplifier	234
8.8.6	System Troubleshooting	234
8.9	Conclusions	240
Problems		240
Project Assignment		240
References		241

Chapter 9 Sensors 242

9.1	Sensor Types	243
9.2	Limit Switches	245
9.3	Optical Switches	246
9.4	Other Sensor Types	251
9.5	Vision Systems	253
9.5.1	Standard Vision Systems	254
9.5.2	Line Scan Vision Systems	255
9.5.3	Vision Approaches and Algorithms	255
9.6	Case Study Number 1: User Input Motion Device	257

Contents

	9.6.1 New Approaches to the Problem	259
	9.6.2 The Leash Leading Method	260
	9.6.3 Conclusions	262
9.7	Case Study Number 2: Pallet Leveling Sensor System	263
	9.7.1 The Industry Need	263
	9.7.2 Self-Leveling Pallet Requirements	264
	9.7.3 Self-Leveling Pallet Concept	267
	9.7.4 Self-Leveling Pallet Prototype	268
	9.7.5 Experimental Findings	272
	9.7.6 Conclusions	272
9.8	Conclusions	273
	Problems	274
	Project Assignment	275
	References	275

Chapter 10 Control   276

10.1	Timing Diagrams	278
10.2	Programmable Logic Controllers	280
	10.2.1 Are PLCs Necessary?	280
	10.2.2 Ladder Logic	282
	10.2.3 Model Train Example 1	284
	10.2.4 Model Train Example 2	285
10.3	Other Programming Options	288
10.4	Case Study Number 1: Agile Automation Control Systems — The Hansford Assembly Flex Project	289
	10.4.1 Agile Control System: Release 1.0	291
	10.4.2 Control System Architecture	296
	10.4.3 A Second Assembly Flex Implementation	296
	10.4.4 Project Conclusions	300
10.5	Case Study Number 2: OMAC Automation Control	300
	10.5.1 OMAC Working Groups	301
	10.5.2 Conclusion: Do You Use OMAC Principles?	304
10.6	Conclusions	304
	Problems	304
	Project Assignment	305
	References	305

Chapter 11 Bringing New Automation to Market   306

11.1	Case Study 1: Precision Automation	306
	11.1.1 The Market Need	306

	11.1.2	The Closed Loop Assembly Micro Positioner (CLAMP) Device	308
	11.1.3	CLAMP Production and Market Acceptance	311
	11.1.4	The Dockbot	313
	11.1.5	Micro Positioners	315
	11.1.6	Dockbot Production and Market Acceptance	316
11.2	Case Study 2: Palletizing		317
	11.2.1	The USPS Mail Tray Project	317
	11.2.2	Commercial Robotic Prototype Implementations	319
	11.2.3	First Implementation: Linear Track System Design	321
	11.2.4	Second Implementation: The Stackbot Design	323
	11.2.5	Multiple Stackbots Solution	328
	11.2.6	Stackbot Production and Market Acceptance	334
11.3	Case Study 3: Pouch Singulation		336
	11.3.1	Standard Parts Feeding	337
	11.3.2	Feeding Pouches	338
	11.3.3	Vacuum Pickup of Pouches	341
	11.3.4	Pouch Feeder from a Modified Bowl	342
	11.3.5	The Trackbot Invention	347
	11.3.6	Initial Test Model	350
	11.3.7	Design Considerations	353
	11.3.8	Trackbot Motion Strategies	356
	11.3.9	Motion Analysis	360
	11.3.10	Distributed Control	363
	11.3.11	Lessons Learned	366
	11.3.12	Trackbot Production and Market Acceptance	369
11.4	Overall Experiences		370
Questions			372
References			372

Chapter 12 System Specifications   374

12.1	Expectations		375
12.2	Other Problems Beyond Specifications		377
12.3	Example 1: Bulk Mail Carrier (BMC) Unloader Specifications		378
	12.3.1	Design Specifications	379
	12.3.2	Comments	381
12.4	Request for Quote		381
	12.4.1	Example 2: BMC Unloader Bid Award Package	382
	12.4.2	Example 2 Conclusions	384

Contents

12.5	Conclusions	385
References		385

Chapter 13 Packaging Machines — 386

13.1	Liquid Filling Machines	386
13.2	Cartoning and Boxes	389
13.3	Labeling	393
13.4	Cases	393
13.5	Palletizing	394
13.6	Forming Pouches	397
13.7	Blister Packs	397
13.8	Bags	399
13.9	Conclusions	400
Reference		400

Appendix A — *401*
Appendix B: Projects — *411*
Index — *419*

1
Introduction

Automation: designing, building and implementing automatic machines. How can such an intriguing concept that has the potential to keep manufacturing located domestically also cause some to be so concerned? The image of intelligent machines producing thousands of quality products for less cost is the dream of many an engineer. But to the current worker at the plant who might be replaced, automation is a potential nightmare.

Depending on one's role in this world, the impact of automation can cause excitement or fear. Let us look at some of these roles to gain an initial view before we start to think about designing and building an automatic machine:

- *Manufacturing Director* — International competition continues to pressure almost all manufacturing operations to reduce production costs. Because labor costs rarely go down, and many workers are operating at reasonably optimal rates, there are no significant gains to be made. Automation, if possible, is a goal of most Manufacturing Directors to remove labor and increase output and quality. Internal manufacturing means greater control of production.
- *Company CEO* — In addition to the concerns of the Manufacturing Director, the CEO is also concerned about employee injuries, Workman's Compensation costs, and the flexibility to raise or lower production outputs if market conditions change. Many a CEO would like to keep production onshore rather than move it to a Third World nation.

- *Stockholders* — To be viable, an investment needs to be placed in a company that has room to grow when markets increase. Adding automation is often easier than finding skilled employees willing to work third shift. So a company that is highly automated can be perceived to be a good long-term investment.
- *Current Company Worker* — They have the most reason to be concerned with the implementing of automation. Some may lose their jobs, while others may be retrained and relocated to maintain the automated production. Fully automated, or "lights out" facilities, are not always cost-effective. However, the jobs that will be replaced are often ones that seem like drudgery and lead to repetitive motion injuries or other physical risks. In society as a whole, there are hopefully better jobs to be done that require the intelligence of a human being, but remaining competitive means staying in business.
- *Sales Representatives* — If a machine is to be sold to many customers, a set of independent representatives are often formed; or it could be your own company's sales force. Your new automation must generate enough cost savings to make the sale. It needs to be a better mousetrap at a good price to be selected over the competition.
- *Consumer* — The consumer wants high-quality goods at a low price. The social concerns of where and how it is made are often left behind when one gets to the checkout line. However, not all products made halfway around the world are of the quality one would expect. Automation, if done properly, can help to lower costs while keeping or improving quality.
- *Environmentalist* — The product and its packaging will have an impact on the world, perhaps for many years after its useful life is over. How it can be recycled is a concern. Additionally, the automatic machine itself needs to be thought about. A machine that will make only a single size and type of product might be obsolete before it has been assembled and debugged, with today's quickly changing markets. Can a machine be designed such that while still being cost-effective, it can live on to make future products not yet conceived? Can it be designed and made in modules that can be reassembled as building blocks? Or will the one-product machine be limited to being a boat anchor in the not so distant future?

Although the engineer responsible for automation may not hold many of these roles, the concerns of these groups do help to define the dynamics that will come into play either directly into the engineer's world, or indirectly behind the scenes. And if the engineer understands all of these issues, one can better anticipate them and be prepared.

Introduction

In this textbook, we will be primarily focused on the traditional engineering aspects of designing and building an automatic machine. The information in the book comes from many sources: companies that manufacture machines, trade show displays and product literature, professional articles and trade magazines, conferences, and the experiences of the author over the past 25 years. These experiences include being a university professor, having a sabbatical at a local automation company, and being the president of two startup companies in automation.

1.1. WHY AUTOMATE?

As the above role descriptions start to mention, there are many reasons to automate a manufacturing process. These include:

- Reducing labor;
- Avoiding labor's sick days, lunch breaks, being late for work;
- Improving quality;
- Reducing waste;
- Enabling production of multiple shifts and weekends;
- Increasing repeatability and quality;
- Increasing Workman's Compensation claims and expenses;
- Keeping production onshore.

Some of these reasons have costs impact savings that are simple to compute. Counting up the X number of employees making $Y per hour can be entered into a spreadsheet. But it is more difficult for a company to put a price tag on improved quality, or to estimate the reduced Workman's Compensation claims. Different accounting practices and methods also create different answers.

Since every manufacturer probably desires to address some, if not all, of these reasons, one could assume that automation is an obvious answer. However, there are not available today automation solutions for every production process. Many production tasks have not yet had a machine designed and made for them. Or the existing machine is not cost-effective compared to the current labor expenses, both direct and indirect.

So the task of automation is often thrust upon a single engineer or engineering team. In some cases there are ready solutions, where the engineer's job is simply to find the best one at the right price. This is similar to what a consumer usually does to buy his or her home washing machine. The consumer rarely designs and builds their own. But in automation, the engineers may need to design and build one from scratch.

If the engineer (or hopefully team) has the skill set and experience, they can do just that, design and build a piece of custom automation. In this case this text is designed to assist the team, either to bring along a novice, or to give more experiences to assist the expert. It may be helpful as a reference to justify the methodology.

Alternatively, the team may be charged to outsource the automation; to be responsible to define the system requirements, write a request for proposal, accept bids, select the right bid, and monitor the development and installation process. In this case the text will assist the engineer to understand the basics, appreciate the range of solutions and the level of difficulty, and aid in the evaluation process so as to select a winning bid. Or it may help the engineer to come to the logical conclusion that the process cannot be economically automated and must remain manual labor for the time being.

As we will see, the field of designing automation is both logical and developed, and as creative and individualistic as fine art. There are often many solutions, no one "right" answer, and sometimes no answer at all. But there is often the better mousetrap waiting to be invented, sometimes by the untrained novice who does not know any better why it had not been done before.

1.2. BOOK TOPICS

We will look at the entire process of designing and building automation (Chapters 2–10), including some experiences of the author, both good and not so good. The topics include the traditional areas, including:

- Steps to Automation;
- Justifying Automation;
- The Automation Design Process;
- Robotics as Automation Tools;
- Workstations;
- Feeders and Conveyors;
- Actuators;
- Sensors;
- Control.

Of particular note is the Steps To Automation chapter, where the understanding of the process to be automated is discussed in great detail. Many a machine was designed and built without a good fundamental investigation of how it was to perform the process, and thus never had a chance of succeeding. Examples of both good designs and "war stories" of limited or complete failures will be addressed. Looking at alternative methods to perform the task may make or break the entire machine development program. For example, no one to date has developed a cost-effective alternative to the dexterity of the human hand.

These chapters of the book have been developed as a mix of examples, case studies, thought-provoking questions, and possible individual or student team projects. It has been found that university students learn these concepts best by

Introduction 5

working with them within the framework of a specific project. Most likely time and cost limitations limit a student to only building simple prototypes of key automation process areas, to further gain more understanding on if and how things might work. The degree to which a project is completed by the reader can be a function of time, education to date, and the ultimate goals of the engineer.

Many example projects are the results of previous engineering student teams. Some of these results are promising, while others might be limited. But the author makes note of any issues and limitations, since much can be learned from looking at not so great solutions as well as the great solutions.

Another area of note is the chapter on Bringing New Automation To Market. This is based on the experiences of the author about his three new inventions, from concept, development, system debugging, and market conditions. A rapidly changing economy makes these efforts troubling at times. Patenting automation is also discussed for its benefits and limitations in the chapter on Justifying Automation.

The chapter on System Specifications will be useful for the engineers who must write a Request For Quote (RFQ), understand the submitted proposals, compare alternate design, and look at project management issues. This is useful to people from both sides of the fence, so as to better understand the partnership.

The last chapter, on Packaging Machines, is an introduction to a large subset of automation machines. Packaging is the placing of the finished product into a bag or box, perhaps also into a larger carton, and maybe finally into a cardboard shipping container or box. Whereas many of the processes to be automated are very specific to a very few industries, the packaging market transcends many industries, and thus has a larger market, leading to a greater number of existing machines available off the shelf.

1.3. CONCEPTUAL DESIGN AND THIS TEXT

This text will focus on the concepts relating to the design of automation. The methodology of how a machine is designed is more important than how many bolts are holding it together. The book will also lead the reader towards many existing bodies of knowledge found in more basic engineering textbooks. It will hopefully remind the reader or point to these principles (such as calculating the rotational moment of inertia for motor sizing) but will not develop these concepts within. It is anticipated that the reader has access to these other textbooks as resources.

The figures in this text have been created as a means to show a generalized concept, not a blueprint to be copied into your CAD design. These figures are meant to show how something happens, since it is often difficult to gain such understanding from detailed photographs. Some figures are generalizations of many different brands of similar automation machines or devices, rather than a reflection of a specific brand of conveyor or robot. In these changing economic times, it seems more prudent to not focus on any particular brand of automation. So there are not a large number of photographs included. The reader is referred to the Web to search on suggested topics for state-of-the-art components. The author has used this method quite successfully for more than a decade of teaching automation students.

2

Steps to Automation

So you want to design and build an automatic machine. Or at least your boss says that you will. Either way, good for you! There is nothing as exciting, enlightening, and sometimes frustrating as going from the overall customer need to seeing some equipment humming along, cranking out product after product. And since this often takes many months, it is not unlike the process of giving birth (although male automation designers would have to take their wife's word on this!!).

2.1. WHAT IS AUTOMATION?

So what exactly is an automatic machine? It could be:

- Automation — by dictionary definition.
- A "black box" that takes supplies and components as input and mysteriously spits out a finished product — perhaps management's view.
- Something intangible such that only an expert knows one when they see one.
- Call it what you will so as to sell it no matter what it does.
- System design of components and control to produce a desired product.

Let us look briefly at all of these.

2.1.1. Automation: Webster's Definition

Merriam-Webster (1972) gives three options in their definition of "Automation":

1. The technique of making an apparatus, a process, or a system operated automatically.
2. The state of being operated automatically.
3. Automatically controlled operation of an apparatus, process, or system by mechanical or electronic devices that take the place of human organs of observation, effort, and decision.

The third choice is probably the closest choice to what engineers design and build, although most do not think about taking the place of human organs (yes, replacing hands for manipulation and eyes for vision, but what would replace the heart or liver?). But this definition is still rather loose in many ways.

2.1.2. Black Box Approach

Figure 2.1 shows a simple block diagram that someone without engineering training or common sense might assume about automatic machines. The simplicity of this diagram may be correct if one is freezing tap water to make ice cubes, and the center box is both a human and a home refrigerator freezer doing the work. But an automatic icemaker in a refrigerator is not a trivial device. Although there are far more complex machines than an icemaker, one should be worried whenever a person (particularly your boss!) draws the automation process similar to Fig. 2.1.

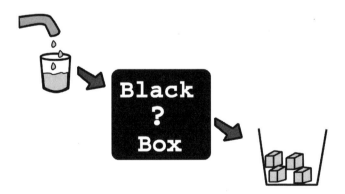

FIG. 2.1 Black box approach to automation: Beware!

Steps to Automation 9

Underestimating the level of effort in time and dollars to produce new automation can lead to the ultimate downfall of a new product coming to market. Experience helps, and this text will try to assist.

2.1.3. The Expert Knows Best

Is a gasoline-powered lawn mower by definition automation? Most automation engineers would say "NO," but the definition is not clear on this. One could argue that taking long grass and converting it to short grass defines a process to be automated. A blender making a milk shake would have a similar argument. Then almost anything with a motor might be classified as automation.

To most engineers in the field, the blender would have to be connected to an infeed supply of milk, ice cream, and flavoring, and an output channel to move away the finished milk shake. Although we will see that automation can be designed to operate in a batch mode, an infeed and outfeed system alone does not distinguish it from other devices. The fact that the standard home blender requires a human to handle the input and output, and additionally it is a batch process, will lead most to call it a home appliance. But this opinion might not hold up in a room full of lawyers!

Is a printing press that published your morning newspaper automation? Most likely the answer is "YES." The continual infeed of paper from a large roll and the outfeed of finished printed and folded papers helps to build the argument that it is automation. Or a machine that takes a measured amount of corn flakes, deposits them into the box a consumer will see on the shelf, and seals the inner liner and box flaps is definitely automation.

Whatever the "expert" wants to call automation, and what is determined not to be automation, is not really important. An "expert" is there to help assess the needs, design, and performance of the desired machine, and if they have the right background, they will be an asset. Some requests to the author of this text possibly looked like automation, so he was called in. And the automation experience helped him to be an asset, even when in reality many might have called the project a home appliance or device.

2.1.4. Call It What You Will

However, one will find that the definition is not as important as the results. If a device or machine can be modeled from previous machines or methods, and the results are satisfactory, then one can sometimes call it whatever one wishes. If calling something automation gets the project funded, and the project has true merit, then roll with it and do not make waves.

A similar situation has occurred over the years about the use of the term "robot" and "robotics" (Chapter 5 will cover more on this roller coaster technology). During the 1970s and 1980s, robotics was a hot term. Growth in robotics was expected to skyrocket. Much of this euphoria seemed to start (in the author's opinion) with the original Star Wars movie in 1979. Although most people over the age of 6 years were told that there was a small person in the R2D2 robot, many were enamored with what technology was supposedly accomplishing. Many assumed that robots could have intelligence, when automation engineers of the day were fighting with industrial robots to simply place a peg into a hole!

But robotics never met this expectation. Many Fortune 500 companies lost tens of millions of dollars. So by the late 1980s, if a company made a press release that they were going into robotics, their stock would drop 5% in a day! Gradually, robotics regained a healthier steady growth expectation, and the term was accepted again in a positive way. The definition of robotics has also changed over the years.

2.1.5. Systematic Design of Components

The primary direction of this text is to look at the entire process, look at the mechanical and electrical components available off the shelf, and those that may need to be custom designed and built, and integrate it with a sensor-based control system to minimize human intervention while in operation. Sometimes this almost "global optimization" during the systematic design process is short circuited by limited time and dollars, but a total systems methodology is the preferred way to approach things.

2.1.6. Generalized Automatic Machine

Lentz (1985) defined the Generalized Automatic Machine. As see in Fig. 2.2, there is functional box for most every part of a classically defined piece of automation. This diagram is a good stepping stone to review in the learning process, but a limitation if one feels compelled to require every new machine to be rigidly designed as such. There are too many options with the machine architectures of today to limit oneself to this design. We shall see other more exciting options in later chapters.

2.2. AUTOMATION DESIGN PROCESS

In many ways the machine design process is similar to the design process for any new product. Depending on which author you have read last, there is somewhere between a five-step and ten-step process to follow. A generic six-step process (Shigley and Mitchel, 1983) is probably the most common (Fig. 2.3). One of the key steps for the engineer is to determine how the product functions, usually internally and not always seen by the customer. This requires the nominal brainstorming and inventiveness tempered by constraints from manufacturing on how it can be made.

Steps to Automation

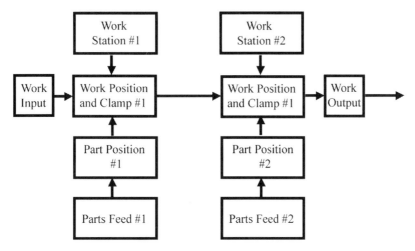

FIG. 2.2 Generalized automatic machine (from Lentz, 1985)

However, automation machines are a different challenge from general new product design. Most likely the machine itself will not be seen by the general masses. The customer may be a team or a single person on the manufacturing floor. The color of the machine may not be a concern (although many companies would like an Occupational Safety & Health Administration (OSHA) recognized color that stands out for safety reasons, and some companies require that all machines in their factories be the same color!!). The automation machine's potential existence springs from a product that needs to get out the door, usually as soon as possible. The machine is one of the boxes or arrows in the parallel generic design process of Fig. 2.3 for the real product the ultimate customers want. And since market forces can change overnight, the need for the machine may disappear before one has finished assembling it!

So with the pressures of a quick time to market, where some production customers during an actively growing economy will pay an expediting fee to get it one or two months faster than the six to 12 month quote you normally propose, there is little time to figure out how the automation is supposed to do its task. So the automation design engineer (or team) is faced with a great challenge. One might do the following, although not necessarily in this order:

1. Look at similar automation processes and machines.
2. Look at how a human does the process.
3. Attempt novel things in seemingly random acts of creativity.
4. Apply a Mechatronics strategy.

Let us now look at each of these.

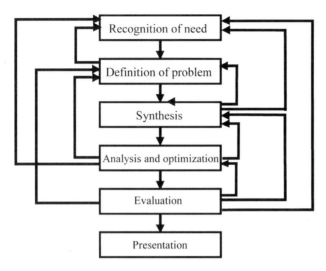

FIG. 2.3 Generic six-step design process

2.2.1. Look at Similar Automation Processes and Machines

One can or should never stop learning, although many people stopped well before they graduated high school. The author strives to learn something every time he goes into another factory or plant. Even during a trip for a possible research project for his university that does not pan out, there is still something to be learned from seeing how others have solved a problem. It is so exciting when after the 5th, 50th, or 100th plant visit, one sees a need in an industry that has been partially or totally solved in a seemingly unrelated field (this is one of the few real benefits of growing old). Multidisciplinary efforts in design come from the cross-fertilization of ideas.

However, one does not have to be almost half a century old to see these things. Specific industries often have very close-knit organizations that, even though they are competition, can learn from each other. Some are even friendly towards each other. The packaging industry (more in Chapter 13) has a professional group of several hundred member automation companies. As a subgroup, almost all of the eight to ten manufacturers of case erecting machines work jointly side by side to gain for the common good. However, this group of competitors may know nothing about the needs of a state library handling audiotapes for the blind, and why should they? But if you are charged to work in this new area, you can apply the knowledge from the case erecting machine to make great gains in a new market.

If one cannot travel into a factory to see such equipment, there are a few other options. Many automation machines will be in operation at the appropriate trade shows, but some of these are held only once a year. Almost all automation companies do have websites with a great deal of information, but few sites have video clips of their machines in operation. It seems to be part of an older, conservative nature of many of these companies. Many of them do, however, have videotapes that they will send you. Their marketing departments usually develop these tapes, so they are filled with audio clips stating that their machine is far better than anyone else's in the known galaxy! So if one can live through the hype (just as one wades through the hype every day on commercial television) one can usually determine if an existing machine has possibilities to solving one's automation problem. Be aware of marketing types overselling technology and capabilities. Often one has to call one of the companies' applications engineers to gain more confidence that the fit is right. But if the machine is not the right fit, and if the concept is not patented, one might look to use some of the strategies in a custom design machine that you design and build.

2.2.2. Looking at Human Performing an Operation

The process of watching a human perform the task, particularly if they have done it for years, can be both enlightening and limiting. Enlightening to see someone's fingers align cut pieces of cloth to sew together a shirt similar to the one you are wearing gives one the challenge to replicate a human hand! Yet maybe the way a human does this task is next to impossible to automate, at least within your budget. Maybe one needs to find a fresh approach to the process to gain success. Do not get biased by the way it is currently done.

How does one tie one's shoelaces? Most people learn from someone when they around five years old. Bow tying can also be applied to fancy gift boxes seen at some high-end jewelers and expensive chocolates. At these companies, all of the bows are tied manually due to a lack of affordable automation. Engineers at these companies attempted to duplicate the motions of the human worker tying these bows, and failed. But a clever engineering designer, Mr. Clay Cooper (personal communication, 1996), looked at this a different way.

When he was stuck at an airport with a four hour layover one day, being potentially bored, he examined the tied laces on his shoe. He decided to look at the finished tied bow, and work backwards. He took the two loops, and pushed them back into the knot, and discovered a novel process. If one created three loops (as in Fig. 2.4a) one could pass the end loops through the middle loop (Fig. 2.4b; first one side then the other), and then pull, creating a perfect bow

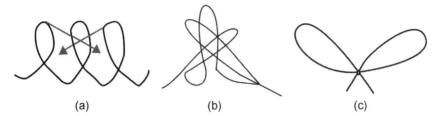

FIG. 2.4 Novel method to tie a bow: (a) three loops; (b) pull outside loops through middle loop; (c) tighten to form bow

with limited passing of the lace (Fig. 2.4c) and no loose ends to handle. A custom designed mandrel (Fig. 2.5) would be required to produce these three loops. But the challenge we find as youth to make the first loop and wrap the other lace around that seemingly flimsy first loop is eliminated.

This works with ribbon for packages just as well. Yes, the resulting process cannot be used to tie your shoes, and the bow is made from a single end of the lace or ribbon. The other end would be aligned by the mandrel, and the end would be held tightly by the bow when the loops are tightened. This is not as structurally strong as a common bow on your shoes, but it would be strong enough for a package, would look just as nice, and might be easier to unwrap since the bow itself could remain after sliding out the other end.

The mandrel would have to be made with parts of its round body able to move for the grippers to grab the ribbon successfully 100% of the time. However, the process now becomes very tractable, consisting of many simpler steps rather than a single set of difficult to impossible tasks.

2.2.3. Try New Things

The dexterity of the human hand and the sensory processing of the human eye and brain are remarkable; so remarkable that researchers continue to try to duplicate the efforts in order to improve on what we have in automation presently.

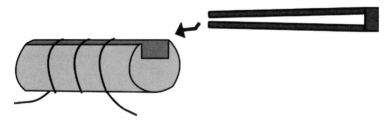

FIG. 2.5 Mandrel for gripper access

Steps to Automation　　　　　　　　　　　　　　　　　　　　　　　　　　**15**

However, the results to date are not close to being cost-effective. So one may have to try to use three, four, or 12 less dexterous simplistic fingers, and manipulate or hold the product in place quite differently from that to which we are accustomed. In automating a sewing operation, perhaps it is more dependable to use six or eight simple actuated fingers to place a pocket onto a shirt front rather than the current method of using one human hand?

There is also the total range of principles from physics that one can explore. Although a human usually does not use vacuum, except to suck something up a straw in order to consume, one can use a vacuum cup to obtain a fantastic grip on a giant sheet of glass. Or one can apply a static charge to hold some sheet of thin film material that would not stand up to the vacuum cup (it could cause some deformation, creating a dimple). One needs to be willing to think "outside of the box."

One challenge a few years ago for Mr. Clay Cooper at his custom automation firm, American Dixie Group, was to pull apart or de-nest paper baking cups from a stack (Cooper, 1994). These paper baking cups are found on some brands of cupcakes and muffins one can purchase, and many home bakers prefer to use them also. Baking cups are made from taking a round thin piece of paper that can stand up to the moisture of the batter and the heat of the oven (who wants flaming muffins, Fig. 2.6c) and are stamped by a processing die that makes the sides of the cup fluted (Fig. 2.6a). Usually 25 cups are stamped at once, and a group of 25 are stacked with other similar size groups (Fig. 2.6b).

Now most humans would peal off one baking cup at a time, from either the top of the bottom of the stack for filling a single muffin pan at home. This relies on friction from your fingers. However, if you are filling 12 pans of 24 muffins each every minute or two, you would rather not be paying to have a team of people all huddled around the row of pans trying to keep up with the never ending flow. So Clay Cooper wanted a new approach. He observed that one could get a good fix on the inside of the top baking cup of the stack with a suction cup, but it

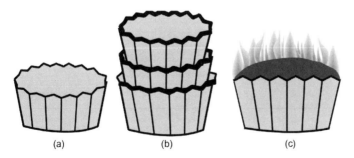

FIG. 2.6 Fluted baking cup: (a) single; (b) stack of multiples of 25 cups; (c) flaming baking cup

puckers the bottom surface and the performance was not repeatable. So he configured a test device out of a 6 inch long, 2 inch diameter piece of PVC pipe with drilled $\frac{1}{4}$ inch holes and some duct tape (Fig. 2.7), and fastened it to his shop vacuum.

This device did not cost much to make in his basement, and he was able to find out some results in a few hours. It did successfully grab the inside cup without deformations, but a second or third cup would come along once in a while. This was because when the 25 cups are die cut and formed into the fluted shape as a group, the edges of one cup can become attached to the next. Now a second cup may not seem a big deal every now and then (it would not cost that much), but a double or triple cup can significantly affect the baking time and consistency of the muffin or cupcake. The muffin pans are baked on conveyors running continuously through an oven. So, unlike baking at home, one does not use a toothpick to see if all the muffins in one pan are done or not. A double or triple cup was therefore not acceptable by the production customer.

So Clay continued to experiment. He found that if he used a jet of air through a nozzle and placed it just right, it would separate the second cup from the snugly held top cup from the stack (Fig. 2.8). But this air jet could not be on continuously, or it would wreak havoc with the stack of cups before the top cup was gripped on the inside by the vacuum. Sometimes the cups were not so tightly packed, and loose cups would go flying around the room.

This experimenting by Clay Cooper, over a few days with a few dollars in his basement, led to the key process of a $175,000 machine, which has been running every day for over 10 years.

2.2.4. Apply a Mechatronics Strategy

Whether you sincerely believe that the Mechatronics approach to engineering is a new creation of the 1990s (Craig, 1992; Derby, 1992), or you more cynically

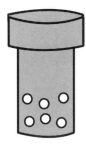

FIG. 2.7 Baking cup de-nester testing device

Steps to Automation

FIG. 2.8 Air jet to separate gripped baking cup from the next one

think that it is just a new buzz word for Systems Engineering (Groover, 2001), a total systems approach is the only way to succeed in designing and building a new piece of automation. As seen in Fig. 2.9, there is the total integration of:

- Mechanical Engineering;
- Electrical Engineering;
- Controls;
- Software;
- Materials and Components.

The key concept here is that one should look at all of the many technologies at one's fingertips, and creatively use any or all of them together. This may sound somewhat obvious, but some custom automation has been built such that the mechanical functions are completely designed, made, and assembled, before

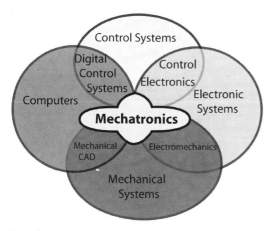

FIG. 2.9 Mechatronics system design

the controls group gets their first look at the machine. And then the control group is supposed to get it to work by the addition of sensors and a controller. If the process is basic and there is nothing new to it, the project still might work, but, in general, this is a recipe for disaster. These are the projects that take months to debug rather than weeks. The extra reworking and redesign to make it work properly eat into the anticipated profits and kill schedules.

Can a single engineer handle this Mechatronics or Systems Engineering approach? Again if the machine task is simple enough, yes. But most likely if it is comparable to most current automation machines, the answer is no. Rarely can one person accumulate all of the required knowledge and have the time to bring the resulting automated product to market. There are not enough hours in a day. And there are often more experienced team members available to assist.

2.3. KNOWING THE PROCESS

With all of this said, what is the most important step in the automation design and build process? Knowing and understanding the process one is to automate! Because there is always a time crunch, there is never any time to go back to the drawing board if the machine fails to work. Yes, there is always major debugging to be done. But if the core process does not work correctly, one has invested many man months of effort and possibly $25,000 to $1 million in creating a potential flop. A revised process may not be applicable to the existing machine architecture, so one may end up starting from scratch. And at these prices, an automation engineer will not be around to see this happen too many times.

Any process that will be potentially automated needs to be stable and well defined. The process of welding a car frame was perfected under human operators' control many years before welding guns were placed on a robot. If there is great variability to the process results, and technicians currently have to test, adjust, and decide which components of a batch of product are good, then a company is most likely not ready to automate.

It would seem that if a company seeking to automate wanted to know the process well before committing development dollars, and if a company had a Research and Development group, one should take advantage of this resource. But this does not guarantee success. The following example is true, but the names have been withheld to avoid embarrassment.

2.3.1. Wound Capacitor Example

Several years ago the author was able to visit a company that makes capacitors for the electric power industry. The capacitor consists of a wound roll of multiple metallic material layers, with a layer of plastic insulation placed between each

Steps to Automation 19

layer. Figure 2.10 shows the simplified roll being formed. These capacitors were wound on decades old machines around a removable central core mandrel, and when completed, were squished to fit into a box that is about 6 in. wide, 12 in. long, and 12 in. deep (Fig. 2.11). Electrical connection wires are added, but are not shown since they were not part of the automation challenge.

The company's Research and Development (R&D) group was asked to study the electrical performance of the capacitor, since the market was always asking for more performance for less price. They discovered that if the capacitor and insulator materials are wound with a constant tension, rather than the variable tension that occurred when the spinning mandrel wound the layers tighter as the roll got bigger, the electrical capacity of the capacitor could be improved! This was quite a find, since for no increase in materials, performance could be improved. All that was required was a better winding system, with controlled tension.

Because there were six supply rolls of metallic and insulator materials, and the central core winding mandrel all need to be controlled to keep the individual and combined tension within parameters, the system needed seven servo motors with coordinated motion control. This is not rocket science, since it is similar to controlling a seven-jointed robot under sensor control. This is do-able but not cheap. But the capacitor company, with the aid of its R&D group, had completed the mechanical, electrical and in their minds most of the control work of the new automation machine when they called the author in to assist with coordinating the motion controllers.

Motion controllers are available for this type of work, but they cannot be a random collection of single and double axis motion controllers hoping to tie them together. They need to be from one manufacturer, designed to handle the specific task. A six-jointed robot uses a single coordinated multiple axis motion controller to perform a specified weld on a car body.

Surprise Problem

Hours after seeing the initial problem (defined by the engineer as a capacitor winding motion controller issue), a great revelation struck the author! Can you

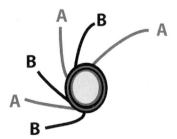

Fig. 2.10 Winding of capacitor layers: A is metal layer; B is plastic layer

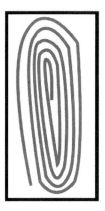

Fig. 2.11 Capacitor in its container

see the fallacy of the above-described procedure? It does not require any expertise in automation, it is simply understanding the process.

The efforts from the R&D group were noble, in finding how to make more performance of the capacitor from seemingly nothing. But the big problem is the finished wound roll. Look at a standard roll of paper towels. What happens if one squishes the roll? The layers gap in irregular fashion. So all of the precise winding of the capacitor is thrown away when it is stuffed into the rectangular cross-sectional container (Fig. 2.12)!

One could argue that this superior capacitor should be packaged into a square or round container to keep the higher performance. But the electric

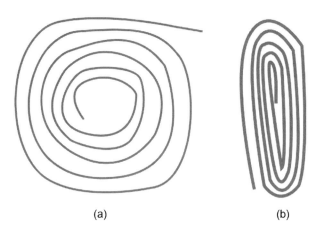

Fig. 2.12 Wound roll: (a) original; (b) squished

power industry has conventions and standards that they need to keep for compatibility reasons. So unless the capacitor could be wound on a long and thin mandrel, the effort was wasted. And no-one knows how to wind such a mandrel at the speeds needed to achieve the required production rates. The capacitor production company had spent almost $1 million at that stage.

The capacitor company and their R&D group did not understand the process. Something that might have been discussed at the first lunch kickoff meeting turned the effort sour. How could this have been avoided? Perhaps by better communication, by making fever assumptions, or by listening to the outside viewers who come in with a new set of eyes and ears. Nothing is foolproof, but this situation was sad and probably avoidable.

2.4. PROCESS EXAMPLE: PEANUT BUTTER CHOCOLATE KISS COOKIES

This section will detail some process options on how one would make peanut butter chocolate kiss cookies with automation. The intent here is to discuss a range of possibilities, but not to state which of them is the best, second best, or worst. These open-ended questions will be left for the reader to experiment with over time. So keep in mind that there are no right answers, or wrong answers. There are just concerns and observations to be made. You the reader will determine what will work for your constraints.

The choice of baked goods as a process example is made because of the relative ease with which the reader can perform some of these processes. There are many other nonfood related processes to automate, but these are difficult for the reader to understand without the proper experiences. The author finds that success in understanding a process only comes from playing with the components to be automated with one's hands. Figures in books or on websites will usually leave one cold and clueless. So please bake some cookies here to gain all of the understanding possible. And when you are done, you will have something to reward yourself with too!

A commonly found home-baked dessert is a peanut butter chocolate kiss cookie. The recipe can be found in many traditional cookbooks, on some packages of chocolate kisses, and is a variation of a peanut butter cookie mix found in supermarkets, both in boxes and stand-up pouches. If we focus on the prepared mix approach, the box or pouch contents is usually a dry mix, where the home baker adds eggs and either cooking oil or a separate pouch of peanut butter included in the mix. When all of the ingredients are mixed, it forms a batter that is rather gooey and yet remains grainy from the dry mix particles. The dough is rolled to obtain a smooth cookie surface after baking.

The process to be automated is as follows:

1. Take a heaping teaspoon of batter and roll it with your hands to form a ball;
2. Place the batter ball in the cookie sheet, spaced as the directions state;
3. Use your thumb to form a depression in the center of each ball;
4. Bake according to directions;
5. Prepare chocolate kisses by unwrapping them while the cookies bake;
6. Remove the tray of cookies when done and immediately place unwrapped kisses on the cookie's center depression
7. Let cookies cool.

Cookies can be eaten as a reward soon afterwards, but if they are to be stored one upon another, will need to cool for an hour or two.

The kisses need to be placed on the cookies in a matter of seconds after the cookies are removed from the oven so as to properly melt the kiss. Over the subsequent 10 minutes after placing the kiss on the hot cookie, the entire kiss will melt. When the kiss is initially pressed into the warm cookie, the melted chocolate and cookie interface will ultimately form a solid bond. The residual heat from the cookie sheet also assists in this process. Often it is useful to wiggle the chocolate kiss when one observes that the bottom of the kiss has melted and the top of the kiss can be manipulated without deformation or the leaving of fingerprints. This bonding is critical if the kisses are not to fall off from normal handling. Another note is that the baking process will lessen the depression you made with your thumb.

2.4.1 Process Possibilities

There are many areas of concern on possible automation processes for making the batch of cookies. We will only look here at what is perhaps the most difficult step, that of taking the batter in step 1, rolling it (Fig. 2.13a), placing it in a single location, and the thumb print indentation in step 3 (Fig. 2.13b). The placing on the pan will be simplified at this time to the placement of a single cookie, in order to focus on one process at a time. Plus there are many existing ways to achieve multiple cookie placements.

Humanly, the batter is dispensed into one's hands by measuring with a spoon, while some cooks will just grab a blob of batter with their fingers to speed up the process. Either way, the amount of batter is not consistent. The rolling in one's hands relies on dexterity to form a fairly well shaped ball, and to sense when the rolling is complete. Not all humanly made cookies are the same size. Directly automating this rolling process is not simple. Here are some options and thoughts for the reader to explore and experiment with:

Steps to Automation

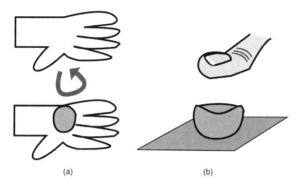

(a)   (b)

FIG. 2.13 Rolling batter into ball with hands; (b) thumb print into ball of batter

Replicate the Human Process

- A bowl of the gooey and grainy cookie batter is usually in a randomly shaped blob. This does not lend itself to easy dispensing of a fixed amount of batter with a spoon. There would also have to be some sensors to find out where the batter is in the bowl initially and after one or more spoonfuls have been taken.
- Since the amount of batter taken with a spoon is not consistent, the batter rolling station would need to adapt to the variable amount. A human does the adjusting without thinking.
- The rolling surfaces might be two flat surfaces, or they might have some curvature. When rolling is complete, it would be great if the ball of batter is always in the same location.
- Either some sensing would be required to determine that the rolling is complete, or the rolling process would be done for an amount of time that guarantees good output.
- A mechanical appendage looking like a human thumb can be configured. It would need either some physical centering of the batter ball or some sensors to find the approximate center.

This is not an impossible set of tasks to perform but some parts are challenging. However, do not limit yourself to only what has been stated above!

Extrusion Process

- Look at the extrusion process for a possible different process (Fig. 2.14). This is how heated aluminum is melted and forced through an extrusion

FIG. 2.14 Extrusion process

die to get constant cross-section of product. There are cookie guns that extrude cookies of various shapes. If the batter is perhaps cooled to the right temperature, the batter will form something closer to a cheese log, and literally can be sliced. This would be similar to the consumer product slice and bake cookies. These types of cookies are usually round in cross-section, and tend to flow enough so as to form a smooth edge and surface.
- If the extrusion process is used, there is no ball into which to place a depression. The chocolate kisses will sit sufficiently well on top.
- To obtain the right viscosity for this process, one might have to adjust the amount of oil or peanut butter to thicken it, or some additional flour would need to be added.

Silicon Dispensing Process

- One might relate to the dispensing of silicon sealant from a calking tube and gun. Or think of a tube of toothpaste. By changing the viscosity of the batter, possibly by warming it up slightly, one can get the batter to flow better. Then the cookie can be formed more like the spreading of frosting as a baker decorates a cake. When dispensed slightly warm, the gooey batter will tend to make a smoother surface, perhaps removing the need to roll altogether.
- After dispensing the cookie blob onto the pan, use another device to make the thumb print depression. Or maybe by dispensing the batter from a height lower than the ultimate diameter of the rolled batter ball, the nozzle forces out the batter into something closer to a pancake rather than a ball or somewhat rounded blob.
- The heating process could be done with normal heat sources like an oven, or perhaps a microwave. In production, warmth could be generated by warm water, or warm air as a byproduct of a cooling tower. In real bakeries, one usually does not have enough floor space to let continuous conveyors run while products cool for one or two hours. Cooling towers are used to speed up the process, and they have an abundance of heat available for other things.

2.4.2. How to Proceed

After a first initial read of these three suggested processes, the reader might think that these are too limited. Maybe there is a significantly better way. The author does not promise that any of these three methods will work successfully, but they do give some areas for exploration.

The reader might also come to the conclusion that one or two of these methods would be superior. Both the extrusion and the silicon methods seem appealing since there is no rolling and the depression seems to be easy or not even required. And the author might bet a nickel that this is the case. However, until some testing is performed on all three process methods, no one should bet the farm! Have fun exploring.

2.5. CONCLUSIONS

When it comes to designing automation machines, creativity is more often a blessing than a curse. It is a blessing since you may be tasked to automate something that does not exist. Others may have tried before you, perhaps for many years. If it was easy, it may be for sale already. It is great when you have the time to try new things, look at other industries to gain insight, or just brainstorm with some teammates. The challenges may be straightforward or they may be significant. If something looks easy to automate, but it has not yet been done, this might be a red flag.

Creativity can be a curse when one feels the need to design a new machine rather than investigate whether there are existing machines available. Sometimes an overzealous engineering student will not do the needed Web searching to find existing segments or entire machines that they can incorporate into the overall design, and they feel the need to design it all. This is usually a luxury in today's economy.

It is vital that the automation designer or design team understand the processes that are to be automated. System requirements, such as cookies that must be rolled, need to be challenged and alternatives presented to the automation customer. To build automation that does it step by step as a human would perform it has been shown in this chapter to not always be the optimal method, but if it is a new product, and the tasks to be automated cannot be performed reliably by humans, then one will need to demand time and support to determine the process parameters before the designing can start.

PROBLEMS

1. Determine the possible processes to take commercially available chocolate kisses (wrapped in thin aluminum foil and that have the thin

white paper label that specifies the producer's brand) and unwrap them for use in the Peanut Butter Chocolate Kiss Cookie machine in Sec. 2.4. Assume that the wrapped kisses are in random position and orientation. The output of your determined process should consist of a single line of unwrapped kisses on a continuous conveyor belt. It is recommended that one purchases a bag of kisses for experimentation. Note: in reality, if one was making such a machine as in Sec. 2.4, one would want to contract with the chocolate kiss producer to supply the kisses in unwrapped condition.
2. Determine the possible processes to make chocolate chip cookies. How would one mix the dry ingredients? Crack the eggs? Mix the batter and know when it is well mixed? Dispense the batter with the embedded chocolate chips? It is recommended that one makes a batch of cookies and take notes of how humans perform these tasks, and then brainstorm on alternate processes.
3. Determine the possible processes to make a sandwich of your choice. How should the bread be manipulated so as to not tear it? How can the lunch meat be pealed from a stack of similar meat. How should something as unconstrained as lettuce be handled? Create a machine concept that can make a sandwich of a single type of meat, and look at the extended needs to use more than one type of meat or cheese, and what would need to be added if there was a general purpose sandwich-making machine?
4. Determine the possible processes to make a pitcher of juice from a can of frozen concentrate. How should the can be opened, traditionally or in a novel way? How should the concentrate be manipulated? Should it be dumped in as one solid block or should something else be done? How and when should the water be added? How should it be mixed, and will it be safe for homes with children? Will it be easy to clean?

PROJECT ASSIGNMENT

Using one of the projects from the Appendix (or any other projects), perform the following:

1. Look at similar automation processes and machines. Use the Web to find related tasks.
2. Look at how a human does the process. Purchase one or more of the processed products if at all possible.
3. Attempt novel things in seemingly random acts of creativity. Have fun trying to use all of the laws of physics at your disposal. Do not feel limited on costs at this stage.

4. Apply a Mechatronics strategy. Concept how the process can be carried out if there are a number of simple sensors connected to a series of simple actuators. Try using simple passive grippers such as clothes pins and binder clips to hold intermediate steps together. You might need to ask someone else to lend you a hand.
5. Keep all of your notes in a bound laboratory notebook, where each page is numbered sequentially. Staple any of your hand-drawn notes onto the notebook's pages. Do not throw out any idea, no matter how crazy it might seem.

REFERENCES

Cooper, C. (1994). *Baking Cup De-Nester.* Patent No. 5,529,210, June 25, 1996.

Craig, K. (1992). Mechatronics at rensselaer: integration through design. In: ASME International Computers in Engineering Conference. San Francisco, CA, August.

Derby, S. (1992). Selective use of mechatronics when designing assembly robots. *Mechatronics Systems Engineering* 1(4):261–268.

Groover, M.P. (2001). *Automation, Production Systems, and Computer-Integrated Manufacturing.* Upper Saddle River: Prentice Hall.

Lentz, K.W. Jr. (1985). *Design of Automatic Machinery.* New York: Van Nostrand Reinhold Company.

Merriam-Webster. (1972). *Seventh New Collegiate Dictionary.* Springfield: Merriam Company.

Shigley, J.E., Mitchel, L.D. (1983). *Mechanical Engineering Design.* New York: McGraw Hill.

3
Justifying Automation

Most likely somebody somewhere in a company needs to justify the expense of purchasing a standard available machine or committing to the creation of a custom piece of automation. They may be the company CEO, the project manager and their boss, or some overworked bean counter in the back room. Often it may involve many if not all of them. Related to this need, the automation machine builder needs to make a profit to survive. There may be purchasing departments and sales forces involved with competitive pricing pressures pushing from every direction. How does this all fit together, and what are the different operational modes with which automation builders tend to align themselves? And who owns the intellectual property, and is getting a patent worth the effort?

These are some of the many different questions to be addressed in this chapter. The level of detail will vary by the relative importance to the automation market. Some of these questions will be sufficiently covered in brief discussions within this chapter, while the complete answers to other questions would require a set of dedicated textbooks. The level of understanding needed by an automation engineer does not have to include an MBA degree.

The topics of this chapter include:

- Traditional project cost justification for a purchase;
- Traditional costing estimating for building and selling automation;
- Win–Win purchasing philosophy;
- Maximum profit cost estimating for building and selling automation;

Justifying Automation

- Justifying flexible automation over hard automation; and
- Intellectual property, patents, and trade secrets.

Embedded in these discussions will be other trade-offs concerning automation builders, including:

- Having one's own sales force vs. a network of independent sales representatives;
- The impact of world events due to the normal sales/purchasing cycle;
- How to deal with a single customer in a specific market vs. all of the competition;
- How does a startup automation company compete?

As we will see, there are some basic facts that assist the automation machine builder to better understand the economic games that get played. Factors like having a person blessed with natural sales abilities can be a large asset not always appreciated by an engineer. Or understanding that the lowest price does not always mean that your machine will be selected. Many purchasers dealing with competitive bids are not legally bound to select the lowest dollar amount, so reputation plays a big part in the selection process. Because the purchaser might be contracting for a machine to be delivered in six to 12 months from the date of issuing a purchase order, there is a long time of waiting that can be sleepless nights if the machine builder looks to be on shaky financial ground.

Starting an automation company has its challenges, too. It is tough to be the new machine builder on the block. Unless your new machine will make the customer millions, being the first to market, few people want to be the first to purchase your designs when economic times are slow. Everyone wants to know who else has bought a machine to minimize the risk. Getting the first few sales is challenging. The customer's warm and fuzzy feelings about the purchase deal carries more weight than one might suspect.

3.1. TRADITIONAL PROJECT COST JUSTIFICATION FOR A PURCHASE

One of the primary reasons a company will investigate automation is to reduce the number of workers. Workers are looked to be a component of a product's cost that can be adjusted. Whereas a car still needs four wheels, the number of workers used to put the four wheels on a car is not fixed in concrete (unless the union says it is). So if one worker can run around at breakneck speed and install all four wheels on a car in 20 seconds flat instead of two or four workers (this worker should probably work in a race car pit crew), the costs of producing the car will drop. Can this super worker keep it up 8 hours a day, 5 days a week? Most likely they cannot, but it sounds attractive to the financial people.

The cost of a worker is more than his or her salary. Sometimes the internal cost to a company can reach close to twice the worker's salary. This complete financial accounting is sometimes referred to as the "burdened rate." The burdened rate may be an internal secret of a company, but one can make some guesses and be close.

Other factors can also have tremendous economic impact, but are traditionally less accepted or more difficult to quantify:

- Product quality — automation can produce a consistently good product.
- Sick or late workers — automation does not call in sick.
- Lunch/breaks — workers need breaks to survive and do their best.
- Ability to meet expanding market — automation can work a third shift on a moment's notice.

These are sometimes referred to as the more intangibles in cost justification. They are all valid points to potentially be discussed with an automation customer, but their answers may differ tremendously depending on their company culture.

Let us look at two workers in two different jobs, and see the relative attractiveness to implementing automation.

3.1.1. Fast Food Worker

The normal worker at a fast food hamburger joint may get paid something between $6 and $10 per hour. They may be stuck on some exciting job such as cooking French fries all shift, and feel like they are soaked in grease by the end of the day. Whether or not breathing in and being covered with grease is healthy for a human being (who can imaging that a doctor would suggest the job for you if given some choices), one could imagine some automation to replace this person.

Now the owners of the restaurant, besides paying the worker the hourly rate, would have to pay:

- Social Security;
- Workman's Compensation Insurance;
- Unemployment Insurance;
- Health Insurance;
- Other assorted payroll taxes and fees.

So the hourly rate to the owners is effectively $3 to $4 per hour greater than what the worker sees in their paycheck (before the personal withholdings). So the total burdened hourly rate is now $9 to $14 per hour. What kind of automation would that start to justify?

The next set of questions to the owners is:

- How many shifts will the worker be replaced?
- Does their French fry person do other tasks?

Justifying Automation

- How many months or years is the owner willing to invest before seeing a return?
- What is the risk factor dealing with the automation being successful?

For the hamburger joint, most likely it is 2 shifts, 7 days a week. So one could take the burdened hourly rate (assume an average rate of $12 per hour) and multiply it by 16 hours per day times the 7 days a week. That produces an amount of $1344 per week being effectively paid out for the French fries employee.

If the owners were willing to invest the amount such that after one year's time the automation has been paid for, the total for the automation would be almost $70,000. (A one year payback period is common in industry, but it can range from 90 days for short-term markets, to two or three years for stable long-term markets.) Added to this are the intangibles, where workers cooking French fries are not always excited about their work, and some are likely to be late or call in sick. The process is well defined, so product quality is not a huge variable.

The author thinks that it might be possible to build and sell automation to this industry at this price, assuming that dozens if not hundreds or thousands of restaurants purchase copies of these machines, so as to spread out engineering design costs and to obtain cost savings by making multiple automation machines in batches. To carryout this scenario, many of today's chains of hamburger joints have many different store configurations, and they may have to do some standardization to make it work at this automation production point.

But do the owners have $70,000 to purchase such a machine at any specific time? Will they have to take out a loan? Can they lease the machine from a leasing firm? Is their credit good? Are they financially strapped as it is? Not knowing how the hamburger chain's finances are positioned, one is left to make an educated guess. The author's guess with his limited knowledge of their internal financing is "maybe."

3.1.2 U.S. Postal Worker

The author has had multiple interactions with the United States Postal Service (USPS) over the last six years. It has been told to him by USPS engineering managers that the internal justification amount per employee to use automation is over $150,000 (J. Weller, personal communication, 1996). This means that the automation to replace an employee can cost $150,000, and with whatever internal accounting process they have, everyone is happy. Many of the USPS jobs are a single shift, with a late afternoon start through very early morning hours. Some of the driving components to this cost are high salaries, fantastic retirement plans, and great health benefits.

Many of the human tasks in a USPS processing center are where people are feeding odd shape envelopes and/or bulk mail (magazines) into current automation sortation machines. The sortation machines do a great job considering the range and quality of mail that is handled every day, but the human operator is basically a slave to the machine, keeping it well fed. The operators risk repetitive motion injuries like carpal tunnel, and many workers there do get long-term disability care that costs the USPS a great amount. Thus the great health benefits are needed and used often.

The $150,000 amount sounds attractive to an automation builder, but the USPS must be in a mode that is financially viable to create such a large purchasing program. There are around 1000 processing centers in the United States, and for the worker replacement program to be cost-effective, a good number of the 1000 sites must be upgraded in a short amount of time. So when the USPS is short on cash as they are in the early 2000s, things move slowly.

3.1.3. Which Market to Approach?

As we have seen, the USPS market seems to be more attractive due to the cost justification numbers of $150,000 vs. the French fries automation for $70,000, but the job of automating French fries might be easier and require less machine complexity and component cost. However, the USPS market would be a single huge contract that might be a windfall if one gets selected, or a massive disappointment if not.

The only suggestion is to work on solutions to both markets on one's free time, and approach them both when given the opportunity. So when either customer can come up with the right amount of dollars to go forward, one is ready to play the competition game.

3.2. TRADITIONAL COSTING ESTIMATING FOR BUILDING AND SELLING AUTOMATION

Developing a good cost estimate for building and selling a new piece of automation is a combination of good financial practices, experiences from past projects, and knowing you have a good method of manufacturing operations that will follow the estimates fairly well. You plan for a slight cost overrun in the form of a contingency, and you plan on making a profit. To understand the core issues here, we will look at a more reader-friendly example.

3.2.1. Room Refinishing Example

Because many readers of this text may not have any experiences with costing out an automation project, let us first focus on a more commonly found problem: refinishing a room in a home or apartment. Let us assume that the goals are to:

- Replace the floor covering;
- Repaint the walls and ceiling;

Justifying Automation

- Replace the ceiling light fixture;
- Add a pair of bookshelves to one wall.

These do not look like overwhelming efforts on first glance. Assuming one has the average person's skill set, or a friend or two who do, the tasks do not look too daunting. A paint sale in the morning paper makes the project seem timely, and you always wanted to get better lighting, so you can cost-justify replacing the old light fixture. The room will be more useful with that nice set of bookshelves, too.

So how long will it take? How much will it cost? What if any expertise will you have to hire out? These are questions that usually are not completely answered by many homeowners or apartment dwellers before they start a project. You see a price for paint at $20 per gallon (you will use maybe three gallons?), carpet for $240, and bookshelves for $100, and you guess the light fixture will be $75. You have a quick total of $475 and your bank account says you could afford $500, so you go ahead with the job. You think it might take you two consecutive weekends if all goes well. Assume four killer days of 10 hours each for a total of 40 hours. (Some people's spouses automatically double their mate's estimates.) Let us see what the impact of a quick but incomplete analysis might be.

Floor Covering

You have an old carpet that is well worn out, so it goes to the trash. But your trash hauler charges you $20 to take it away. The new carpet is $15 per square yard, and your 12 ft by 12 ft room requires 16 square yards. But you would like it to be a wall-to-wall installation, and you need 48 ft of the thin wood pieces that have the embedded nails at the correct angle to hold it in place, another $40 (Fig. 3.1). You might want it installed professionally to remove any wrinkles, but we will go on a limb and say that you have the right friend with the right tool and they owe you a favor.

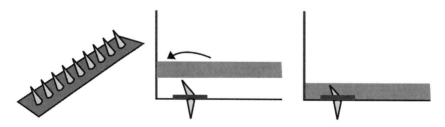

FIG. 3.1 Wall-to-wall carpet strips

Painting the Walls and Ceiling

The walls require two coats of paint so you get two gallons at $20 each. The ceiling can be done with one gallon at $20. But you need at least one paintbrush, one roller and roller pan at $20. You do have a stepladder, so no need to purchase one, but there are some nail holes that need patching and sanding first. So the patch material, putty knife, and sand paper cost you another $15 you were not planning to spend.

New Light Fixture

For some reason unknown to the author, light fixtures always seem to cost about twice the amount you plan for. Maybe it is the custom use of glass that makes even a moderate fixture somewhat expensive. So the initial guess was off, and you spend $150. We will say that you can rewire the light without giving yourself an electrical shock by the proper selection of the circuit breaker to turn off and good common sense, but you do not have a roll of electrical tape and connector nuts, so you spend $5 more. The old light fixture can hopefully be thrown out in normal trash after some effort to dismantle it.

Adding Pair of Bookshelves

You purchase the bookshelves at a discount home improvement store for $100. They are nothing fancy, but you are a practical engineer. You would like to find the studs in the wall so the shelves can handle all of the textbooks you have accumulated from your school days, or you might have to settle for some of the wall anchors that came with the shelves. The wall studs might be too hard to find, or they are located in poor locations for good room layout efficiency. The wall anchors look adequate to do the job, but do not give you the sense that they would handle all of your books. So you buy stronger wall anchors for an additional $20 and you are more content (Fig. 3.2).

Total Costs and Time Required

Remember your initial cost estimate was $475, and this amount was within your budget, so it energized you to do the project. Also remember that you and your friends' labor costs are free (except that you probably will have to feed them a few times), but if we focus only on hard purchase costs and the normal underestimates and surprises, we find that we have spent $665, an additional $190 or 40% overrun. This does not include the cost of food and beverage to keep the "free" workforce happy.

So you have to find the additional $190 from your personal budget. If this happens once every two years or so, you look upon your refinished room with pleasure and hopefully just make do. You say you will try harder next time to have a more realistic budget. You convince yourself that the free labor was worth far more than the additional $190 and you move on to something else.

Justifying Automation

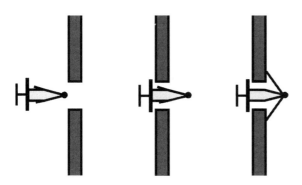

FIG. 3.2 Superior wall anchors

The time estimate of two weekends was not even addressed in the previous discussion. Can or will it be done only on two weekends and no additional evenings? It is harder to estimate how much talent you and your friends have here. But let us assume that you spend a few evening hours making purchases, getting things prepared for or cleaned up from the prime weekend time, for an additional 10 hours. This is a time overrun of 25%, but of no real consequence to a homeowner.

3.2.2. Impact on a Machine Builder

If you were running an automation building company, and you constantly have material cost overruns of 40% and time overruns of 25% of your labor, you would be out of business very quickly. When all of the accounting is done on a successful project, the machine builder might make a profit of 10–12%, so there is little room to absorb errors of this magnitude.

One might argue that the room refinishing project is the kind of job you do so infrequently you forget the lessons learned from the last time. If you did remember these lessons, you might look at your budget and decide to wait a few more months until you save more money. But the impact of your room decision is probably not going to make you bankrupt.

A company that makes all of its sales by making proposals and giving budget costs it must later work within must make better estimating efforts. We can learn from the example, however, where many of the estimating mistakes can occur. We can:

- Be swayed by the advertised price and not consider hidden costs (trash hauling);
- Use data from memory that is not very accurate (light fixture);

- Not perform enough critical calculations (wall anchor needs);
- Not define the customer specifications completely (carpet being wall to wall install);
- Underestimate processes (wall preparation before painting);
- Underestimate time to do all of the tasks (shopping, preparations, cleaning).

So our quick little example manages to hit many of the key issues project costing will grapple with, whether automation or most any project. One needs to know the project specifications and requirements involved, and to minimize the number of surprises.

3.2.3. Cost Structure for an Automation Builder

An automation machine builder may have a line of standard available products, or they may design and build custom machines, or they may do both. Reputation and word of mouth can help to initiate the sales process, but this is usually not enough. So there is a need for an internal sales force and/or a network of external sales representatives. External reps are usually independent agents who work for a commission. These commissions are paid when the sales are complete, and are in the 5–25% range, usually 10% for automation machines.

External reps may seem at first glance to be the best way to approach sales. They cost you relatively little, and you only pay them the commission when sales occur, but these reps are most likely handling a range of other products and automation machines. When they walk into the door of a potential customer, they have a large three-ring binder filled with product literature that includes your machine and maybe 10–30 other machines made by other manufacturers. So they are listening to the customer, and looking for the easiest sale combined with the biggest commission check. Some reps will be great for you, others disappointing. It takes time to select and train a good network of reps that will potentially cover the entire United States and beyond.

Cost Components Within an Automation Project

Many of the components of an automation project will probably roll off your tongue. Others are less obvious. Some can be allocated to a specific project, others are for the well-being of the company and no-one really knows to which account to charge it. Good management practice just knows it needs to be done sometime somehow. The following is a list of expenses when one deals with automation:

- Purchased components;
- Engineering design time;
- Fabrication time;
- Inventory and Work In Progress (WIP) financing;
- Profit (10–15% hopefully);

- Administrative costs;
- Fringe benefits;
- Workers compensation insurance;
- Unemployment insurance;
- Travel to see the customer before and after the sale;
- Installation costs (depending on contract agreement);
- Warrantee costs (depending on contract agreement);
- Proposal writing;
- Commissions to product reps if used;
- General marketing expenses (trade shows, literature, websites);
- Real estate related overhead;
- Contingency funds.

Some of these are worthy of note. Warrantee costs are for the normal one year parts and usually labor expenses, which might run to 3–7% depending on the machine. Travel to see the customer and write the proposal need to be included, and might appear as an administrative cost, since customers do not want to think that they are or have paid for the initial contract development work. However, it does need to get paid for from some account, and if one ignores this for multiple projects, it will eat up the profits.

The contingency funds approach to automation development is sometimes used when there are uncertain issues or true development needed, since this is a novel or one-of-a-kind machine. When both the customer and the builder acknowledge the risk factors, a range of costs needs to be covered in the proposal if things do not go optimally. So as to not create a proposal with a lot of excess built in, a contingency amount is specified. This can be handled in several ways. One way is to state that the amount (perhaps 10–20% of the total amount) will not be used without the customer signing off at an agreed-upon milestone. Or the amount might be approved, but the builder agrees to provide documentation to prove the need, or is compelled to give a rebate at the project's conclusion.

Groups Involved in Automation Development and Payments Timing

There are many small and larger variations to the norm, but here is one possible scenario on how automation is developed and paid for:

1. The internal sales force and/or external sales reps bring in a Request For Proposal (RFP).
2. The engineering and management teams use experience and financial data to make a step-by-step assessment and develop the proposal.
3. The proposal is submitted to the potential customer by the sales force or reps.

4. The proposal is hopefully selected from competition: 5–30% success depending on market, customer type, and the purchase cycle timeframe.
5. Depending on the customer's budget cycle, the selected proposal may get approval in a few days or a few months. Most proposals have an "effective until" date.
6. The customer issues a Purchase Order and may be required to place a down payment (20–30%, negotiable).
7. The management team dusts off the proposal and implements the project timeline.
8. Engineering design and development produces CAD drawings. They generate a list of off-the-shelf components and subcontracts for custom parts for the Purchasing Department.
9. Manufacturing produces any in-house components.
10. Management determines timing for purchases deliveries and internal production and schedules assembly. It notifies the customer if some component lead time will impact delivery.
11. Assembly builds the machine.
12. Engineering works on the debugging procedure.
13. Optional Progress Payment may be required depending on contract language (20–30%, negotiable).
14. Quality Assurance meanwhile inspects internally made components and any critical purchased components where needed.
15. The customer comes to visit the machine for a performance review.
16. Major Payment is made, bringing things to the 85–90% paid level.
17. The approved machine is shipped to the customer site and installed.
18. A final check is issued (10%) when the customer is happy or begrudgingly agrees to pay.

There are often changes to the automation machine design because this is usually such a long process (4–12 months). A machine in early testing might show some performance that was originally thought to be great, but is now no longer desired. Or the product to be assembled is modified. Authorization needs to be given to deviate from the original specifications. So an Engineering Change Order (ECO) is issued, and depending on who issued it and why, some additional payment may be required. During the first development of a machine, whether it is $50,000 or a multimillion dollar line, if the change orders from the customer comprise less than 10%, things are proceeding normally.

3.3. WIN–WIN PURCHASING PHILOSOPHY

What is another approach one could take towards costing and sales? Instead of just using cost plus 10%, one could know what the project is worth to the

Justifying Automation

customer. If the customer shares enough information with the machine builder, they can work together to see what are the most important needs to be addressed by each side, and recognize that each side should benefit. This has been referred to as a "Win–Win" situation. Here, the customer states how they will save money by replacing labor and reducing the other less tangible costs. The machine builder shares enough about his pricing and component costs that both sides can agree on a price where a profit can be made but the customer does not feel that they are being gouged.

This is a wonderful situation when it can occur. It does rely on both sides being open, and this sometimes can be quite a game. One of the required pieces of information is how many people will the automation be replacing. How much does the customer's accounting method attribute to each person being replaced? What are the current Workman's Compensation expenses for each worker? What timeframe does the company require for a project to be effectively paid back?

Answers to these questions can easily be put into a spreadsheet to determine the potential savings. Then the automation builder can determine what the value is to the customer. If everyone is up front, a price point where both sides make or save money is the "Win–Win" situation. The builder will prosper enough that they will survive and be able to support the currently purchased machine for years, make additional machines for future sales, and all is good in the world.

However, the games can begin if the customer wants to keep more of the potential savings. They may "stretch the truth" about how many workers currently are performing the job. They may not be forthcoming about a second or third shift. Finding the correct numbers to enter into the spreadsheet can be difficult. The author has seen this happen several times personally. A true "Win–Win" needs more honesty than this. It needs bilateral trust.

3.4. MAXIMUM PROFIT COST ESTIMATING FOR BUILDING AND SELLING AUTOMATION

Some automation builder companies are aggressive towards every dollar they can get. They are not concerned about "Win–Win" as much as growing their balance sheet. They know how many workers their machine will replace. They know they have the only solution available. They have filed for a provisional patent on their concept and will aggressively defend it. They know that having the leg up on the competition does not last for decades, so one must go for the brass ring with every sale.

So, with this strategy, the price for the automation is just short of the maximum that the customer will be able to justify. They know that there will be repeat sales, since they have their customers over a barrel. The first machine designed and built costs more than the second and third one. Then who cares if building a second machine is only $15,000 when the repeat customer can cost justify it for $65,000. The extra $50,000 is all profit and the machine builder wins big time! But some customers will begin to realize the game that is being played here, and sometimes will try to find other sources to make future copies of your machine.

Another clever strategy that may not seem as outwardly aggressive can be found in the packaging area. A popular new product package has a strong set of international patents for protection. The automation machine and the required metal backed paper used to make the product are both covered by these patents. The machine builder will lease the machine for something unbelievable like $1 a day, but requires users to buy their special paper at an outlandish price. And because no-one else can supply this special paper, they have a monopoly. World acceptance of the packaged product helps to drive the monopoly towards great profits, yet their machines are seen as cheap. This ends up being extremely expensive once the total costs are calculated.

These profits are seen as so great that other forms of competing packaging are being thrust upon the consumers. Some potential automation packaging customers refuse to buckle to the high costs of the monopoly. The new competing product is now making headway, but the owners of the monopoly are still in a good financial situation.

So these aggressive companies, following no matter which strategy, are looking to make as much money as possible, for as long as it lasts. In their defense, the consumer market is so flighty at times that one cannot blame them. They may be out of business in 1–5 years if they do not expand their product line, but that is not the main goal — making money is.

3.5. JUSTIFYING FLEXIBLE AUTOMATION OVER HARD AUTOMATION

Many of the machines described in this text can be classified as "hard" automation; that is, they cannot be easily modified and reprogrammed to do another task. A hard automation machine that assembles car stereo faceplates will not be able to assemble a car generator. Hopefully the hard automation has been optimized to crank out the product faster than what is referred to as "flexible" automation. Flexible automation usually means that there are one or more industrial robot arms as the key elements in the workcell.

As one can imagine, there are financial scenarios when either type of automation can shine over the other. Part of this can be seen as the reuse of a significant part of the automation cell vs. the option of creating a potential boat anchor when one creates a single-task machine, but if that single task is to be performed

Justifying Automation 41

for many years (such as the assembly of spark plugs), then the flexible automation may or may not be able to keep pace.

Brian Carlisle, CEO of Adept Technologies, presented some business research (Carlisle, 1999) he conducted in the late 1990s. Adept is a leading U.S. manufacturer of assembly robots, so it was in his best interest to show how flexible automation can be the best choice. He also compares automation to manual production, but raises some good points no matter which side one is trying to take. Figures 3.3–3.6 shows some of his findings dealing with financial justification. His key observations are:

- Most automation engineers do not have a readily available financial comparison tool;
- Quantifying flexibility had not been done before his research;
- Depreciation was often used as a hindrance factor by accountants, but does not have much impact;
- Rarely can a company make up financially for lost time if changeover takes days or weeks.

Potential customers have to know how a company carries out internal finances, what it wants out of automation, and what the range of possible products to be manufactured will be in the next few years. Without this information, it is tough to justify anything.

3.6. INTELLECTUAL PROPERTY, PATENTS, AND TRADE SECRETS

Intellectual property is the term used for the collective group of patents, trademarks, copyrights, and trade secrets. Depending on a company's size and sales volume and the market conditions, intellectual property can be one of the strongest assets a company can own. These assets can be created, purchased, or sold. They are the benchmark of many startup companies and their investor funding potential. They can also keep a room full of lawyers employed for decades. Many times when an employee starts work at a company, they are required to sign away any or all of their intellectual property rights. Check with your

---

- Explain Important Financial Metrics
- Provide Tool to Automation Engineers
- Compare Flexible and Hard Automation to Manual
- Help Sell Proposal

FIG. 3.3 Purpose of justification analysis (from Carlisle, 1999)

- Quantify Value of Flexibility
- Changeover Cost Analysis
- Incremental Capacity
- Show Relative Sensitivity of Key Assumptions

FIG. 3.4 Generate justification automatically (from Carlisle, 1999)

company if you are currently employed. Or if you have been employed, your past company may hold certain rights for some time after you leave.

Each of these terms will be briefly discussed, with focus on what is common to automation companies and their customers.

3.6.1. Patents

The granting of a patent from the United States Patent and Trademark Office (USPTO) can be one of your career's highlights. It means that you have invented something new and novel that no one else has thought about before. It gives the assignee (you or your company) specific rights in ownership of the idea in bringing the concept to market. If some other person or company tries to use your idea in the protected timeframe, you can go after them in a court of law. There are a significant number of rules that will not be addressed here, so the reader is referred to the USPTO website (www.uspto.gov).

The key point in an invention is that it must be something that is not currently available in the market, that has not been patented before, and is not an obvious extension of these existing works to one who is a practicing engineer in the field. This means if someone takes a good current product and simply paints it red, one cannot get patent.

- Changeover Scenario
- Changeover Times and Ramp-up Rates
- Is Revenue Lost During Changeover Recoverable?
- Special Packaging for Automation
- Cost of Quality
- Financial Impact of Errors
- Cost of Labor
- Utilization

FIG. 3.5 Most important assumptions (from Carlisle, 1999)

Justifying Automation

- Depreciation Life!!
- Depends on Profits and Tax Rate
- Cost of Utilities
- Cost of Floor Space
- Cost of Maintenance

FIG. 3.6 Least important assumptions (from Carlisle, 1999)

Patents do not cost a great amount strictly from the USPTO fees structure, but to properly interface with the USPTO office and to fulfill all of the regulations and rules on style and types of drawings, etc., one often relies on a patent attorney. This normally brings the cost of getting a U.S. patent to something between $5000 and $20,000 on average. International patents can range from $3000 to $20,000 for each country in which one thinks that there is a viable market. So if one foolishly thinks that a new patent on space heaters will sell well in the Amazon region, then one can waste the dollars getting a patent there.

However, for every U.S. patent issued, perhaps only 5–10% of them actually generate any income. The patent process takes time, energy, and money. Inventors always want each of their inventions to be patented for a chance to get rich, but the market is fickle, and other people come up with a better method, a better machine, and therefore a better patent before one has recouped the costs of one's own patent. There are a tremendous number of great ideas one can see by searching the USPO website database. It is free of charge. And it will tell you if a patent has expired or has been abandoned. It will not tell you if the idea ever came to market, however.

Patents that are issued have a renewal fee every three to four years. So if a patent is not profitable, it may be left to expire by the inventor or his company. This leaves it open to the general public to use free of royalties or other financial obligations. It makes for interesting surfing of the net to see what useful and crazy machines have been designed, but possible never built! One does not have to build a machine to get a patent on it.

3.6.2. Trademarks

Trademarks are the name or term that a product is marketed under. It helps tremendously to establish brand recognition, and to assure repeat customers. It is something that everyone needs to respect so as to not violate the trademark laws, but this is something usually left to the marketing department. At the level most automation engineers work at and conduct their business with customers, there is little to be concerned about trademarks. That is the job of Marketing.

3.6.3. Copyrights

Copyrights have to do with written documents, music, photos, and similar products. On the surface, they would seem to have little to do with automation, except when one gets to the point to write the user manual for the machine one has built. However, there are several tricks sometimes used that have not been necessarily proven to be useful, at least not yet.

Some companies will copyright the proposals they submit to a potential customer for protection. It is not unheard of that a shady customer likes the concept given in a proposal, but does not like the pricing. The customer might try to shop around the proposal concept to other manufacturers to see if someone else can make the same machine for a lower price. Or the customer decides to make the proposed machine internally. Both of these practices are very unethical, but unfortunately not unusual. It is common enough to make proposal writers very anxious. So the proposal is marked with a "Copyright" designation, in the hopes that if some other entity tries to use this proposal concept, it would be infringing. This approach seems to have limited legal effect other than to possibly scare away the potential new manufacturer, who might have been duped into trying to quote someone else's concepts.

3.6.4. Trade Secrets

When a company has a machine that represents a large commitment of time and dollars, and it is believed to be vastly superior to commercially available machines, that company may not try to patent it at all. It will just keep it as a trade secret. That is because one must present all of the technology in the patent disclosure, and it is then available for the competition to review and perhaps improve enough to warrant a new patent. With Trade Secret policies, employees are bound by agreements not to disclose the technology under penalty. Visitors are not allowed to see the machine. No-one takes photographs. It is simply a well-kept secret.

This strategy is most likely less expensive than obtaining and defending U.S. and international patents. One small company known to the author budgets $50,000 per year to move a patent along in the international process (which takes more time than the U.S. process) and to defend the patent when it is awarded. This is for a single invention! Some people say a patent has not really proven itself until it has been defended from another party in court, and patent litigation is very expensive.

3.6.5. Intellectual Property for Automation

With all of the expense of obtaining and defending a patent, it may not be the right thing for all machines. First, if it is in an area that is very mature, the core patents may have expired 20–30 years ago. To make a better machine means you have to try some radical new way to do it. Secondly, one has to generate enough profit to pay for the patent work. This most likely will not happen if you are making a single special purpose machine, and therefore unable to spread out your engineering design and development costs. So if you have a product that your automation building company is developing, and you plan on selling dozens to hundreds of them, then it may be fruitful to proceed with a patent.

In today's climate for contracts and business negotiations, often the customer who is contracting for a piece of custom automation will want the intellectual property, including the patent rights. Some customers might let you retain the patent rights if they get their machine for a cheaper price, but this is usually in a situation where if their competition gets a copy of the same machine, the first customer feels like they will still have the superior product. Or they may allow you to sell into markets that do not compete with their defined product line.

3.7. CONCLUSIONS

Whether the current automation challenge before you is the first or the 100th machine to design, one should look at all of the costing components in Sec. 3.2.3. Many of these 18 steps will become obvious after a few times down this road, but if one changes employment between the previous automation design and the current project, a designer can be surprised by how many of these steps can be different due to company policies and culture.

When it comes time to plug some numbers into these categories, it is always best to have someone else review the suggested numerical entries. Even on your 100th machine you may forget to include some sensor or actuator that just seems to have slipped your attention. This process of costing a project can turn a fun and potentially profitable time into a nightmare that seems to last for years!

PROBLEMS

1. Create a cost and time estimate for:
    a) A five-course dinner.
    b) A party for 10–12 people.
    c) Refinishing a room.
    d) Some other event.
    Keep notes on costs, time, and unfulfilled expectations. Look at your estimating time as a function of the total time. How did you handle any cost overruns?

2. Search the USPTO website database (www.uspto.gov) and try to find a machine that:
   a) Mows the lawn.
   b) Washes the dishes.
   c) Does laundry.
   Which of these correspond to consumer products than you have seen? How do these differ from commercial products?
3. Types of sales forces:
   a) How do you deal with commission-based sales people, like car dealers or some large appliance dealers?
   b) How do you react with full-time sales staff, such as a waiter in a restaurant?
   c) Who does the better job and why in your opinion?
4. You have just invented a new automation machine that everyone will need in their home. It peels and slices fruit with no human intervention. What sales method would you use and why?
   a) Dedicated sales force going door to door.
   b) Commission-based sales force in shopping mall kiosks.
   c) Phone operators and television infomercials.
   d) Internet-based sales provider.
   What advertising would you need to support your selected sales method? What would make the everyday homeowner want to part with $100 for your device?

PROJECT ASSIGNMENT

Using one of the projects from the Appendix (or any other projects), perform the following:

1. Estimate the current human operator cost of processing the product. Use the following labor rates (These include estimated company overhead rates, for unemployment insurance, worker's compensation, social security, etc. These are not the take-home pay rates of the fictitious employee):

Food Service	$10.00/hr;
Light Assembly	$14.00/hr;
Machine Operator	$20.00/hr.

   To do this assignment, you most likely do not have enough hard facts to be totally accurate. Perform a mock processing of the tasks and have someone record the time needed.

2. Create a list of probable reasons why one should automate this process. Concerns should include the relative likelihood that manually there would be:

    Improvement of quality;
    Reduction of repetitive motion injuries;
    Improved productivity;
    Ability to meet expanding market.

3. By looking at the probable steps you developed in Chapter 2, make an educated guess on what the parts and labor will be to make such a machine. There is not sufficient information given in this text in Chapters 2 and 3 so far to do this accurately, but view this task as an interesting estimate that you will look at again after you are done with the total design. It is interesting to see how sometimes one can be fairly close to the actual cost with limited knowledge, and yet other times are off by an order of magnitude.

REFERENCE

Carlisle, B. (1999). Financial justification analysis for flexible automation. Adept Technology System Integrators Conference. Monterey, CA, January.

4

The Automation Design Process

In Chapter 2 we looked at the Steps to Automation, where we needed to discover the appropriate process that was to be implemented by the automation machinery yet to be designed. These steps helped to establish the best way to do something, but not how the machine should be configured. One could envision a totally novel machine configuration to solve such a problem, but this is a great undertaking. It may be better to first look at commonly found configurations, and investigate the best available choices before becoming very creative. There are many components of these commonly found layouts that are readily available, and should be more cost-effective and timelier compared to starting from scratch.

In Sec. 3.2.3, we looked at the 18 steps commonly found with "Groups Involved in Automation Development and Payments Timing." In Step 2, the engineering and management team use experience and financial data to make a step-by-step assessment and develop the proposal. This also includes the need to understand the process. It is interesting that this work gets done without any dollars coming in from the potential customer. The team has no guarantee at this point that they will get the work. So this planning work may be done half heartedly or as a pipe dream if another company is seen as a lock (and it is surprising when the lock does not win and you do get the job once in a while).

So it is understandable that many new automation machines are not developed beyond what has been done before. When a company does get the

The Automation Design Process

successful bid, it might look at changing or improving the machine layout, but sometimes the customer is locked into the design that was pitched in the proposal, if only for emotional reasons. Thus, the proposal writing stage is very critical, usually underfunded at the automation company, and often is not given the time it deserves.

4.1. SYSTEM SPECIFICATIONS

When the Request for Proposal (RFP) or Request for Quote (RFQ) comes to the engineering and management team, someone somewhere must set some systems specifications, or there will be great disappointment when the machine is turned on. Someone needs to establish things like:

- Throughput rates;
- The processed product's size and weight;
- How to determine out-of-tolerance products and what to do with them;
- Who is responsible to guarantee that the incoming product is within tolerance;
- Available electrical and compressed air facilities;
- Floor capacity — in weight per square foot;
- Floor space available and ceiling height;
- Pillars and other obstacles to be negotiated;
- Noise levels;
- Available entry passage to desired machine site;
- Expected machine life — Mean Time to Repair;
- Safety procedures at the desired site including relevant ANSI standards and local codes.

The absence of any of these specifications can haunt an engineer, lead to lost profits, and make one's life miserable. Imagine, for example, that the customer has a limited entry way into their factory. Perhaps the only accessible doorway is just 5 ft wide, and one's machine is 6 ft wide! If known before hand, it might have been possible to make the machine narrower, or to make it in modules that unbolt for travel and are later reassembled during installation. Or if not possible, the customer might have made a temporary opening and/or installed a bigger doorway if they had several months to think about it and implement it. But if your new machine is on the truck waiting to get into the factory, and then someone notices that the doorway is too small, everyone gets angry and starts pointing fingers at each other. It is too late then to think about a narrower machine or a different mechanical architecture. It is not unusual for a machine to take a specific form due to it needing to be virtually wedged into an existing building, possible around a pillar or two.

If the customer will not provide you with system specification, you can take one of three paths:

- Ask enough questions of the customer to write them yourself and include them with the proposal.
- Make intelligent guesses on your part so as to include them with the proposal and get the customer to sign.
- Walk away from the RFP.

You would prefer that you copy the list the customer has provided in the RFP as an appendix to your proposal. The third choice may sound unreasonable in a competitive business world, but your potential profits may evaporate, and an unhappy customer can create significant negative marketing that may take years to overcome. We will look at specifications in more detail in Chapter 12.

So you never want to submit a proposal without a list of system specifications. You do not want to start designing or building a possible machine configuration without a system specification either. However, life is not always this simple. Sometimes situations place the cart before the horse and you must adapt if you want to keep your job.

4.2. BRAINSTORMING

It is hard initially to differentiate between the discussions in Chapter 2, where we try to understand the process to be automated, and brainstorming on how the machine architecture should be laid out. The process to be automated will define some operations, the required local motion, and fixturing needs. It may also lead to one or more likely candidates from the range of traditional machine configurations possible. But a predetermined machine layout may limit the methods investigated in understanding the process, and this action might be fatal in terms of project success. For example, to state beforehand that a desired solution must use an industrial robot similar to what one sees welding car bodies in television commercials is most likely not wise, but major automation users such as the USPS have done just that in the late 1990s due to upper management edicts (Derby, 2000).

Brainstorming about new automation machines heavily relies on one's experiences and the review of existing machines, and from resources such as this text. However, as has been stated for various reasons above, very few really "new" sections of these machines are designed if at all possible. It would be desirable that a machine builder with great experience looks at their CAD database and borrow a good chunk of many previous machines to get a reasonably optimal performance. The true optimal may be a luxury. They do not want to perform what is usually referred to as "research" as found at many universities. It is too risky and takes too long. So the brainstorming done is usually on the limited

The Automation Design Process

side compared to what is often found in new product development. These totally new creative developments have their high costs, where the first unit development costs range from three to 10 times unit production costs.

Brainstorming does happen when a machine builder sees a market need that cannot be addressed with a traditional machine (since many builders often seem to clone each others work anyway). The builder sees the opportunity as a potential for great profits if an investment is made from their internal funds. Or the new concept may come from an independent inventor or entrepreneur who is looking for a partner. We will see more of this in Chapter 11, "Bringing New Automation to Market."

4.3. MACHINE CLASSIFICATION BY FUNCTION

Traditional automation machine classification is often based either on its primary function or by the form of the material handling system at its foundation. We will review the most common forms here, add some insight into their usefulness over the past 20–30 years, and comment on their potential for the future.

There are a wide number of functions that a machine can perform, but if one looks at the history of what has been built to date, and one was to try to classify the significant groupings that would result, several major classifications would emerge:

- Assembly;
- Inspection;
- Test;
- Packaging;
- Computer Numerical Control (CNC) Machine.

The concept of the last classification, CNC Machine, does not seem to fit well into the general theme of this text. And with good reason, it does not, but it is included in this list for completeness because many of our working definitions are not very legalistic in nature, and the CNC Machine can get included, if even by mistake.

The term CNC stands for computer numerical control, as opposed to manual operation. The grouping includes milling machines, lathes, nibbling machines, surfaces grinders, and many more. Within the term CNC, many include internal automatic tool changers and material handling systems to bring in raw parts and remove finished goods. The traditional concepts used today do have some overlap with the design of automatic machinery, and this text can still be an aide, but the machine tool industry is more structured in its choices of controllers, languages and intense knowledge of product feeds and material removal rates that need to be addressed by additional technical resources.

4.3.1. Assembly Machines

Assembly Machines as a group can range from the production of a high-volume part such as a spark plug or a piece of home kitchen cabinet hardware, to the construction of a cell phone. Throughput rates and product flexibility expectations can vary. The machine to assemble a J8 model sparkplug most likely can operate for 10–20 years without any required changeover. J8 sparkplugs will live on forever. But the market novelty for a particular model of a cell phone can be six to nine months. Changeover is required if the automation is to be kept from being a boat anchor. And it most likely is not cost justified in such a short timeframe with today's competition.

Figure 4.1 shows a generic Assembly Machine. There is the base part of the product brought into the workcell in some fashion. Parts feeders are used for the components to be added to the base part. And another method of transferring out the completed assembly is usually required. If the assembled part is a hinge for your home kitchen cabinets, then the output could be simply dumping them into a bin. (We shall see later that this may be a costly error, losing the part's position and orientation, only to have to redetermine it again.)

4.3.2. Inspection Machines

Inspection Machines (Fig. 4.2) are significantly fewer in number than Assembly Machines. Although in-process inspection is currently desired even more, inspection is often performed as an integral operation within the Assembly Machine. Computer vision systems and dimensional measurements are two of the commonly found inspections performed within the Assembly Machine.

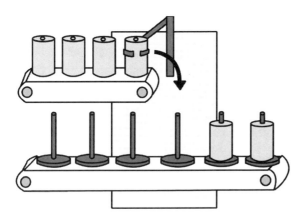

Fig. 4.1 Generic assembly machine

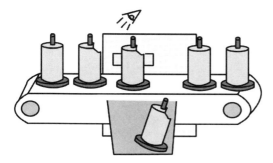

FIG. 4.2 Inspection machine

However, stand-alone Inspection Machines for checking a packaged product for the correct weight (check-weighers) and making sure no metal filing from all of the food processing machines fall into your box of corn flakes (metal detectors) do exist and have meaningful niche markets.

Most of these Inspection Machines generically have a product inflow, a checking station, and two outflows. One of these outflows is the good product (hopefully 99.9% or better) and the other outflow is for defective products. Depending on the product and its defect, the defective product may still be sold. If it is a food product and is only underweight, it can be eaten, possibly showing up in a factory seconds store. Obviously, products with metal filings are recycled, burned for heat value, or thrown away.

The inspection process in one of these machines (or within another machine) can include:

- Checking one or more dimensions with mechanical gauging or electrical sensor.
- Checking one or more dimensions or features using a vision system.
- Checking weight for correct amount.
- Checking a liquid's volume by weight or level.
- Checking a filled Stand up Pouch (SUP) for leaks.
- Checking a product for metal filings, etc.
- Checking a cereal box for the free prize inside.

The results from all of these can range from health risks (metal filings) to disappointed customers (leaky SUPs). Perhaps the most dangerous situation is when a child gets to the bottom of a box of their favorite cereal and the free prize is missing!

4.3.3. Test Machines

Machines that conduct some performance check on the filled, assembled, or processed product are sometimes referred to as Test Machines. Although some might argue that testing is part of inspection, the distinguishing feature is often the cycling of the product in some or all of its designed operation. In other words, an Inspection Machine functions by either a noncontact mode, or with a simple contact where some measurement or property is determined. A Test Machine makes the product do some action or work, such as cycling a spray head from a hand-powered misting bottle. The test is carried out on either a random basis, or on every spray head if trouble has been observed in the past, but the spray head is either passed as working properly, or is rejected.

As opposed to Inspection Machines being potentially integrated into an Assembly Machine, most Test Machines are separate from the assembly process. Test machines are often highly specialized to the product being assembled or processed, and the devices used to perform the test and to judge the results cannot be easily integrated into the other machines. Figure 4.3 shows a Test Machine that is checking the previously assembled widget to see if it will hold together or whether it will fall apart. Following the example of testing the spray head, there would need to be devices to move a single spray head into the test station in the correct orientation, an actuator to perform the test, a device to advance spray heads that pass the test, and another device to dump a rejected spray head into a hopper. The controller may need to be smart enough to allow the spray head to be actuated a variable number of times, so as not to reject heads that are good but not the best performers. Remember here, we are talking about something that goes on top of a bottle of kitchen cleaner, not a $250 cell phone. So if a spray head does work with one or two extra pumps, the company's marketing group may say to keep it, but if it works in fewer pumps, do not waste time and pass it.

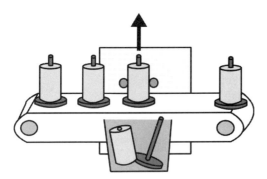

FIG. 4.3 Test machine

The Automation Design Process 55

Another option, if the process takes significant time and the product throughput rate is significant, is to test a set of products as a group or gang at one time. Perhaps the same pump spray actuator can be used for six spray heads at a time. However, six separate sensors will be needed to judge each spray head's individual performance, and six different actuators are needed to dump the bad product selectively. No-one can afford to dump all six spray heads if only one is bad.

4.3.4. Packaging Machines

Any finished consumer product of any value gets packaged in one of many different types of packages. It can be bags, boxes, cartons, SUPs, aseptic boxes, and more. None of these packages significantly improves the performance of the product inside, but the packaging does help the consumer understand the product, differentiate the product from the competition, and improve sales dramatically. One packaging user's group even had the 2001 slogan, "Packaging Matters."

Commonly found Packaging Machines include:

- Closing filled corrugated cardboard boxes;
- Filling bottles with liquids;
- Filling bags with dry products;
- Placing products into cartons;
- Weighing products for accuracy;
- Metal detection for safe consumption by consumer.

As you will note, the last two machines have been discussed in the previous section. Trying to segment the market and production of automation is a gray area.

Packaging Machines are such a large and important group that they will be covered in some detail in Chapter 13.

4.4. MACHINE CLASSIFICATION BY TRANSFER METHOD

Two of the chief concerns with any new automation design are how many copies of the important processing devices, fixtures, etc., can one afford to build, and what is the cycle time. This will become more apparent as we look at the different configurations and trade-offs.

4.4.1. Linear Indexing

Indexing motion is where a line of products in a similar state is moved into a location to have a process carried out. The spacing of the products is at a constant interval, and if an empty space occurs from a missing product, appropriate

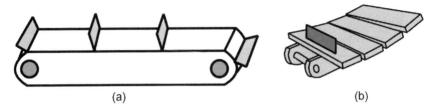

FIG. 4.4 Flighted linear conveyor: (a) continuous belt; (b) segmented

measures may need to be taken. For example, if a bottle is missing from an indexing conveyor, automation should not attempt to dispense 20 ounces of beverage, resulting in a mess on the transport and the floor.

The two commonly linear indexing configurations are:

- Conveyors;
- Walking beams.

Conveyors can comprise a continuous belt or be made up of many sections or segments similar to a tank tread, but either design needs to have an attached bar or set of features to assure that the transported product is located at the appropriate place when the process occurs. Figure 4.4 shows both types. These features on a conveyor will classify it as a "flighted" conveyor. Some automation producers will refer to them by the alternate spelling — "flited." A flighted conveyor is also useful in transferring product up and down inclines without the product's location and orientation becoming unknown. Some segmented flighted conveyors have customer-specific designed features to facilitate positive locating of a particular product to be transferred.

There are probably 1000 different types of conveyors and options available. We will look at them in more detail in Chapter 7.

As with most continuous conveyors, the belt or series of segments form a contiguous chain that requires a return path to form the total loop. Most often the return path is directly under the working surface. This is usually satisfactory, but does mean that more than twice the length of flighted belt is required to be designed and purchased. For belt conveyors, the rollers on the ends are often tapered so that the belt is likely to stay centered and will not drift to one side or the other. These rollers usually have adjusting screws to keep the centering correct for belt wear.

Conveyors do not work in all applications. Sometimes the expense of the customized flighted belt is too high, or, more likely, the surrounding process or system obstacles do not support a standard belt design. Possible belt conveyor issues include:

The Automation Design Process

- The rollers at the ends of the conveyor where the belt reverses direction can be a space issue.
- The rollers may be in the way as product is loaded at the start of the conveyor or unloaded at the end.
- The segments are large, so the rollers would have to be extremely large and costly to flip the segments around.

Thus there is the need for the Walking Beam. A Walking Beam will transfer a part from one set location to the next location along the index path, and can work right up to the end of the machine. There are no return rollers, so transferring larger parts is not an issue. Often there are some registration features to keep the parts in known positions when the Walking Beam is not in contact over a small part of its cycle. It is usually during this noncontact part of the cycle that a useful process or subsequent transferring occurs.

In Fig. 4.5 the part transferring is performed by the lifting devices that are represented by the shape similar to the capital letter "T". After replacing the parts (shown in the lowest row of the sequence), the lifting device then loops back to their starting position in order to repeat the cycle.

A side concern with an index machine and its implied coordinated or synchronous motion is the proper timing of all of the operations above the con-

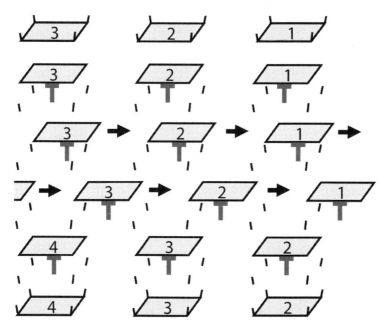

FIG. 4.5 Walking beam motion sequence: lift, shift, and place

veyor or Walking Beam. Some machines have avoided the use of sensors and multiple controllers by resorting to a common drive shaft for internal motion devices (Fig. 4.6). This is a somewhat historical approach since this dates back to waterwheels driven by a local river in the 1700s and 1800s. Today, a single large motor drives a large driveshaft that has gears, chains, and timing belts and that powers all coordinated motion. This is still a viable answer for the right application, but does lend itself to the potential of tremendous frictional and safety shielding problems.

As one can image, the use of a common driveshaft can either produce a machine that can last for years, or one that is a perpetual headache. The ability to adjust for a slightly smaller batch of product may not be available with a common drive shaft. Any adjustment in timing and action may be limited. There is little flexibility compared to what is available from good robot software programming.

4.4.2. Linear Continuous

As the name implies, a linear continuous conveyor transfers the product. This is often used for rigid products that can accumulate and are not fragile if they bump into one another, or it can be designed to handle a multitude of products while they are kept separate from each other. It can be one of the lower cost implementations of the many configurations discussed in this text. There are several different modes worth exploring here.

Product Guided by Rails

This method works great for rigid objects, such as empty bottles, that can be awaiting the filling operation. The conveyor belt is usually "slippery" compared to belts found in supermarket checkouts. The product will move reliably on the belt until a movable stop comes into place to register a product's location. Figure 4.7 shows such a situation. The other bottles coming into the filling area are constrained by the guide rails, but their spacing is not regulated, and is not of any concern. The next in line to be filled bottle may be pressing right up against the filled bottle, if this causes no difficulty. Or if it somehow is not

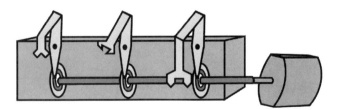

FIG. 4.6 Common drive shaft

The Automation Design Process

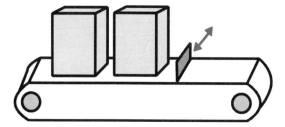

FIG. 4.7 Movable stop on continuously moving conveyor belt

satisfactory (such as it would disturb the weighing process), a second stop can be located previous to the filling stop, to regulate the traffic (Fig. 4.8).

The system of rails makes it easy to adjust from one size bottle to the next. Most often this adjustment is done cost effectively with a technician using a wrench. This assumes that product changeover is every few days or perhaps every shift. When product size changes more often than once a shift, systems with motorized adjustable rails can be cost justified due to less changeover time.

Continuous Product Without Position Registration

A simpler version is a standard conveyor belt system, where the product is spread out in two dimensions to undergo a process step. A chain-style belt could be used to transfer pretzels into a waterfall of melted chocolate (Fig. 4.9). The chain holes allow the excess chocolate to pour through. The covered pretzels would then be transported to a quick chilling tunnel so as to harden them. The location of each pretzel is not very important, as long as any two pretzels are not touching or are so close together such that the chocolate coating process would produce a double (yet very tasty) treat.

The key to loading this conveyor system is to keep a reasonable rate of pretzels flowing onto the conveyor, while guaranteeing them from being too close. For some automation systems, one could use a computer vision system to assure reasonable locations, perhaps for chrome plating parts, or for something else expensive. For something like a chocolate covered pretzels, one might find an empirical method to have success more cost effectively, and live with the once every five to 10 minute problem of a double-coated treat.

Vision Guided Product Processing

The most difficult of the Linear Continuous methods is probably the most exciting to watch, when it has been implemented well. Imagine a higher priced sandwich cookie, where on the conveyor belt one half of the cookie outside is lying randomly on a continuous conveyor, outside up. A second cookie half, outside down, has had a layer of chocolate and raspberry filling placed on it, and it is

Fig. 4.8 Two movable stops to regulate traffic flow

still warm and gooey. A robot system with a computer vision system guiding it, matches a gooey half to a top half, and places the top half on the gooey half, all while the conveyor and cookies are in motion.

This task used to be done by humans, but has been automated since the late 1980s in some factories (Fig. 4.10). However, there are still many food and other product-related factories not automated since the operation is more difficult than matching cookie halves, or the automation costs are too high.

Vision guided assembly while the conveyor is in motion is far more difficult than lining up two cookie halves. The tolerances for cookies are fairly sloppy compared to cell phone component assembly. Most precision operations are therefore not recommended to be approached using this method.

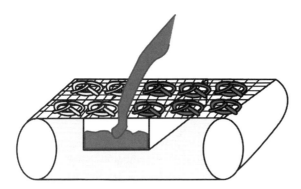

Fig. 4.9 Continuous products: pretzels being coated by melted chocolate waterfall

The Automation Design Process

FIG. 4.10 Vision guided robots matching cookie halves

Before committing to this approach, since the expense and technology is relatively high, one should ask the question, "Can it be done in batches while stationary?" The answer may be that this batch approach will imply a very fast indexing of the next batch to be processed. However, in some cases people just seem enamored with the technology when there is no logical need for it. In a similar situation, when industrial robots were developing, every manufacturer wanted the robot to track the motion of a product on the moving conveyor. In a very few instances (like welding car bodies moving at a fixed rate) this does make total sense, but as robots have evolved, most robots no longer have the tracking option. The other ways to do the tasks are more reliable and cost-effective!

4.4.3. Linear Asynchronous

If the product being processed is somewhat fragile, and the operation is more precise, or the product needs to be processed while standing on its end (and is therefore dynamically unstable), a pallet or puck system can be used (Figs. 4.11 and 4.12).

The Pucks or Pallets can be machined from plastic or metal. Depending on the processing tasks required precision, they can be injection molded. They are propelled along the conveyor by either a single "slippery" belt or a pair of them. Pallets usually have some precision features that allow for movable pins connected to the conveyor frame to engage into the pallet, and lock it into a proper location. The belt(s) keep moving, allowing the pallet to slip, or in some designs, the pins actually lift the pallet off the belt(s). There are many different versions, many of them patented.

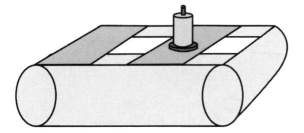

FIG. 4.11 Conveyor with an empty pallet and a pallet with product on it

Pucks hold product for filling or other material handling requirements. Many calculator and hearing aid batteries are "activated" when a foil pull-tab is removed from a hole on one side (Fig. 4.13). Air is involved in the chemical reaction that generates electricity. If the foil tab were damaged in the packaging process, one would potentially be purchasing a dead battery before it ever gets into your calculator. So some manufacturers use small pucks while processing in their factories to guarantee that one battery does not displace a neighboring battery's foil tab. Then the battery is pushed out of the puck only when grabbed by a suction cup in the final packaging machine.

4.4.4. Rotary Indexing

One of the oldest and still popular automation configurations is the Rotary Indexing machine, often called a "Dial Machine" (Fig. 4.14). It consists of a round plate of steel anywhere from 18 in. to 10 ft in diameter. The wheel has identical stations for fixturing a part's base, and as the wheel is indexed in a motion increment so as one station moves to the next processing location, an additional part is added to it, or a testing process occurs. The number of stations on the wheel is

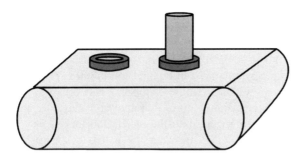

FIG. 4.12 Conveyor with an empty puck and an empty tube to be filled in a puck

The Automation Design Process

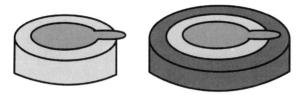

Fig. 4.13 (a) Calculator battery; (b) calculator battery in a puck

usually an even number, since most suppliers of the wheel base and its internal gearbox have standardized on even numbers. Anywhere from four stations to 16 and 20 are available, depending on the supplier. Since the incremental rotary motion needs to be quite precise, an average automation firm does not want to waste time creating such a precise and large component as the wheel and gearbox. So there are several popular models available, many of which have been around for decades.

Each station on the wheel needs to be aligned consistently, from number 1 to the possible number 20. So demanding machining is required, and, if need be, individual alignment to an external reference point. Externally mounted processing devices will interact (assemble, test, load, unload) with the station next to it when the dial wheel comes to rest. The last thing an automation engineer wants to see is a forklift back into his Dial Machine! Realignment is very time consuming.

The motion of the dial wheel is important, in terms of its dynamics and the sufficient holding of the base part, as is the time it takes to move with respect to the available rest or dwell time. The slowest process will determine the limiting cycle time of the machine. If one process is consistently longer than every other process, then one should consider if the process should be offloaded from the dial for that sequence, or in some cases the processing tool will operate on the dial,

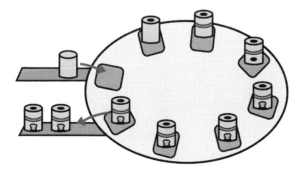

Fig. 4.14 Rotary index or dial machine

being situated at every station on the wheel (a costly proposition). This also means that the wheel diameter and number of stations increase. So without trying to explain all of the possibilities, one can see that if the time to perform the processing steps is very disproportionate from station to station, there will be an imbalance to the system for which the customer would need to allow. Watching many stations just sit for 95% of the cycle time can be frustrating when throughput rate needs to be increased.

Rotary Motion Issues

One of the limiting issues with a Dial Machine is that the wheel is always rotating in the same direction, hour after hour, day after day. The system never rewinds itself. This perhaps silly concern is very challenging in the fact that any compressed air, electrical power lines, or control signals need to be coming from the machine's base, resting on the floor. Any cables or hoses cannot just simply be allowed to twist as it spins. This is similar to your normal homeowner garden hose reel. There is a rotary joint in the hose reel, and as most homeowners know, it does not take too much sun, cold, or other abuse before that rotary joint starts to leak! A single pneumatic rotary joint functions similarly to the hose reel. It just has better tolerances and seals.

Commercially hardened pneumatic (compressed air) joint and electrical couplings (usually called "slip rings") are available for just this application. They can work fine, but the number of lines, both for air and electrical, are quite expensive. So if on your Dial Machine you have compressed air clamps to hold down the base part at the 12 stations (for example) one has the choice of running 12 separate air lines through a very costly rotary fitting, or one air line and enough electrical signals to trigger the 12 valves located on the dial.

Figure 4.15 shows an electrical slip ring. The power is supplied by two batteries, and the voltage is connected to three brushes held in contact by compression springs. The brushes contact continuous metallic bands that are fixed to the rotating cylindrical surface. Thus the light bulb and electric motor can operate while the cylinder spins. The brush contacts can get dirty from wear and the factory environment, and can need to be replaced every so often. Dirty contacts can lead to a noisy electrical connection that may not be robust enough for control signals.

As one looks at the options of different machine configurations, these trade-offs start to imply escalating costs and demands, depending on the strengths and weaknesses of each system. A controller based on the fixed part of the machine would need to both control all of the actuators and get input from all of the sensors on the rotating part of the machine. To battle the dial system rotary joint limitation, another answer is to locate a smaller processor on the dial itself, so it can rotate along with all of the operating componentry on the dial. Then the smaller processor can talk to the overall machine processor. Using state of the art

The Automation Design Process

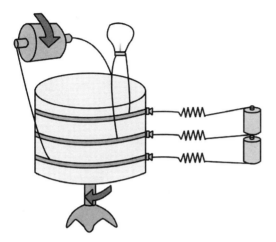

FIG. 4.15 Electrical slip ring

tools, one could even configure a wireless radio frequency (RF) ethernet connection. Then the needs at the rotary joint would be a simpler slip ring that supplies only DC power to everything on the moving dial.

4.4.5. Rotary Continuous

Many bottle filling machines are configured with a rotary unit that continuously moves. The design is rotary since there is usually a tank in the middle, and the bottles can be filled starting from first contact up to about 270° of the motion. The system has a serpentine conveyor belt bring in the bottles (Fig. 4.16) and some gating device to align one bottle under every filling nozzle. A star wheel or timing screw is used to pull away a single bottle from an incoming line, where often the bottles are in a linear continuous conveyor with guide rails, and are in contact with each other side by side.

Similar to the Rotary Indexing systems, there is a need for compressed air and electrical signals going to and coming from the rotary filling system. Often under each bottle as it is being filled is a load cell to weigh the amount of product. These load cells can:

- detect an empty bottle that is underweight, and therefore a likely defect;
- detect a bottle that is not increasing in weight appropriately, and therefore has a leak;
- withstand the chemical being filled if it does leak out; and
- withstand a washing down once a day for food cleanliness reasons.

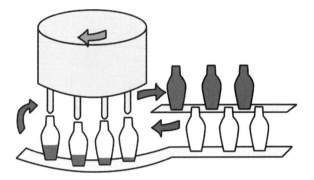

FIG. 4.16 Rotary continuous configuration with bottles on a conveyor system

And again similar to your garden hose reel, the bottle filler might have a rotary joint to transfer in more liquid to be filled.

4.4.6. Robot Centered

If one is thinking about using a robot as the main machine configuration, one needs to generally select the robot to be used. This may sound like circular reasoning, but is mostly true. If the robot type to be used has a central base, and the robot can reach generally in a cylindrical envelope, then the feeders and reach locations must be within this cylinder. Or if the robot is a large gantry type, similar to an overhead crane found in some factories to move products around, then the layout of locations may be more optimal if laid out in a linear fashion. But how does one know which type to use to start this analysis?

It is best to segment all commercially available industrial robots into several groups. This grouping may eventually limit one's thinking, but is a good starting point for the novice. Smaller robots for moving 10 pounds or less are either a SCARA design (cylindrically designed) or a six-jointed arm (spherically designed) looking similar to a large human arm. There will be much more discussion on Robotics in Chapter 5. Figure 4.17 shows the simplified working volume of each.

So if the robot is the main device to move things within the automation configuration, the choice of either of these two highly influences the layout. There are also a great number of six-jointed robots that can move up to hundreds of pounds if one needs the strength and extended reach, and can afford them.

A more recent option is to fixture the robot base on the upper surface of a workcell, effectively placing it on the ceiling. One needs to check with a robot vendor if this is permissible. But Staubli Automation recently has created its Modular Automation Workcell, using its six-jointed robot in this manner, so as

The Automation Design Process

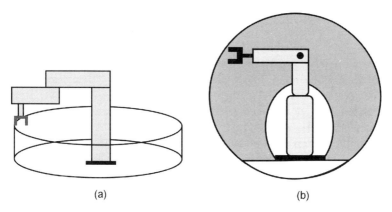

FIG. 4.17 (a) SCARA working volume; (b) six-jointed robot working volume

not to waste the space the robot base traditionally occupies. A six-jointed robot mounted to a work table can normally access the space directly above it, but it is impractical to use.

The third general configuration mentioned before was the gantry robot system. Figure 4.18 represents this rectangular work area. This may look appealing since there seems to be more accessible volume than the SCARA or six-jointed robots, but having access from above to reach things without interference with other items in the volume may not be trivial. Humans should generally not be located in these volumes while the robot is in operation, so the smallest volume that gets the job done is sometimes the easiest in which to implement safety precautions.

There are some other special purpose robots with other configurations, but they will be left for detail in Chapter 5. No matter which general robot configur-

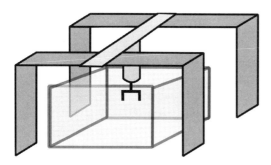

FIG. 4.18 Rectangular gantry robot work volume

ation you choose, you often look to move multiple pieces of equipment to maximize effectiveness of the automation cell. In some processing operations, a single robot can service two or more CNC machining centers if the processing time allows. On the other hand, if one tries to make a robot do more operations than is possible in the allowable cycle time, the work may then need to be split between two robot centered cells, not just one.

This determining of what and when a robot can do in the automation workcell can get complex. It is often useful to get a second or even third set of eyes and ears to look at the system requirements, and see what other possible options can be determined. Sometimes by asking the right questions of "why" and "does this really have to happen in this order," can a potential set of robot centered workcells be combined into a single cell.

4.4.7. Dedicated Custom Design

The automation device that does the real work is the center of attention. All feeders, conveyors, and so on are secondary, and no standard configuration will work for all problems. A good example of this is found in Sec. 2.2.2, on the tying of bows on high-priced jewelry or candy. Trying to create a Rotary Indexing or Dial Machine would require that the boxes to be wrapped be fed into an input station, and that there would be four, six, or eight special bow tying assisting mandrels made, and that the bow tying would be processed at several stations or stops along the rotary indexing motion.

However, it would probably make more sense to have the box and bow tying assisting mandrel stationary. A human operator might manually load the box under the mandrel, and even adjust it for the proper height. Remember that in the jewelry market, individual custom orders are bow tied when requested. We are not filling boxes with corn flakes as in a true assembly line here.

An industrial robot could be used as the motion device, gripping the loops under and through the center loop, but since the first loop needs to be held when the second loop has been passed through the center loop, two robot arms might be in order. Two robots could be purchased and integrated to do this task, but the cost might be too high. The motion of each loop gripping and pulling might be more cost-effectively achieved by a set of motors driving a cam and follower to do the motion. The bow tying itself will be done the same way no matter what size box is tied. The robot's flexibility and cost might not be justified if the cam system is perhaps 20% of the cost in comparison.

Some of this is economy of scale, and how many bow-tying workcells one can sell. The author on many occasions has used a robot with a prototype custom gripper as a Proof of Principle Model (POPM). This is when and if the potential jewelry customer would like to see how the process is achieved, and perhaps to focus more on the mandrel design. The robot is a very flexible system and highly programmable, which is helpful since the required motions are not well defined early on. But once the POPM has proven the task worthy, and if the customer wants many bow-tying workcells, then the motions achieved by the robot might be cost-effectively turned into specific cams, actuators, and a dedicated automation device.

4.4.8. Modular/Flexible Design

As product lifecycles become shorter, industry is placing more emphasis on agile manufacturing systems — systems with unparalleled flexibility. Recent years have shown a shift in paradigms from attempting to exploit economies of scale to economies of scope. The ability to quickly adapt to changing market forces, new product designs, and changing technologies is now key to an organization's survival.

At the manufacturing floor level, this translates into being able to deal effectively with short product life cycles, frequent part changeovers, and small production lot sizes quickly and efficiently. Flexible, Agile, or Modular (the terms are often used interchangeably) Automation is the response to that need. But what is flexible automation? The ideal system uses easily reconfigurable equipment to enable simultaneous production of different part types with zero online setup time and costs. This includes system software, including the goal of an Agile Automation Control System (AAGS) (Durante, 1999), which will be covered in more detail in Chapter 10 on Controls.

Modular Automation Goals

A list of Modular Automation Goals has been developed (Derby and Cooper, 1998). The 11 items below are not ranked as to importance. For example, safety is listed here last, when in fact safety should be of the highest concern when designing and fabricating automation. Safety was just not the first goal in many people's mind as a reason to use Modular Automation.

The goals of Modular Automation are:

- Interface standardization;
- Different size automation products;
- Flexibility;
- Singular process;
- Less floor space;
- Reduce down time;
- Product life cycle;
- More off-the-shelf parts;
- Technical simplicity;
- Quick construction;
- Safety.

Interface Standardization

This perhaps is one of the biggest problems to today's automation builder. When a company is given specifications for the automation subsystem they are to build, they are relying on the written specifications to be honored by the subsystem above stream and below stream to themselves. When the subsystem is built and delivered, and the mechanical interfaces are not compatible, there is often a lot of finger pointing going on. A fraction of an inch may be just as bad as 6 in. when it comes to incompatibility. Someone somewhere has either missed or misread a piece of information on how the subsystems are to be joined. The same can be said for electrical interfaces and overall control system timing.

Different Sized Automation Products

Can the same approach be used to make different sized modules that perform similar tasks? This will allow for the smartest use of engineering time and talent. It is needed to be profitable in today's competition, and CAD models are one of the vehicles to make this happen. Components of multiple sizes and shapes are modeled in AutoCAD and other software readily available from company websites.

Flexibility

Can the module work for a range of product types? Can the same singulation conveyor that handled pretzels be used to sort packages for the shipping industry? Can a donut loader be used to pack pouches? This is one of the biggest challenges for automation firms that have specialty areas. They are now looking at other new markets for their existing product, and sometimes a goldmine is waiting right under their noses. Consultants who know many fields can yield interesting results.

Singular Process

It is best to create modules that have a single dedicated task, then the modules are true building blocks, where all possible functions will be in use. But it is not as simple as it sounds. Every module you construct may have some overhead, such as a frame, a sensory system, local controller, and so on. So making an automatic machine into too many modules can drive the cost higher if not done with great planning. This is similar to the economic justification problems that general purpose robots have faced.

Less Floor Space

Rarely will an alternative concept be considered if it will take more floor space than the traditional methods. Possibilities include thinking in three dimensions,

The Automation Design Process

not just the traditional two dimensions. Can product path loop back over the original path at a higher elevation to reduce space?

Reduce Down Time

Down time is the enemy of all automation. The robotics industry suffered when insufficient engineering did not cover enough contingencies during the first wave of applications. What happens when a module fails, and it will eventually! Can a human step into a work area safely and fill in for a short time so production keeps going? This is often a requirement, but hard to do in real life when worker safety might be compromised.

Product Life Cycle

Can all or most of the modules' design be individually updated without the need to update every other module in stock? Can the control system be updated appropriately, and does it have enough capacity for expansion? Are your component vendors state of the art, or in the dark ages.

More Off-The-Shelf Parts

It does not usually pay to produce custom parts when commercial parts are available. This also helps when creating a replacement parts list for product support. Is there a backup supplier if some vendor ceases to trade. Conservative customers will stick with known quantities, such as demanding a specific brand of PLC controller, since they have a good track record. Or the customer may not want a specific component since they are known as the cheaper, low-cost device that fails too often. Customers may specify this by stating number of cycle or mean time between failures that are required.

Technical Simplicity

Each module should be a clever engineering or application of existing technology. This may sound too conservative, but if the equipment is to run for many months or years without unscheduled maintenance, it is a must. There is inherent risk in a new technology, but also great rewards in performing better than your competition. Many customers do not have advanced engineering degrees. Each customer has a level of comfort to this problem.

Quick Construction

Quick response to today's financial operations is a must. Expectations are bordering on the impossible when economics can range into the millions of dollars on an automation project. Everyone wants it yesterday, at a cost lower than before.

Safety

Any automation machine produced has to comply with existing safety standards, as well as the general rules of safe design. If any module is not used, and humans are performing a set of operations, will overall safety be compromised? This will be addressed more in Chapter 6.

Why Use Modular Automation Concepts?

If an automation company is looking at the long-term prospects of its future, Modular Automation approaches make a great amount of sense, both from a good use of engineering resources, and from an overall economic viewpoint, but it does take some "fiscal courage" to do this. Why call this courage? Modular Automation:

- may cost more than making one product at a time;
- may increase development time for the first product;
- does not know if a sister product will be developed or not;
- may not be obvious to financial bean counters!

Also, if all automation ever built by every vendor in the world was mechanically and electrically plug compatible, then an automation company would run the risk that some competitor could produce any or all of their machine module product line and put them out of business. Trying to keep high performance and cost effectiveness in a plug compatible world is a tricky business. Just ask the vendors in the PC market!

So if one company was a true leader, and perhaps even a dominant producer of automation, they could set the interface standards for the world. And then they could say, "Follow us or good luck," almost like Microsoft's domination in PC operating systems. But to date, no one company is in a position to attempt such domination, and very few mechanical interfaces have been established.

4.5. MACHINE CONFIGURATION TRADE-OFFS

There are good, sound reasons why one machine configuration is preferable to another, and then there are the reasons due to politics or emotions. And many times, when a process to be automated is considered, the best answer is not obvious. But nothing stirs the pot with a group of automation engineers more than this topic.

Some will either swear by or swear at industrial robot arms. These systems are so powerful currently, and the controllers can often handle everything that needs electrical interfacing, that the job seems more than half done when the robot is simply uncrated, but others have been forced to use robots when

The Automation Design Process

the required cycle time has been reduced after first implementation, and they just cannot get any more speed out of the robot without some part flying out of the gripper, or the proper part alignment suffering.

Meanwhile, other engineers will start with a rotary dial system, a walking beam, or whatever, and sing the system's praises or gripe about its shortcomings. All of the evaluation matrices for nonsubjective reasoning will mean nothing. These engineers all may have been burned by one configuration, and achieved success on another. So if your boss insists on one particular design for some either substantial or unknown reason, you most likely can find some automation company who will agree with that view if you look hard enough. Do not always cave into your boss.

Evaluation matrices comparing:

- costs,
- time to completion and in production,
- ability to get support and replacement parts,
- ability for the system to upgrade/improve throughput,
- risks and unknowns, and
- vendor references

are all viable quantities to understand. However, it is interesting to the author that just about when it seems that one of the configurations is possibly obsolete (say for example, the walking beam), then state-of-the-art product processing requirements will almost demand that configuration as the obvious choice.

So the final answer is that it all depends on who one talks to (previously burned engineer, vendor of a specific product line) on what recommendations you will get. Do not be afraid to carry several configurations forward at this point in the machine development cycle, since some future surprise may change one of your prime assumptions. A comparison matrix of some of the trade-offs is given in Table 4.1.

Some specific definitions are needed to help understand these terms:

- Costs — for the same throughput rate;
- Time — to design, build, and debug the system;

TABLE 4.1 Comparison Matrix of Trade-Offs

	Pallet system	Rotary index	Robot centered
Costs	Moderate	Moderate	Moderate/higher
Time	Moderate	Longer	Shorter
Support	Depends on builder	Depends on builder	Good
Risk	Moderate	Moderate	Lower

- Support — builder can range from excellent to poor, if they go bankrupt. Robot centered cells rely on the robot controller, which most robot suppliers can usually get you through in an emergency.

The application of current technology within a system is an issue for all three options. Any system that uses significantly older technology might have no available replacement parts in a very few years.

4.6. MECHANISMS TOOLBOX

Traditional automation designers are usually somewhat conservative, and often do not have enough time to think about options. They are pressed most days to balance the design of a new machine, debugging a second machine they designed last year, and assisting to quote on three new machines that marketing does not have the warm and fuzzy feeling that they will ever get to design. They probably are a degreed engineer, but sometimes are a converted draftsman from the old school, having some good but possibly limited practical experiences.

A four-bar mechanism offers a range of complex yet possibly more useful motions than a simple rotary or linear motion can accomplish, but this requires more analysis than the automation machine builder often wants to use. Figure 4.19 shows a generic four-bar, with the links pivoting about the fixed trapezoidal bases, and the resulting action being the dotted curved line on the right. Their toolbox is usually a CAD system, and if you force them, they might run some finite element analysis code to predict bending and failure properties, but many of these designers would rather not have anything to do with these tools that just seem to slow them down.

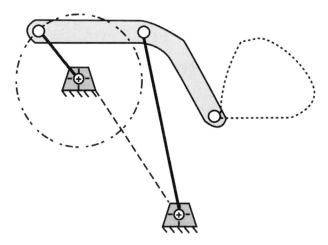

FIG. 4.19 Four-bar motion from single rotary input

The Automation Design Process

This is unfortunate, since Mechanisms is a well established area of technology. Some types of mechanisms have been used for hunderds and even thousands of years. Many books like that by Chironis (1991) list case study after case study of potential solutions, devices that are useful for quick return motion, when the effective output of the device can be maximized with just a little insight. Some well-seasoned engineers know of these options, and guard these case study books with their lives.

Perhaps an inexperienced automation engineer thinks that they need a two-dimensional servo control system. It may fall well into the Mechatronics approach that can usually be so helpful, but a simple four-bar device using a single rotary motor that can be timed to other machine motions may be all that is needed. The challenge is to find the correct ratio of link lengths to achieve the desired motion. Computer programs like LINCAGES (Erdman and Sandor, 1991) and SYNTHETICA (McCarthy, 1998) are a quick method to explore these design spaces. Older practices included the making of cardboard links and using straight pins as pivot joints to test out a four-bar mechanisms. However, what might have taken weeks or months of exploring with dozens of cardboard links and more pins than one could count can be duplicated in a matter of hours on specialized CAD programs like LINCAGES.

The topic of Mechanisms includes:

- Cams;
- Gear trains;
- Linkages — four-bar, six-bar, and others;
- Specialty devices such as Geneva Mechanisms, Scissor Jacks, etc.;

All of these have their strengths and weaknesses. Some of them are very mature (gear trains and cams), while others like four-bars are viewed as only used by university types.

4.7. TBBL AUTOMATION PROJECT*

This section will define some of the background requirements of several case studies within this text. The overall project and the case studies are from a four year effort at RPI, documented thoroughly by Mr. Matthew Simon (2000a,b,c). Hundreds of pages of reports, CAD drawings, and analysis will be reduced to a few key points for useful examples, but the reader will undoubtedly find a few terms or process steps not completely defined. That would be impossible to do in this text. However, this example does represent some highly custom automation operations and machines that were built and are in operation today.

---

*This case study supplied by Mr. Matthew Simon.

4.7.1. The Need

The New York State Talking Book and Braille Library (NYSL/TBBL) distributes free material to the blind and handicapped as part of the National Library Service (NLS), in particular "talking books" on cassette. However, a limited staff, an increasing demand by its expanding list of patrons, and documented cases of repetitive motion injuries from work in its entirely manual processing system led them to investigate automation alternatives. The NYSL/TBBL group approached the author and the New York State Center for Automation Technologies (CAT) at RPI. The effort to provide a solution was conducted in a multiphase series of projects.

The first phase concerned the characterization of the NYSL/TBBL mission, problem, and process, and required a system-level design of a series of automation modules to assist the NYSL/TBBL staff in meeting their projected patron demands, moving from 3000 to 7000 books per day. Like so many other situations, this one began with little detailed knowledge. Phase I was concerned with creating this knowledge base within the CAT project staff, by immersion in the Library process, discussion with Library staff and administration, and observations. With this established, the problem and constraints were understood to the most complete extent, and only from this point could the "meat" of the effort, the design of an automation system for the Library, commence. The plan for the remaining phases was to look at the possible automation solutions in various groupings or modules, so as to facilitate the state funding limitations.

4.7.2. The Product

Of the 30 million Americans over 65 years of age, many are temporarily or permanently unable to read standard print, and most elderly people lack the tactile sensitivity required to learn Braille. Thus the talking-book program was established, under the jurisdiction of the Library of Congress.

Cassettes are shipped in a green plastic case designed by NLS (Fig. 4.20). The case has undergone several minor design changes in the 20 years of use. These cases come in two sizes, one that holds four cassettes (the more common) and one that holds six cassettes. Two latches hold a case closed for mailing.

The System Goals were established at the onset. They were mostly predictable, but their impact on the final results was totally unknown.

4.7.3. System Goals

1. Reduce repetitive motion injury;
2. Reduce costs associated with worker compensation claims;
3. Reduce other work-related injuries;
4. Improve process flow;

The Automation Design Process

FIG. 4.20 Two sizes of talking books cassette cases

5. Increase productivity;
6. Increase throughput and capacity of TBBL;
7. Improve employee job satisfaction;
8. Increase flexibility to handle variable system demands;
9. Meet long-term throughput projections (7000 books/day).

Several students and staff spent a good number of days watching the TBBL workforce perform their tasks, asking them many questions, and often getting the answer "because we always have done it that way." Many industrial engineers and automation specialists have found this similar situation in other plants wanting to automate. If there ever was a great reason for a task in question, the person who really understood why was no longer around.

And since many times in the design process you find too many constraints, it is useful for the Decision Criteria to be ranked to the relative importance. TBBL staff input produced the ranking in Table 4.2. Not on this list, but a long-term goal of the TBBL director, was the desire to create a modular automation system that addressed some of the operations in groupings or clusters. The NYSL/TBBL facility services only part of upstate New York. There are 75 similar libraries covering the entire United States, though some of them were much smaller, and only some of the automation functions might be cost effective. So if the resulting automation was in smaller clusters, the NYSL/TBBL could re-licence the technology in which they were investing.

The books on tape are delivered in standard US Postal Service Bulk Mail Carriers (BMC). These are 63 in. high by 43 in. wide by 63 in. long, weighing hundreds of pounds. Other forms of media in different style cases are mixed with the green plastic cases of the books on tape. The 1998 manual operation task list is given here. Of note is the rewind and inspection operation. It was assumed from the onset that this process would most likely not be automated, since the clientele often would mix up tapes if they had several green cases at the same time, some-

TABLE 4.2 Ranking of Decision Criteria

Criteria	Rank
Reduces repetitive motion injuries/Ergonomic design	1
Cost (ranks #6 if whole system is forecast at more than $250,000)	10
Noise level	2
Adaptability to changes in staff size	4
Throughput and capacity	5
Maintainability/Reliability/Durability/Availability	3
Easy to learn/operate	8
Graceful degradation	6
Degree of modularity/ability to implement incrementally	7
Availability of off-the-shelf systems	9

times they would inadvertently damage a tape, and once in a while stick some foreign object in, such as their dentures! So these operations that use so much of the skill set of humans were deemed to have to continue for the foreseeable future.

The P-label noted is an alphanumeric, device readable text format label used by the TBBL and all of the 75 other similar libraries. Technology-wise, it was a precursor to the barcode systems commonly found on many products.

Many patrons would not be specifying a specific book, but rather a book of a certain genre (say a Mystery) that they had not read before. So instead of re-shelving every book on tape that was returned, these tapes sat on "quick turn-around shelves" for a week to reduce the amount of running around for the staff.

The operational task list shown in Table 4.3 was generated by RPI students and staff after they spent much time and went through several rounds of document review. This diligence is often not found in this great detail in most industrial automation applications, however.

As stated at the beginning of this section, many operations will not be detailed here, but from the onset of this project, the author knew that this was not a simple robot centered configuration or the like. There were too many specialized operations, and at the time of Phase I, not enough internal system requirements were known. For example, no one at the TBBL had quantified the force required to open a case, and how much variation there was from case to case, and style to style.

Despite the seemingly unique product to automate, the RPI team conducted a complete patent search. Using the USPTO and www.uspto.gov more than a dozen somewhat relevant patents were found. These included a total automation system to be used in a branch videotape rental store. Interestingly enough, however, no-one could find such an automation system in actual operation. None of these patents really addressed the specific goals and requirements.

The Automation Design Process

TABLE 4.3 Operational Task List

TBBL operation task list

I. Bulk mail carrier (BMC) handling
   A. Roll BMC from loading dock area to lobby area
      1. Push BMC from behind to avoid accidents
   B. Unload BMC into cloth bins
      1. New tapes
         a. No P-label, go into separate bin
      2. Other mail
         a. Separate bin
      3. Library mail
         a. Separate bin
      4. Flexible disks
         a. Separate bin
      5. Rigid disks
         a. Separate bin
      6. Braille books
         a. Separate bin
      7. Video tapes
         a. Uncommon
         b. Separate bin
      8. Garbage
         a. Trash
      9. Tapes
         a. Sometimes bagged
         b. Separate bin
II. Stacking and opening of tape cases
   A. Check P-label and mailing label
      1. If P-label missing or not identical to others, belongs to other library
         a. P-label is unique to NYS Library
      2. Put in box for further examination
      3. If yellow label present on mailing label, put in basket for office processing
   B. Check case
      1. Rubber band(s) around case is OK
      2. Rubber band(s)/string(s) means discard
   C. Stack in groups of 10
      1. No more than one large case
   D. Remove mailing label
      1. Place in box for discard
   E. Re-stack in groups of 10 with latches up
   F. Use tool to unlock latches
   G. Stack in groups of 10 with open latches up
III. Inspection of cases and tapes
   A. Remove any "extra" material
   B. Check number of cassettes enclosed with number of cassettes specified on the label

(continued)

TABLE 4.3 *Continued*

TBBL operation task list

        1. Remove any additional/wrong tapes that are included
        2. If any are missing, discard the case
    C. Check enclosed cassette titles with title specified on the label
        1. Remove any additional/wrong tapes that are included
        2. If any are missing, discard the case
    D. Check pressure pads and tape casing condition and whether tapes are rewound
        1. Discard if pressure pads are mangled or worn or casing is in inoperable condition
    E. Rewind if necessary
        1. If tapes won't rewind, "twist and whack" and attempt rewind again
        2. If tapes won't rewind, discard the entire case
    F. Place tapes back in case in order
    G. Close latches
    H. Look at case
        1. If one strap is bad, keep
        2. If one hinge is bad, discard
        3. If case condition is extremely dirty (i.e., food caked in), discard
    I. Check book number for A, B, and C ending
        1. If present, place these in separate, multi-volume bin
    J. Place in alternating stacks of 10 and 13
    K. Move stacks to table to awaiting pick up for re-shelving
IV. Re-stacking books
    A. Take card to inspection room, put stacks on cart
        1. Place two rows of 23 on each shelf
            a. Latches down and P-labels out
    B. Move cart to stacks area wanding station
    C. Enter row numbers of quick turnaround shelves into system
    D. "Wand in" the cases one by one
        1. Books can be wanded more that once without detriment to process, but not after another book has been wanded
        2. If multiple beep sounds, check reason on computer
        3. If still cannot be wanded, replace on cart and set aside
    E. Place stack of 23 on appropriate quick turnaround shelf
V. Checking out books
    A. Get pile of labels
        1. Log in which labels were taken
    B. Find book on quick turnaround shelf (or stacks), insert label
        1. Place in cloth bin
        2. Continue until cloth bin is filled
    C. Bring full cloth bin to wanding station
    D. Scan account number and book number
        1. If multiple beeps sound, check cases and label scanning procedures
    E. Make small pile of wanded cases
    F. Place in BMC for egress
VI. Move BMC to loading dock

The Automation Design Process

After some brainstorming, the Operational Task List was reduced to the most likely clustering or module of automation. They were:

1. Bulk mail container unloading;
2. Separation of cassette cases from other materials;
3. Opening of latches;
4. Removal of shipping labels;
5. Inspection/rewinding of cassettes;
6. Closing of latches;
7. Re-inventorying/de-inventorying and associated shelving processes.

It is interesting to note that after many months of looking at the overall need of the automation system, and trying to go beyond the process of the current human powered approach, some of these steps were reordered to improve efficiency. Also, floor space limitations (walls and beams) had tremendous impact. No-one would have predicted this at the onset, however.

4.7.4. Case Study Number 1: Case Opening*

One of the simplest tasks, yet one of the most stressful on a person's wrists, was the opening of a case. A TBBL client could usually unsnap the two latches by hand, but they most likely would not be unsnapping one case every second for 10 minutes straight. So the TBBL staff came up with a pair of modified gardening tools (Fig. 4.21) that would catch both snaps and latches in a single motion. Of key concern is the proper orientation of the green case since there were four possibilities even if the cases were constrained along a conveyor with side rails (latch up leading, latch up trailing, latch down leading, latch down trailing). It only made sense to select one of these (latch up leading) for automation purposes.

Functional Specifications

This module reorients, as required, all cases to a single desired orientation, and opens the latches of the talking book case.

- This module must correctly orient 99.99% of talking book cases and open both latches of 99.9% of cases.
- The module must detect improper orientation of the case before latch opening is attempted, notify an operator, and take the appropriate corrective action.

Investigated Methods

Various methods of operation were explored. These included:

---

*This case study supplied by Mr. Matthew Simon.

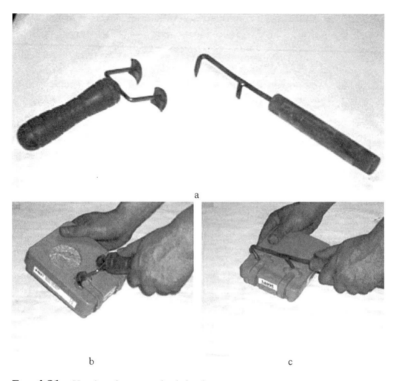

FIG. 4.21 Hand tool to open both latches

- Trying to manipulate the modified garden hand tool. This included attaching it to a robot, and creating a four-bar linkage mechanism.
- Creating a linkage that moved quickly that first opened one latch and then the second latch.
- Rolling devices with claws on them.

Each of these had their limitations. The green cases needed to be held down firmly and in a very repeatable position since the lip under the latch is not too large. And some of the older versions in service seemed to have warped slightly, or the case's original manufacturing tolerances were not too closely held, giving some variation. In the manual operation days, this would not have been a problem, but automation needs repeatability.

The method selected is shown in Fig. 4.22. It grabs both latches at the same time, and from two opposing sides. This tends to assure a good grab and lift every time, and the production unit works in this mode. Since the travel of the finger is small, adequate tolerances in the construction are vital to effective operation.

The Automation Design Process

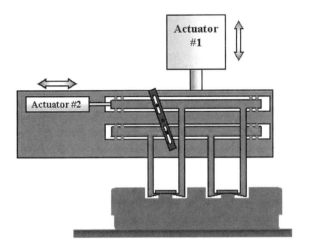

FIG. 4.22 Unlatching mechanism

Additionally, control must coordinate the timing of the descent, finger closing, ascent, and finger opening with the arrival of a cassette case. Whether it is best to open the latches when the case is moving or stationary was not yet explored.

System Specifications

The latch opening module was constructed to the following specification:

- The equipment footprint is approximately 3 ft wide × 6 ft long.
- The functional components are at the typical working level of a human in a standing position.
- The case is stopped at an explicitly known position for assessing the orientation and for opening the latches.
- Four sensors are placed to determine the orientation of each case. These are placed $1\frac{3}{8}$ in. from the front of the case, and $\frac{11}{16}$ in. from the side.
- The desirable case orientation has the latches leading (in the direction of travel) and up.
- The sensing and orienting system must be able to accommodate both case sizes.
- Fingers must be designed without sharp edges to prevent wear on the case latches.
- A means must be provided to position the case securely, and hold it down, while the latches are being opened.

- Improper case orientation must be sensed and appropriate corrective action taken.

Again, the green cases would have different thicknesses depending on whether they contained four or six audiotapes. To help address this, when the module was actually built, it was found to be easier to lift the case up to the pair of gripping fingers, and then grip the latches and lift against a hard stop (Fig. 4.23). Then there was no sensing or controlled adjustment needed for the two different case thicknesses.

In operation, sensors determine if the latches are not oriented correctly, or if the unlatching was not accomplished. If either of these conditions occurs, the case is simply diverted into a cardboard box. Since this does not happen often, operators can manually inspect if a latch has become defective or not.

Configuration Determination

Since this module was to work either as a stand-alone unit for other smaller libraries, or as a section of the overall system for the TBBL site, a Dial Machine was ruled out immediately. The cases themselves were quite rugged, so no gain from using a pallet could be seen. The required motion was quite limited, and the operations did not require the dexterity of a robot, so that was ruled out.

The system was seen to be linear, so the only question was whether it should be continuous or indexing. A conveyor with raised intermittent lips or flights would be useful if we needed to specifically keep a case in a timed mode. However, the overall TBBL system was designed to handle one case at a time, and when a module was done with the current one, it would hand it off to the next module and request an unprocessed one from the previous module. A conveyor belt with the right amount of slip would suffice best for this product. When the case was lifted off the belt as in Fig. 4.23, there was no friction issue. Yet to be opened cases would be gated and would sit still and slip while the belt continued to run.

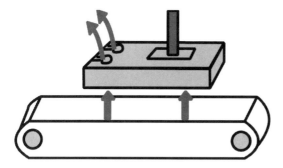

Fig. 4.23 Lifting case against hard stop and then unlatching

The Automation Design Process

It should be noted that if the flighted belt were selected, this would have been correct for the later part of the machine not discussed in this text, where the case needs to be elevated three feet to deposit in an outgoing BMC. The flights are needed to assure that the case will make it up the steep climb. No amount of friction from a rubber or other kind of belt would ever make the case travel up three feet at a 60° incline!

4.7.5. Case Study Number 2: Label Insertion and Printing*

This case study is a good example of how duplicating what is manually done is not always the best automation approach. For the manual approach, the cases of books on tape were found from a list and placed in that order. A computer printed out a stack of mailing labels, and then the operator took the top label and inserted it into the top case. The first reaction in automation concept development was to duplicate this task. However, the molded plastic guides or lips found on three sides of the mailing label area are often warped (Fig. 4.24). This means that the staff person needs to use their fingernails or something like a screwdriver

FIG. 4.24 Case's three guides or lips for holding label

---

*This case study supplied by Mr. Matthew Simon.

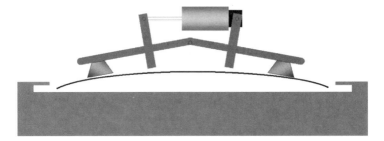

FIG. 4.25 Initial concept for label insertion

to pry the lips up enough to insert the label properly. So if the TBBL staff is trying to stuff a label in every few seconds, this gets even more time consuming. These lips do need to hold the label firmly through the rigorous sorting operations found in the U.S. Postal Service, but it is worth noting that the blind client also must remove the label, reverse it, and then pry with their fingers to place the return label side up properly. Some clients do get frustrated with this task.

The initial concept for label insertion is shown in Fig. 4.25. Although it looks simple enough, when human hand operated suction cups were attempted and an early experiment to achieve the operation developed, the results were disastrous, with a less than 50% success rate. It helped us to realize how far the process was from the specifications, and helped to modify them.

The manual TBBL method of inserting a printed label into a case was eventually achieved every time, even if the staff person had to fight with the case for many seconds or even minutes. In the world of automation, if a label was not properly inserted, it should be recycled and a new label attempted, but if the labels were all preprinted as had been done in the past, then someone would have to handle all of these bent or destroyed labels as exceptions, as well as the associated cases, without losing track of which case was to go to whom.

So it was determined that the labels should be blank when they are inserted, and then printed while attached to a case. A bent or destroyed label is worth a fraction of a penny, so the investment of time and materials was negligible. Commercial ink jet printing heads were selected that could print on the already inserted labels, and the automation concerns could focus on "simply" inserting a label.

Functional Specifications

The Shipping Label Insertion module requires insertion of a blank shipping label into the talking book cases identified as fit for sending to patrons, such that:

The Automation Design Process

- Insertion of 99.99% of labels into cases must occur without error.
- Sensors to detect failed insertion are required. An operator is alerted when this occurs, so that proper corrective action is taken.
- With respect to content, names and addresses must be printed to an accuracy of 99.999%; additionally 99.999% of the printed addresses must fall within the printable area of the inserted labels.
- This module must be equipped with a barcode scanner capable of 99.99% accuracy in scanning legible, properly affixed barcode labels. This confirms a label to a selected book.
- This module must interface with the overall computer system to
  — send the barcode scanned from the talking book cases to the computer, and
  — receive the names and addresses of the recipients from the computer.
- The medium used for imprinting the names and addresses on the shipping labels must be durable and "permanent" (i.e., nonsmearing and waterproof).

Investigated Methods

Early attempts to lift the three lips or guides were made with suction cups. Two small suction cups were configured to lift each lip. Figure 4.26 shows one trial. The results were not satisfactory, since the lips were often warped, sometimes so much so that the suction cups coming down parallel to the case surface were not aligned enough to make dependable suction contact. So a more dependable and forceful method was devised. A set of three fingers, each with a notch at the end (Fig. 4.27), was forced into the lip at a location where warping was minimized. Then the finger was slid along with the notch under the flap to a location where inserting the label was most likely to work effectively. This was the

FIG. 4.26 Suction cup lifting of the lips: (a) alignment; (b) lifting

FIG. 4.27 A single finger with a notched fingertip

method selected for use. The label insertion was then a more reliable process. Several methods of sliding or forcing the label were tried, but they all reduced to the concept in Fig. 4.28.

Label insertion does work in practice at the TBBL currently, but is probably the module with the least repeatability due to the warped case lips.

Configuration Determination

Similar to the Case Opening module, a dial system was ruled out immediately, as were the pallets. A robot was considered briefly to insert the label, but sliding in a piece of thick paper with some sensory feedback was not available on the robots at RPI during the investigation time period. Most paper handling systems like copiers function with dedicated paper handling devices (although most copiers still seem to jam too often).

The system again was seen to be linear, but the two processes within the module have different demands. The label insertion needs the case to come to a complete stop, with great registration. This is similar to the case opening. However, the ink jet printer head prints a single array of dots and requires

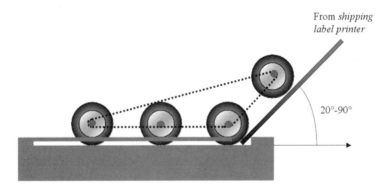

FIG. 4.28 Label insertion device

The Automation Design Process

constant motion to achieve quality lettering for the address label printing. So a continuous belt was chosen again. This time, besides the stop for the label insertion, there is the continuous motion of the case with the blank label in it for the print head to properly function.

4.7.6. Case Study Number 3: Crossed Four-Bar BMC Unloader*

Perhaps the most physically laborious task in the current manual process is unloading the Bulk Mail Containers (BMCs) received from the post office. Each BMC contains between 800 and 1200 talking book cases, and an assortment of other media, machines, random mail, and the occasional odd item (including "coffin-sized" boxes). The TBBL staff commented that there is "always something new and unexpected" in the BMCs.

The BMCs are delivered by a USPS truck to the loading dock and rolled to the TBBL. In the manual process, the green cases are removed and placed in Canvas Mail Carts (CMCs). Other items are separated out. Sometimes staff climbs into the BMCs to scoop out all the talking books, but at the very least they must bend and stretch into the BMC to reach and remove all the items. This is a process that, even at its quickest, may take the better part of an hour.

The BMC are approximately 63 in. × 43 in. × 63 in., and sit 7.5 in. off the floor on recessed 8 in. diameter wheels (depending on the wear on the casters) (Fig. 4.29). Each BMC has a full-sized door on one side, hinging on the top, and a half-sized door on the other side, hinged at half the BMC height. The top is open.

The definitive problem was that, given the use and abuse of the BMCs, none of the above features could be relied upon to function, from the doors opening or closing, to the brakes working, to the wheels pivoting. During the brainstorming session, there was a suggestion that the BMC be rolled into a space between two conveyors, the doors opened, and the items raked onto the conveyors. Given the unreliability of the doors functioning, this concept was dismissed. Similarly, a conveyor with shelves to scoop items out of the BMC also required functioning doors. Another concept was to pivot the BMC about a point on its front. However, this seemed to require more height than available in the State Library building.

Since the BMC is a USPS item, it made sense to see how the USPS unloaded BMCs. A tour of a Postal Distribution Center revealed that the BMC unloader utilized by this and many other postal distribution facilities is the one produced by Lockheed Martin Postal Systems in Owego, NY. These unloaders are a reasonable approach to the brute force, overengineered method of unloading

---

*This case study supplied by Mr. Matthew Simon.

FIG. 4.29 BMC: (a) small door side and back view; (b) small door side and front view

BMCs. The BMC is lifted into the air and inverted, requiring a 15 ft ceiling and two sets of actuators (for each motion). An advantage, however, is that the BMC is unloaded through its open top, obviously the most reliable location out of which its contents should spill out.

The question for concept generation became "how could this solution be improved upon?" especially considering the constraints of the TBBL's facility? Certainly emptying through the top was reliable, but requiring such a ceiling height was prohibitive. The entirety of the mass of the BMC and contents were lifted. Was this necessary?

An invention was recollected that offered a solution to these questions. A rocking chair, utilizing a crossed four-bar linkage, had been invented in the 1980s. Rather than moving the body through an arc, with the rotation point being the contact point of the chair with the ground, this chair rotated the person around some point near their mass center, making for an "interesting" rocking experience. The merit of this as applied to unloading is apparent: moving a load near the center of mass minimizes the required force and torque necessary for the unloading motion. Also, rotation inverts the load without lifting, requiring a relatively smaller actuator and less clearance height.

Thus the crossed four-bar linkage unloader concept was born. This very closely resembles the Chebyshev straight-line linkage, where each body link is 2.5 times the coupler length, and the coupler is half the fixed pivot separation. The midpoint of the coupler link moves approximately in a straight line. Models, utilizing Tinker Toys and Erector kits, demonstrated the concept. Although practical problems would mean deviating from the set formula and models, these were the foundations for the concept.

The Automation Design Process

Functional Specifications

The operator rolls the BMC from the loading dock to this module. Once secured, this module empties the BMC into a receptacle, with no human effort except control. The functional specifications were:

- This module must lift a minimum load of 1500 lbs, which consists of the BMC and its contents.
- The unloading motion must be smooth.
- The BMC unloader must operate so that all objects within the BMC are emptied out.
- The process of maneuvering the BMC into the module must consist only of operations in which the BMC rolls. Neither lifting nor pushing the BMC up an incline is an acceptable loading procedure.
- The process of securing the BMC in the unloader must be simple and require no more than one minute.
- The operator must have control over the unloading motions of the module at all times. These controls must incorporate a "key lockout" type control with a dead man's switch, which, upon release, holds the BMC in position. From the control position, the operator field of view must encompass the entirety of the unloading area.
- Module control must include a feature whereby the module may make a quick return to the starting position once unloading is complete.
- The module requires a safety enclosure, with a safety gate that prohibits operation while open.
- Guarding must be provided to ensure that all materials are directed onto the sorting conveyor.
- The module must be capable of completely emptying its contents in no more than two minutes.

Testing

Testing dictated that the dumping should take place with the narrow end of the BMC to the front, for two reasons. The first reason is that the BMC has a ridge on the long top edges that could stop some items from escaping. Secondly, in order to easily steer the BMC into the unloader and not complicate the linkage layout given the position of the nonfixed casters, either narrow end would need to be positioned to the front, and thus dumped over.

Simulation

The design of the unloader began with a linkage design tool, SphinxPC (Fig. 4.30). Developed at UC Irvine (McCarthy, 1998), SphinxPC (now it is called SYNTHETICA) has you enter desired positions as input, and it creates

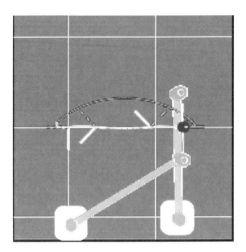

FIG. 4.30 SphinxPC simulation solution

a map of possible linkage that satisfy those positions (and associated orientations). The software graphically accomplishes linkage design (which can be animated in the software) that in the past could only be represented mathematically or by sketching points, drawing perpendicular bisectors, and so on.

There was definite progress toward a workable solution. The linkage proportion information in the task windows was scaled up and brought into a less analysis, more design-oriented program, Working Model 2-D. Working Model 2-D (MSC.Software, Redwood City, CA 94063) is, as the name implies, a two-dimensional motion simulation software. Rigid two-dimensional bodies can be created and interconnected with a variety of joints, including rotational and prismatic, with actuators, anchors, and accessories. Motions and collisions can be modeled, and forces, velocities, torques, and accelerations can be measured. Although now made somewhat antiquated by its CAD compatible three-dimensional successor, the software was ideal for realistic development of the unloader mechanism.

The final linkage proportions had nearly straight-line motion. Most of the mass was moved around the center of gravity. The design was compact vertically and required a reasonable amount of force. Indeed only one pair of actuators was necessary rather than two pairs as in the commercial unloaders (Fig. 4.31).

The resulting geometry was then modeled within the Solid Works CAD program. The final CAD model is shown in Fig. 4.32. This information was given to the list of potential automation machine builders, and will be detailed more in Chapter 12, System Specifications. The system in operation does work great.

The Automation Design Process

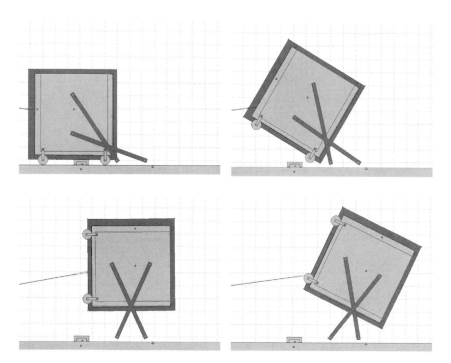

FIG. 4.31 Crossed four-bar mechanism BMC unloader

Configuration Generation

Because the operation of the BMC unloader was the first module in operation of the entire system, and because the contents were unloaded over many minutes rather than the several seconds per each green case, the time frame did not have to synchronize tightly. The speed of changing a BMC was very slow, and could not justify a dedicated large conveyor or material handling system to move BMCs in and out of the unloader. So this module's configuration is basically a Dedicated Custom Design. Although much larger than the bow tying system described earlier in the text, it is similar in many ways.

4.8. CONCLUSIONS

There is more than one way to skin a cat. And there is usually more than one machine configuration that will work for an automation process. So during your brainstorming, concept development, and evaluation comparison matrices do not throw out any intermediate work. Sometime in the near future during

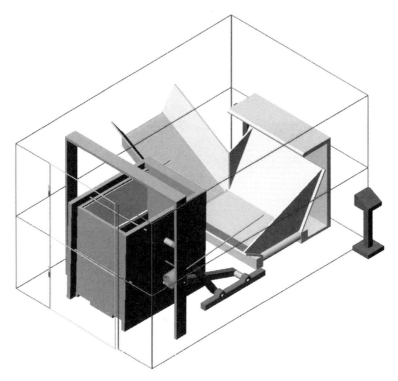

FIG. 4.32 Isometric view of the CAD model of the BMC unloader with supporting hardware

some detail phase of designing process, some road block may occur when you least expect it. And later, even when the machine is half built or has been on the market for several years, out of the blue some patent infringement situation comes up that you thought that you had covered. A different machine configuration might be the key to avoiding the infringement problem; one never knows.

So even if an expert tells you that your concept is second to their concept, keep it in your desk drawer for a while. You might be the hero yet.

PROBLEMS

1. Select one of the projects listed in the Appendix and:
    a) Brainstorm several methods to achieve the process;
    b) Determine several possible machine configurations;
    c) Create a matrix to list the pros and cons of each configuration;

d) Select the best configuration by making any assumptions you must make.
2. Do a Web search on any of the machine configurations of your choice. What off-the-shelf components, machine bases, and so on, can you find?
3. Check out the Robot Industries Association (RIA) website for possible robots to use (www.roboticsonline.com).

PROJECT ASSIGNMENT

Using one of the projects from the Appendix (or any other projects), perform the following:

1. Review the given system specification. Write down any other assumptions you feel you need to state to better define the problem. Any reasonable assumptions will work satisfactorily for the remainder of the text assignments as long as they are not too constraining. You may need to relax some of them as you proceed so as to be able to finish all of the assignments. Remember in real life, potential customers can and do give overconstrained automation specifications, and the resulting answer is that automation cannot be performed. That is not the goal of this text's assignments; you are to be able to complete a reasonable machine design!
2. Review all of the chapter's machine configurations, and determine the impact each has for your project. Create hand sketches of all possible designs.
3. Investigate commercially available workstations. If you by chance happen to find a machine that does your entire list of processes, assume that your project constraints do not allow for that choice (for example, it costs too much, is too fast, is too big, etc.). There is almost always a better mousetrap to be built.

REFERENCES

Chironis, N. (1991). *Mechanisms & Mechanical Devices Sourcebook.* New York: McGraw Hill.
Derby, S. (2000). Novel Palletizer Design. Warehouse of the Future Conference. Atlanta, GA, May.
Derby, S., Cooper, C. (1998). Modular Automation Concepts: Opportunities and Challenges in Industry. ASME Design Technical Conference, Conference CD-ROM Paper DETC98/FLEX-6016, Sept.
Durante, R. (1999). *A Design Methodology for Developing Agile Automation Control Systems Architectures.* Ph.D. thesis, Rensselaer Polytechnic Institute, May.
Erdman, A., Sandor, G. (1991). *Mechanism Design.* Upper Saddle River: Prentice Hall.

McCarthy, M. (1998). SphinxPC Computer Program. Now it is called SYNTHETICA at http://synthetica.eng.uci.edu/~mccarthy/.

Simon, M. (2000a). *A Modular Automated Handling System for the New York State Library: Investigation, Design, and Implementation*. MS thesis, Rensselaer Polytechnic Institute.

Simon, M. (2000b). State Library Materials Handling System Design, Part I. ASME 2000 Design Engineering Technical Conferences, DETC2000/FLEX-14043, Sept.

Simon, M. (2000c). State Library Materials Handling System Design, Part II. ASME 2000 Design Engineering Technical Conferences, DETC2000/FLEX-14044, Sept.

5

Industrial Robots

Here are some questions a TV reporter might ask the person on the street about robotics:

- What does the term "robot" mean to you?
- How do you determine if something is a robot or is automation?
- How smart is today's robot right out of the shipping crate?
- How smart can a robot be with the right engineering?

For some of these questions people might have an answer, for other questions only a wild guess. Most people could develop a response concerning a robot, but they may be far from the automation world. If the term "robot" means something from the Star Wars movies, then one needs to scale one's expectations back quite a bit. In automation, most robots are either Industrial Robot Arms, or they are Automated Guided Vehicles (AGVs), which run around a factory floor, transporting goods while following a wire embedded in the floor. Such AGVs will be ignored for now, even though they are a useful material handling device, since they are usually not the device performing a required process.

Industrial robot arms do look something like a mechanical version of a human arm; sometimes it is human scale, sometimes 3–10 times larger. But the rest of the torso is ignored. There is no head or legs. There is no effort made to look like a human. That would only add to the cost. No practical humanoid robot is likely to be found in a normal automation role in the near future. Humanlike robots performing applications in outer space or other nasty environments are the exception, not the rule.

The common definition of a robot is based on its capability to perform multiple tasks and its ease in reprogramming. So all robots are automation, but not all automation machines are robots. There are practical exceptions to this when it can benefit someone. Sometimes by misusing this term one can get an automation project approved by one's boss, or one might sell a machine to a stubborn customer who insists that they need a robot. So misusing the term robot does happen once and a while.

For example, the machine that folded and assembled the morning paper is automation. It needs to be highly programmable to handle that day's number of pages and sections. It needs to be able to accommodate the Sunday comics and all of those sale flyers, but it cannot be used to make spark plugs or cell phones. It lacks the capability to perform multiple tasks.

Now just because one implements a robot to make a cell phone, does not guarantee that this particular robot will ever get used to do some other task. But it could. Automation, if not modular in design to anticipate other uses, might be useless and therefore a potential boat anchor (the author has never actually seen a machine used this way — most often old machines are sold for scrap).

With this simple argument, one might think that everyone should only purchase robots, so as to minimize any scrap and lost investments, but what if one's newspaper was folded and assembled by a Selective Compliant Articulated Robot Arm (SCARA) robot, or even a series of SCARA robots. It most likely would be too slow, or the associated material handling issues, like constraining pages so as not to get improper folds and creases would demand tremendous external equipment beyond the simple robot out of the shipping crate. Publishers would no longer be able to compete financially.

So when a dedicated process needs to happen at a very high rate, often the use of robotics is not a viable choice. This can be argued another way. A robot was designed and built to be useful in performing a wide range of tasks. For some of these tasks they can work very well, but just like humans, it is difficult to be an expert in all vocations. One is often paying for flexibility that one may never use when one uses a robot in the same application for a long time.

5.1. HANDLING OF PARTS WITH ROBOTICS AND AUTOMATION

Many of the material handling concerns discussed in this section are true for both robots and for automation as a whole. It is just that since the man on the street's expectations for robotics is greater than for automation (more intelligence, reasoning, etc.) there are more disappointments if everything is not wonderful, but the following discussions are equally useful for both types of devices.

Industrial Robots 99

Imagine that one is to load batteries into a standard two-cell screw-on endcap flashlight, and that one had no tactile abilities. One might try to emulate this by wrapping their hands with multiple layers of aluminum foil before trying to unscrew the cap off the end of the flashlight. How sure would one be that the hand has unscrewed the cap? How can one be sure that one even touched the cap at all? As silly as the last statement may seem, many a robot technician has grimaced during a demonstration for guests while the robot continues to perform a detailed program even though it did not grip the first part in the sequence.

Robots can be fairly intelligent, but only in handling situations the engineer and/or robot programmer has anticipated. One has to look at the planned motions and tasks, determine what allowable deviations and exceptions are likely to occur, and then add a multitude of sensors and appropriate robot code to handle these situations. Robot code content can range from 50–80% for error handling, yet if something happens that the sensors cannot detect, or for which the robot code has not been written, then the robot will malfunction and will not be perceived to have much intelligence at all.

Some companies insist on robotic factories running in a lights out condition, able to handle virtually all possible error situations without operator attendance. To do this, the parts to be assembled have been appropriately designed, there are sensors integrated everywhere, and many segments of robot code have been written. Robots are then seemingly intelligent. Other companies have determined that creating that level of automation perfection is not cost effective. If a key part such as a torsional spring is likely to get entangled in its feeder once every few hours no matter what an engineer does, having an operator who is monitoring a reasonably large production area and who can untangle the group of springs can be more practical. This is not the robot's fault, but it is still looked upon as less intelligent.

Robots and automation transfer devices need hands called grippers or end effectors, but when one uncrates most robots, the robot stops at the wrist plate. No gripper is supplied. That is because anyone's robot application may be such that a somewhat generic two-finger parallel gripper would not suffice, and they would have wished that they had not paid the money to purchase the gripper in the first place. So one must select or design the appropriate gripper yourself.

Since the human hand is not currently cost effective to replicate, hundreds of simpler robot grippers are available from dozens of suppliers. They are often vacuum cups or two finger grippers (Fig. 5.1). Some have multiple fingers, and a few have sensors integrated into them as supplied by the vendors, but not many. And since each gripper is somewhat dedicated to a task or series of tasks, if the robot in a workcell is targeted to perform additional tasks, there are removable wrist joints so as to change grippers (Fig. 5.2). This allows for easy connection and disconnection of electrical wires for motors and sensors, and air lines to power some gripper actuators. Tapered pins assist in the proper alignment of the gripper to the robot endplate when the gripper is attached. Even then, as an

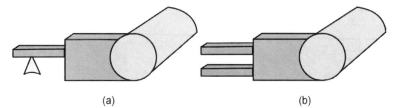

FIG. 5.1 Robot grippers: (a) vacuum cup; (b) two-finger gripper

automation engineer, it is not uncommon that after searching the Web for a day or so, one comes to the conclusion that none of these grippers will work satisfactorily. One must either combine several of them to a custom implementation, or start from scratch. The reason there are not any sensors attached to a gripper is that the applications vary so much. There is no universal sensor everyone can use that is cost effective.

5.1.1. Hold Onto a Part as Long as Possible

An old robot adage is, "Once one has determined a part's orientation, never lose it." This is because it took fixturing, vision systems, and money to determine the orientation of each and every part coming down the production line. It would be a shame to replicate this within the same robot cell or a neighboring cell. Again, this adage is true for all types of automation. It sounds simple enough, but sadly is not always done.

At an ink jet cartridge supplier currently out of business (and one might wonder why?), a Dial Machine was built to assemble the first 40% of a standard ink jet cartridge, one type found in many computer printers. Multiple feeders were used to align component parts and assist the assembly devices in the production. A second Dial Machine 20 feet away then took the first subassembly, and added a few more parts, all aligned by feeders. Both were an engineer's dream to watch. The seemingly crazy part was that the output of the first dial system, the completed subassemblies, was dumped into a tote (a plastic reusable

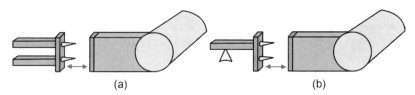

FIG. 5.2 Robot wrist gripper exchanger

open top box). These filled totes were humanly carried 20 ft, and then the subassemblies were taken out of the tote by hand and aligned on a conveyor to be the starting point of the second dial assembly system. The parts orientation was lost as it was dumped by the first Dial Machine into the tote, and an operator's salary was almost entirely being spent in re-establishing the part's orientation. A simple conveyor system could have been installed for a one time cost that was surely equivalent to only a few months salary of the operator. Yet, no-one seemed to care about this situation!

Just like that 1983 James Bond movie, "Never Say Never Again," there are almost always exceptions to rules like these when it comes to practical life. If the first part of the total product is made in Wisconsin, such as the packets of cheese from a box of macaroni and cheese, and the noodles are made in the farm belt, the packets may not be economically stacked to hold their orientation during the transport from state to state. By moving the cheese packets in bulk in large totes, the shipping costs could be significantly cheaper, and even if at the final packaging plant in the farm belt there needs to be either employees sorting out the packets one by one, or automation to perform the sorting process, it may be the economical thing to do.

5.2. SELECTING A ROBOT ARM

When exploring robot arms as the core device of an automation machine, several questions will arise:

- What kinds or types of robots are there?
- What are their speeds?
- What are their cycle times?
- What are their payloads?
- What are their costs?
- How intelligent are they?
- How are they different from automation?

We will explore the first question in the next section. The next four questions can be related to the consumer comparing automobiles and asking the car salesperson for information, but there are hidden issues, just as with cars.

The payload question seems simple enough. Either a robot can lift X pounds or it cannot, but one will find that the robot manufacturer will list a maximum payload that corresponds to the robot being able to operate under effective control of its electronics, 24 hours a day for 7 days a week. A robot listed for 10 lbs sounds like a weakling compared to the average human. One might compare it to a child. However, this rating is for continuous operation. If one misused the robot, it might lift 50 or 100 lbs before something broke. This is why for safety reasons, a novice thinking a robot is only as strong is as a child draws a false sense of security. In an uncontrollable motion, the robot night be able to swing a 50 lb weight at a tremendous speed and either knock one out cold or perhaps even kill.

The payload question is also coupled to the lack of a gripper out of the shipping crate. If one decides upon a robot with a 10 lbs payload, and the required gripper is 5 lbs, one only has 5 lbs of capacity to lift something! So it is important to scope out the tasks, and the probable robot gripper, before selecting the right robot.

As robots evolved over the 1970s, 1980s, and 1990s, their speed and cycle times were advertised similar to a car going from 0 to 60 mph. Faster speeds and quicker cycle times seemed an obvious goal, and in many operations this is basically true. The baseline cycle time was defined by a standard motion as shown in Fig. 5.3. It assumes a cycle where the object is grabbed and lifted 1 in., moved 12 in., and placed 1 in. below. The robot and gripper must also return to the starting grabbing location to complete the cycle. The cycle time for many years hovered around 1 sec. Then some SCARA manufacturers and others were able to reach 0.5 sec and less.

Now if this were a reduction of time for 0–60 mph, race car enthusiasts would be drooling. It has its merits for automation also, but there are other issues to consider. If one uses high school physics, one will determine that the 0.5 sec time and given displacement conditions produce accelerations of close to 10 times gravity. Other issues are:

- Can the gripper close and open almost instantaneously?
- If it does close instantaneously, will the reaction forces on the gripped part be OK?
- Would the part be misaligned if the gripper closes this fast?
- If a suction cup is used, will a vacuum be achieved in time?
- Will the part slip during the 12 in. motion and become a projectile?

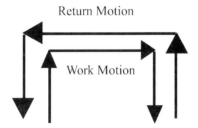

FIG. 5.3 Standard robot cycle

Industrial Robots

Many of these issues cannot be answered without some experimental testing on the specific parts one has to assemble. Many robot vendors have customers testing robots at various locations around the country just for these purposes. One may not get all of the answers to 100% satisfaction, but many of the risk factors can be removed.

The next two questions on robot costs and intelligence are somewhat linked. Usually robots right out of the shipping crate are not well outfitted as mentioned before, but one needs to explore if the candidate robot has the appropriate controller processing capabilities if one does add sensors and external devices. Most robot controllers today have these features, but they range from limited to powerful. One needs to discuss the details of a pending automation project with robot vendors, well beyond payload and cycle time. Nothing is more frustrating than someone giving a project team a 15-year-old robot that seems to be able to perform all of the required motions without trouble, yet the controller does not have the capacity to handle the types and flows of sensory information.

As for how a robot differs from automation, one needs to look at the components found in most robots:

- Structural members and bearings;
- Electrical motors;
- Optional gearboxes (depending on motor type);
- Position encoders;
- Wire cabling for motor power and feedback from sensors;
- A computer/controller;
- Amplifiers to boost computer level output signals to high amperage power signals.

All of these components are most likely to be found in automation. It is just that as an automation designer, one would be selecting these individually from catalogs, not simply selecting the assembled robot as one item. And the robot's software is usually far more versatile than what one uses in automation. So this is where robots get their advantage in reprogrammability and ready availability.

5.3. GENERIC ROBOT TYPES

The author performed a review of available robot geometry in the 1990s (Derby and Cooper, 1997). The result of this review showed that current robot designs are limited to a few generic designs. Nearly every general robotic textbook around confirms this. Groover et al. (1986) is an excellent resource and confirms this point. Potential robot customers were sometimes being sold robot solutions for problems that sometimes do not match well with the robot's kinematics or usable workspace. One problem that often arises is the need to rework factory floor layouts to suit robots, which can be costly. Not much is different in today's robot geometry in the marketplace. Very few technical advances have been accomplished in the last 20 years.

Early industrial robot arms were generally developed along the direction of existing geometric systems. The first three or four moving members of the robot were often:

- Cartesian;
- Cylindrical;
- Spherical;
- Jointed spherical.

These simplified base geometries (Fig. 5.4) have appeared in almost every robotics text written. A key to the choosing of these geometries was the fact that the geometric expressions for the joint angles and slider lengths (collectively called joint variables) could be calculated using simple geometry. These expressions needed to be calculated every few milliseconds, and the computer controllers of the day were very limited when compared to the Pentium® computers found on most engineers' desks today.

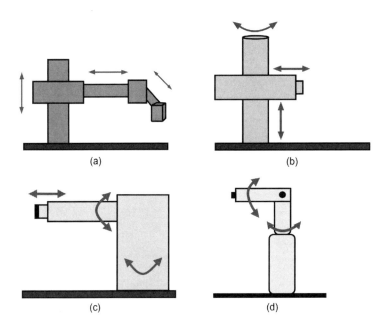

FIG. 5.4 Four standard robot base types: (a) Cartesian; (b) Cylindrical; (c) Spherical; (d) Jointed spherical

Industrial Robots 105

There was a massive entry into the robotics market by many of the Fortune 500 companies in the early 1980s. Except for the introduction of the SCARA-type high-speed assembly (or pick and place) robots in the mid 1980s, and a few Gantry type robots, most of the robot manufacturers were content to copy each other's products. They were creating general solutions and trying to market them to every application, in a sense a solution looking for a problem. It should be noted that the discussion in Sec. 4.4.6 presented the most popular robot base geometries in recent years. The Gantry type is an evolution of the Cartesian in Fig. 5.4. However, the Cylindrical and Spherical robot bases have virtually disappeared from the market today.

5.3.1. Kinematic Solutions

The university academics addressing robotic kinematic problems were led by many dozens of educated scholars from around the world. Most researchers were trying to find the required values of the robot's joint variables for the robot's gripper to be located at the desired position and orientation. It would be too difficult to list them all here (and the author would run the risk of insulting those who were not listed). A complete bibliography of all robot kinematic papers would most likely reach 1000 entries!

The required geometric expressions for the values of the robot's joint variables, most often referred to as inverse kinematics, were approached from two different directions. The first direction was the deriving of exact solutions, where the joint variable was solved for using a closed form expression. However, some of these closed form expressions required finding the roots of polynomial equations, which generally used iterative computer solution techniques to find the answers anyway.

A second direction was to not use a closed form solution at all, but to allow for some iterative or adaptive method to find the inverse kinematics. There were many comparisons to performance time based on different computers and computational techniques. More recent methods include Neural Networks and Genetic Algorithms. There were early attempts by some in industry to create generic robot controllers using some of these general methods, but most of these early efforts were not practical in industry.

A problem is that the majority (maybe 98%) of the inverse kinematic solutions never made the transition to industry. Yes, the kinematicians had solved every reasonable problem they could uncover, and many times there were anywhere from 2 to 12 methods derived to find the same identical answers, but the added versatility that these more complex robot arms could address has yet to be seen as worth the cost to actually implement.

Most industrial robot arms built today use inverse kinematics that can be solved using first year college geometry. Only a few special-purpose service robots have used the more advanced knowledge.

5.4. ROBOT WORKSPACE ANALYSIS

After the inverse kinematic field started to get filled with researchers, industry did ask a relevant question. It concerned the relative merits of choosing one type of existing industrial robot over another. Industrial engineers wanted to know when they should select a Cartesian robot over a Cylindrical one. Many researchers joined in this effort, some limiting their work to planar robots, and some to the three-dimensional world. The author lists a single reference (Derby, 1981) as an illustrative example.

The referred paper deals with the maximum reach of a six revolute jointed (or rotating joint, as opposed to sliding joint) robot arm of general geometry (i.e., all of the robot's fixed angles are not 0° or multiples of 90°). This research effort created a simple recursive formula to determine where this robot could reach. However, a practical flaw to this problem statement was that the results would fix the robot's gripper, or hand, at an arbitrary orientation (Fig. 5.5). So if the robot was holding a glass of water, it was likely that the glass was empty while the robot was at this magical maximum position (Fig. 5.6)! The practical maximum reach of a robot is based not only on the robot geometry, but also on the gripper type used, and the task to be performed.

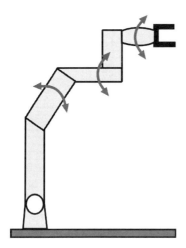

FIG. 5.5 Maximum reach of six-jointed robot

Industrial Robots

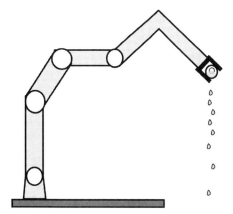

FIG. 5.6 Robot at maximum reach and gripper holding a glass with water

Drastically different robot arm configurations would be possible using these general geometry robot designs that the kinematicians had solved for the joint variables, but these robots usually required a much larger working envelope of unoccupied space in order to perform the same tasks. None of these possibilities has yet been brought to market. So again, many of the results from countless well-educated and well-meaning researchers have let down the industrial workforce.

Industrial engineers today can plan their robot application using computer graphical simulations. This is a positive step forward, but in reality, the simulation is generally a quicker and cheaper method of a simple trial and error process. It allows the actual robot to keep performing the first task, while it is being reprogrammed for a second task.

On first inspection, the floor space claims for a robot consisting of all revolute joints appears to be much less demanding than a large slider track base, but as will be shown in this chapter's case studies, this assumption can be proven false for some types of applications.

5.5. ROBOT MECHANICAL ACTUATORS

To make a general statement about the types of robot motion actuators, the overall trend has been to use revolute (or turning) joints. It is usually easier to support and contain all of the mechanical and electrical related components into a joint that bends like a human elbow, as opposed to a sliding joint. Track surfaces must be kept clean and orderly to handle all of the wires. Movable cable chains serve to control wires and pneumatic air lines at most sliding joints (Fig. 5.7).

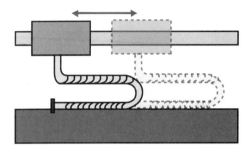

FIG. 5.7 Cable chain containing wires and air hoses in two positions

Friction and lubrication are not trivial. Robot costs can generally be lower using revolute joints.

Many of the early medium and larger robots of the 1960s and 1970s were powered by high-pressure hydraulic systems. These have virtually all been replaced in the 1980s and 1990s by electric servomotors, many with novel designed gearboxes such as the harmonic drive (Fig. 5.8). A harmonic drive can produce a gear ratio in the order of 100:1, without a series of step down gears. Using a flexible thin band gear in its internal workings, it achieves what is geometrically impossible from rigid planetary gear calculations.

The three major components to the harmonic drive (Fig. 5.8) are the fixed outer gear that has teeth on the inside surface, the smooth oval connected to the input shaft (usually the motor, connected to the front of the figure, not shown), and the flexible thin band gear with outwardly facing teeth. The flexible thin band gear is connected to the output of the harmonic drive at the rear of the figure (not shown), and rides on the smooth oval. Bearing on the oval surface can also be used. The thin band gear has two less teeth than the outer gear, so if the oval moves clockwise starting with the large arrow pointed upwards, the smaller arrow shows the various locations of a particular tooth on the flexible band. After one rotation of the oval, the smaller arrow tooth has moved over two outer gear teeth. In practice there are many more teeth than one shown in Fig. 5.8.

Many of the sliding joints were originally hydraulic pistons by design. The electrical version of today usually has an expensive lead screw driven by a rotational motor. On systems not requiring high accuracy, a chain drive can be used.

5.5.1. Robotic Linear Modules

Several robot companies have recently produced robotic control single axis of motion units, which have an integrated controller attached (Fig. 5.9). One

Industrial Robots

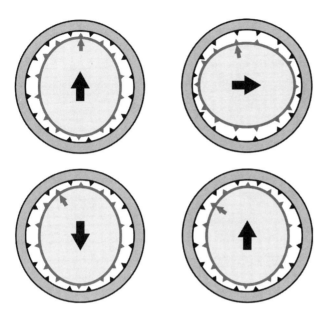

FIG. 5.8 Harmonic gear drive elements: oval turns clockwise

could then bolt them together in a myriad of possible configurations, using the different length of travel units available, and their different attachment mounting locations. A PC or other controller could handle the overall programming by the user and the playback in automation mode. Many of the possible configurations

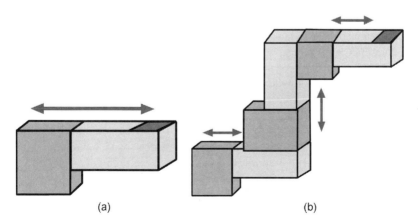

FIG. 5.9 Linear robotic modules: (a) single (b) several chained together

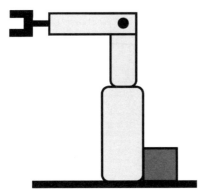

FIG. 5.10 Jointed Spherical robot with integrated controller

look something like the Cartesian robot base in Fig. 5.4. Some rotary motion modules are becoming available from these suppliers, but this motion is only for the last module, or in other words, the module closest to the gripper. There are presently not modules strong enough to rotate additional linear modules through space, similar to the Cylindrical-based robot in Fig. 5.4.

Some of these same suppliers are integrating the same technology into their SCARA and Jointed Spherical assembly robots. This is extremely convenient for manufacturing floor space concerns, but the overall functioning of the robot is the same as if the controller is 20 ft from the robot's base (Fig. 5.10). There is just more available floor space and less clutter from the cabling and hoses.

5.6. INDUSTRIAL ROBOT APPLICATIONS

The area differentiating robotics from automation is a very gray one. Almost all of the applications presented here can be carried out with dedicated custom designed automation, but because each area represents an application niche that the robot producers and integrators have been trying to fill over the years, there are certainly advantages of implementing robotic workcells that are essentially ready to run when they are installed. This may look advantageous compared to the development time associated with creating a totally new machine. Alternatively, one of these niche applications may only constitute a portion of the overall automation needs, and they would still be a welcome addition to one's developed automation.

A list of familiar robot applications is given is the following section.

5.6.1. Welding: Spot and Arc

The most commonly known robotic welding application is seen in many automobile manufacturers' television commercials (Fig. 5.11). This task, when

Industrial Robots

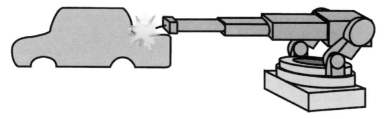

Fig. 5.11 Welding robot

carried out manually, was a very laborious job prone to worker injuries. Robots in this niche require significant payloads, in the 100–150 lb range, to carry the large welding end effector.

5.6.2. Spray Painting

Spray painting of products, ranging from cars to refrigerators to garden tillers, is a nasty job. The spray paint is not very good for one's health, and even in an electrostatically charged paint booth, the paint still seems to go everywhere. Painting robots are often a special breed of robots, specifically designed to only paint, and can do nothing else. A human operator, hopefully the most skilled craftsperson, guides the balanced robot while spraying paint on a sample part. The robot remembers the style and flow of the expert's motion, and will repeat it endlessly as new unpainted products come along (Fig. 5.12). This type of robot needs to be designed with the concept of required human interaction, since most robots need to be designed and installed to guarantee minimal human contact for safety concerns, listed in the Robotic Industry Association (RIA) safety standards (www.roboticsonline.com).

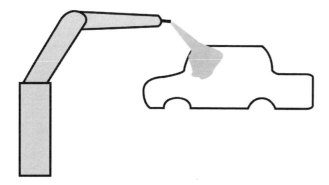

Fig. 5.12 Robotic spray painting

5.6.3. Dispensing

Dispensing of adhesives at home can be a challenging job when one is trying to seal a window or door leak with a tube of silicone rubber. The silicone always seems to be too thick to start the flow, and then does not seem to stop soon enough to prevent it from continuing to ooze out and deposit where it is not really desired. The automated dispensing is an art in itself. Some automation firms will place their dispensing end effectors on a series of available robots, while other firms will make their own specific dispensing machine, usually a gantry system.

Figure 5.13 shows the dispensing of adhesive for an automobile windshield. Most car owners will agree that the original windshield of a car is always installed with better quality compared to a replacement windshield. This is because the replacement is installed by hand application of the adhesive, and the car frame surface it is applied to is never perfectly cleaned of the original adhesive, no matter how hard one tries.

5.6.4. Assembly/Material Handling

The difference between Material Handling and Assembly is usually the level of contact between the manipulated product components. Assembly often has a connotation of precision mating or threading of two parts. This drives the precision requirements of the automation, and often leads to a SCARA-type robot.

The Selective Compliant Articulated Robot Arm (SCARA) was designed for the higher precision requirements of assembly. The use of six-jointed Spherical robot arms had proven to be insufficient for the markets assembly robots were trying to exploit. The novel improvement of the SCARA design was to keep the two horizontal arm members always parallel to the floor. This meant that the mechanical design could be optimized for gravity always pulling the total robot arm in the same direction. The changing effect of gravity and its pulling of members to the limits of its tolerances did add tremendous challenges to the

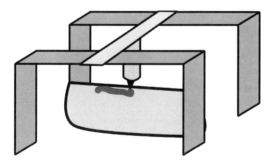

FIG. 5.13 Robotic application of adhesive to automobile windshield

Industrial Robots

six-jointed Spherical design. The loads on the actuators are also highly improved with the SCARA design. To see a six-jointed Spherical robot have different performance capabilities whether it is in one orientation or another is frustrating, but totally reasonable due to gravity effects.

When a part is unloaded from a stack as in Fig. 5.14, it is usually referred to as Material Handling. In this figure, the top gasket is most likely placed upon a base part, and if the gasket needs to be aligned properly, then some alignment features will guide the gasket into place. However, one person's Assembly is another person's Material Handling.

5.6.5. Packaging

Many packaging automation tasks are focused around an industrial robot arm, so as to bring it to market quicker and to limit the risk in development. Figure 5.15 shows a Jointed Spherical robot loading product into its individual carton. There are established cartoning machines available, and the market is quite competitive on which design is more productive. Packaging will be discussed in greater detail in Chapter 13.

5.6.6. Food Processing

Many food processing tasks are either automated, or their manufacturing directors want them to be. If a product is somewhat firm, like chocolate candies, they can be placed in a box by the use of a robot and a vision system. A suction cup is often used as the robot gripper, so as to not damage the candy.

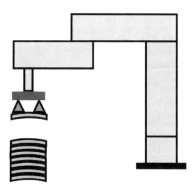

FIG. 5.14 SCARA robot transferring a gasket from a stack

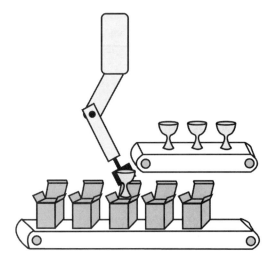

FIG. 5.15 Robotic cartoning

5.6.7. Drug Discovery

Laboratory robots are a niche product that dispense chemicals into small trays that contain hundreds or thousands of tiny wells. These wells act as arrays of very tiny test tubes, which are easier to handle and store as a group than thousands of individual test tubes. Much of drug discovery is trying to find the right chemicals or compounds to address a specific disease or problem. The process is sometimes viewed as a giant shotgun approach, where almost every chemical in the stockroom is tried to see what works. The wells are very small, which is also good since for reasons of cost one does not want to use up the stockroom supplies too quickly.

Laboratory robots also perform the material handling to move the trays from various testing machines and other processes, such as ovens and detectors. This form of business is not of the size and speed associated with traditional manufacturing, but it is a growing market than may be integrated with more equipment as time goes on.

5.6.8. Deburring and Polishing

Many parts, after molding, casting, or forging, require some rough edges to be removed and the entire product polished. Robots are used to both load and unload dedicated machines, and are also used as the manipulation device to bring all surfaces to the processing location. Figure 5.16 shows a robot polishing a golf club.

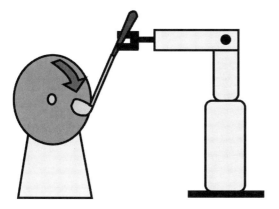

Fig. 5.16 Robotic polishing of a golf club

5.6.9. Machine Loading/Unloading

There are many manufacturing processes where the unfinished product needs to be loaded into a machine, and when processed, the finished part removed. Some of these applications are:

- forging;
- press;
- heat treatment;
- foundry processes;
- machine tools;
- glass manufacturing.

Many of these can benefit from a double robotic gripper (Fig. 5.17). The robot brings along an unfinished part in one gripper while it uses the second gripper to remove the finished part. The time and motion needed to swap these two parts is quite small compared to making the complete trip to remove the finished one, and then get a new one and to load it (Fig. 5.18). The cost of the second gripper is usually paid for in a matter of weeks, if not days. The idle time when the machine is not working is also greatly reduced, assisting with productivity calculations.

The steps for the loading sequence are as follows:

1. The machine has a finished product (shown as light surface color) and the left robot gripper has an unfinished part (shown as dark surface color), and moves to align its right gripper in front of the finished part.

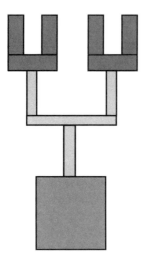

Fig. 5.17 Double robot gripper

2. The robot moves in and grabs the finished part.
3. The robot retracts and holds both parts.
4. The robot shifts to the right so as to align the left gripper with the machine location.

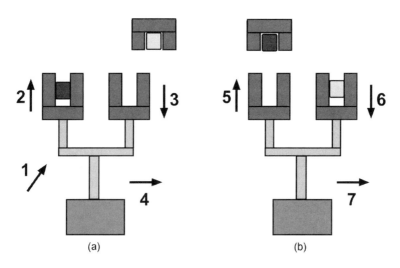

Fig. 5.18 Double robot gripper loading sequence

Industrial Robots

5. The robot moves in to place the unfinished part into the machine and releases.
6. The robot retracts from the machine.
7. The robot moves away to place the finished part in its desired location.

Other processes like plastic injection molding simply use the robot to unload the finished part. There is no loading in these cases. And in some markets, like injection molding, dedicated unloading robots are available.

5.7. CASE STUDY NUMBER 1: MACHINE LOADING/UNLOADING

This case study was part of the author's experiences of one of his many industrial visits. It has been generalized to remove any confidential information. Thus, the figures are simple blocks. But this situation can still be transferred to many other real-life applications.

A company machining and processing mechanical parts is currently loading and unloading various lathes and milling stations, or as a group, called machining centers. Some newer machining centers have built-in parts loading features, but many of the existing machining centers are loaded manually. Because there is a great investment represented by these older machines, and the fact that they still function quite well, automation solutions are being explored.

The company director thinks he needs a "robotic" solution. Various robotic vendors inspect the factory, and list the best robot they sell for each potential application. Almost all of the robots are of Pedestal design (the standard, non-Cartesian designs) from Fig. 5.4, usually Jointed Spherical in nature. Figure 5.19 shows a view from the top, with a simplified robot and its pedestal base. Because most machining cycle times are reasonably longer than

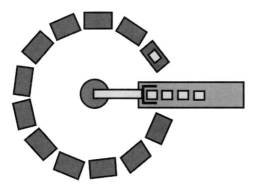

FIG. 5.19 Pedestal robot in a workcell

a prospective load/unload cycle, a single robot is projected to service two machining centers. This is often needed to cost justify the workcell, since usually one person would have been performing this same job manually.

Because most machining centers are rectangular, and they are often much longer than they are wide, the floor space geometry can become a problem. Figure 5.20 shows two machining centers, with the inflow of raw parts on the right, two parts being processed within the machining centers, and a finished part on the left. Available floor space is usually limited to a safe working path for a human operator. However, the installation of a pedestal robot often requires the slim linear workspace between the centers to be enlarged so that the Jointed Spherical type robot may move reasonably and safely for all required trajectories and approach paths. One reason is that the center's load/unload location is not located in the middle of the machine, so the pair's configuration as per Fig. 5.20 is not directly opposed to each other.

Moving the large machining center equipment is not a trivial task, or a trivial cost, and some proposed redesigned floor plans by robot vendors require significant increases in square footage (more costs) and a layout that is not rectangular (often cylindrical is required). Very few industrial plants would want to waste space by having effectively many circles within their rectangular grid of pathways.

By looking at the customer's needs, and trying to find a reasonable solution, as opposed to being limited to having a list of potential products that one is trying to sell, one may turn to custom automation. And after some inspection of the custom automation concepts, one sees that many solutions can be considered as custom designed robots. That is the case in the proposed solution to the case study presented here.

In the single factory briefly described in this case study, there were at least seven sites for either a single or dual arm tracked robot system. Chosen was a vari-

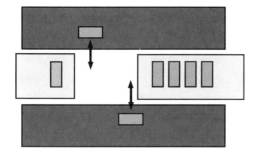

Fig. 5.20 Machining center layout: two different machine models

Industrial Robots

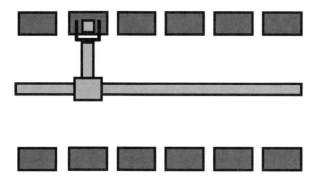

FIG. 5.21 Linear Tracked robot concept

ation of the basic robot workcell shown in Fig. 5.21. The requirements for all seven sites ranged to a reasonable degree, but the speeds and capacities were similar.

The gripper would have to be custom for each application. A double gripper on a single robot arm would allow the robot to hold the raw part nearby, as the finished one is unloaded. A single arm version with a SCARA arm, or a single linear arm with a horizontal rotational degree of freedom could also have been used.

The overwhelming number of linear motions required by the machining centers load/unload process seemed to demand a tracked system. To be able to install a robot, without moving any of the equipment and its supporting electrical and environmental hookups, would be very attractive.

5.7.1. Where Did Things Go Astray

The root problem is that there are many robot sale persons trying to find applications for their products. This case study shows that an available robot is not the best answer, with respect to the costs at least. So a custom automation solution, which looks something like an industrial robot, looks to be an attractive answer.

A truly creative CAD analysis tool that uses the problem specifications and geometric workspace demands to design a set of custom robots needs to be developed. If one of the custom answers is close to an existing robot, then it should be highly considered. Otherwise, the custom answer should be developed as the better solution to the real problem. This will create market niches of robots that most likely do not exist today.

The most important lesson is that to best address the automation challenge, one has to look at the underlying geometry and workspace considerations. It is important to not be limited to a predetermined list of robot geometries, potentially forcing a solution to the problem.

5.8. CASE STUDY NUMBER 2: PANTS PRESSING ROBOT

In the early 1990s the United States Defense Logistic Agency (DLA) was concerned with the production of uniforms for the military. There were a myriad of issues involved, but one key issue was that it was becoming less economically attractive to have uniforms made in the United States. Robotics was seen as a savior for many branches of production, so the tasks involved in making a pair of military dress pants were reviewed. One of the nastiest jobs for humans was the operation of a clamshell pants press to place the original crease into the pant legs. Figure 5.22 shows such a press. One will find similar models in dry cleaning establishments.

The steps to be automated were as follows:

- Take the pants from a hanger from a transport system (Fig. 5.23).
- Place pants on clamshell press.
- Inspect for any wrinkles over 0.25 in. in height.
- Smooth out wrinkles if necessary.
- Signal press to close and perform pressing cycle.
- Remove pressed pants from clamshell.

Many approaches to this automation problem were explored. The research team (Heines, 1991) visited several commercial uniform manufacturers to see if their workers had any undocumented tricks they incorporated, but there was little to be learned. The researchers spent hours in local dry cleaners, and even obtained on loan their own clamshell press with which to practice.

Significant time was spent on wrinkle detection. Because the pants are a woven fabric, and depending on how the person or robot placed the pants on

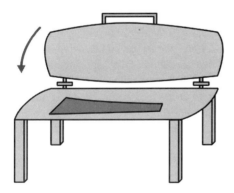

FIG. 5.22 Clamshell press

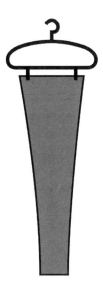

FIG. 5.23 Pants on hanger

the press, the direction, size, and location of any possible wrinkles were difficult to anticipate. Moving optical sensing devices not unlike a flatbed scanner were attempted. Vision algorithms were compared. All of this was before a particular robot was to be selected. The researchers needed to know how many robots, different operations, and grippers would be required before selection could start, but early concept discussions were leading to multiple robots and quick exchange wrists.

However, in what can best be referred to as either a brainstorm of an idea, or that the team finally woke up, was the realization that trying to mimic the human operator was a flawed concept. All of the effort to detect and smooth wrinkles could be eliminated if the pants were grabbed at all four corners as shown in Fig. 5.24. A robot end effector would have four simple hands, not much more complicated than a spring operated clothes pin, and the proper tension would be applied both along the pant's leg and across the leg. Then there would never be a wrinkle, so why detect them or try to smooth them out. The pants would be held in place while the pressing occurred, so the challenge of having to regrip the pants was eliminated too. Because humans do not have four hands, this approach was not easily forthcoming.

In production the clamshell press body would have to be slightly modified to allow the four hands to be situated inside and the press surfaces to close properly about the pants. This was not seen by the press manufacturer to be a big issue, as long as multiple presses would be purchased.

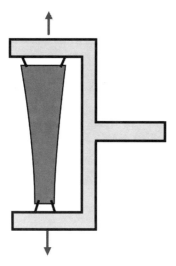

Fig. 5.24 Four hand gripper holding pants

5.8.1. Project Assessment

Now that the process was well defined, the number of tasks for the robot was greatly reduced. The payload for the robot was about 30 lbs, 90% of it being the four handed gripper with tension control. But the trade-off was very acceptable. The reach of the robot needed to be about a 5 ft radius circle, with most of the motions taking place 3–4 ft above the floor. A medium-sized six-jointed robot was selected to do the job. The automation task was successful, but was never continued due to DLA program changes.

5.9. CONCLUSIONS

Robotics is a very viable solution for many automation projects, and has matured tremendously over the last 40 years. Software has become reasonably easy to learn, and sensors are assisting to create some intelligence. There are many significant automation machines being built today that have one or more robots as integral devices, and the falling prices are attractive.

So, do strongly consider robotics as a major part of your automation machine, or simply as part of a flexible part feeder, but be warned that hidden among the customer base in the world are some people who have been burned

Industrial Robots

big time by robotics in the past. It is always wise to float the idea of using a robot to your customer before investing a great amount of time. However, if all they talk about is Star Wars robots, then perhaps a different approach is in order!

PROBLEMS

1. Investigate using the Web the available robot arm simulation programs. Compare features for your educational or working environment.
2. Explore the current robot manufacturers who are members of the Robot Industries Association (www.roboticsonline.com) and note the model types available.
3. Take an assembly task, such as the stacking of several items and determine a robotic workcell layout assuming that you are the robot. How does this layout differ if you use one arm vs. two arms? What is the impact of performing the task with your eyes open (modeling machine vision) or with your eyes closed (modeling the use of tactile sensors only)? What general trends do you see after you have performed this process multiple times?
4. Take a nut and bolt, and thread the two using your two hands. Write a list of concerns you would have as an automation engineer if your boss suggest that you use two robots to perform the same task. Investigate on the Web if any company suggests that their industrial robot arm can perform two arm assembly, and then see what research organization is performing such work.
5. Emulate a robot/automated machine by trying the following exercise. Create a model of some devices you have in your office/room, but place a pair of heavy socks over your hands. Or use a pair of pliers to emulate a two finger gripper. Try performing some of the steps with your eyes closed. How do you verify when things are correct? What is your comfort level that you are complete?
6. Search the Web to see what commercially available industrial robot would be best to install next to your kitchen sink to manipulate dirty dishes and to wash them. Design a suitable gripper, and any auxiliary devices for cleaning required. Can this system handle pots and pans as well as everyday dishes?
7. Investigate robot delivery times for several industrial robot arms, and compare to the time to design, build, and test a new workcell device. Can you compete with the costs? What training level would you need from the robot supplier?

8. What additional concerns would you need to be investigating to make a humanoid robot?

PROJECT ASSIGNMENT

Using one of the projects from the Appendix (or any other projects), perform the following:

1. Review your work from Chapter 4, and see which of your machine configurations can be handled by a commercially available industrial robot. State how much complexity would a robot remove.
2. If the entire project can be performed by a robot, determine the process step trade-offs compared to hard automation. Most likely you will not want to choose the option of complete robotic implementation for the sake of performing the other assignment steps in the remaining chapters. However, in real life this may be an option.
3. Estimate the cycle time for both the robotic operation and the hard automation solutions. Have someone record the time you pretend to perform the various tasks to get a reasonable estimate. Confirm any operation in question using simple physics of motion.
4. Determine how many robotic grippers you will need, and if you will need a wrist exchange to handle them all. Will you need more than one robot to have a reasonable cycle time?

REFERENCES

Derby, S. (1981). The maximum reach of a revolute jointed manipulator. *Mechanisms and Machine Theory Journal* (International Federation for the Theory of Machines and Mechanism), 16:255–262.

Derby, S., Cooper, C. (1997). The evolution of robot geometry and its impact on industry. Fifth National Applied Mechanisms & Robotics Conference. University of Cincinnati, Cincinnati, OH, paper AMR97-056, Oct.

Groover, M., Weiss, M., Nagel, R., Odrey, N. (1986). *Industrial Robotics: Technology, Programming, and Applications*. New York: McGraw Hill.

Heines, R. (1991). Automated Handling of Garments for Pressing. MS Thesis, Rensselaer Polytechnic Institute.

6
Workstations

So far in this text we have discussed some of the overarching design approaches and traditional automation formats. We have looked at cost justification methods, and the impact of estimating costs. However, we have not looked at any real level of detail in the design of our automation. Because every automatic machine can be different from the next one, one will need to look at the discussions in this chapter as a series of examples, while trying not to force your current viable concept into a rigid mold.

We will use the term "Workstations" as a catch all to some of these missing details. A workstation is where the automation does something, whether by a dedicated machine, an industrial robot, or even a human operator if it is a process step that cannot be successfully handled any other way. Workstations are a logical grouping of commercially available components, integrated with a series of custom design parts and/or electronics. They is more than a single motor or an air cylinder. Workstations are usually connected either by conveyors, a rotary index system, or any of the other options discussed in Chapters 4 and 5.

To best understand the specifics of a workstation, one could argue that the topics in later chapters on actuators, sensors, and controllers need to be covered first. In general, the order of the topics in this and the next four chapters are similar to the old statement, "which came first, the chicken or the egg." What one finds in designing automation is that the process is not linear. Solution concepts sometimes come coupled to a new actuator, or an implied control strategy. Students using this text may find themselves iterating through these topics several times until it all synchronizes together. With this said, there will be many small points on actuators and sensors that will be mentioned in this chapter that will hopefully make more sense when their details are covered later.

It should be noted that not everyone in the automation field uses the term "workstation." If one searches the Web for workstations, one will either find an efficiently laid out workbench for a human to process some manufacturing task, or they will find some laboratory automation workcells used in the pharmaceutical market for drug discovery. One needs to search the Web for the specific function required at that workstation. However, it is logical to approach the next phase of automation design by grouping technology using this somewhat arbitrary naming convention.

There are many commercial workstations available. Many can be found by determining the specific function one wants to incorporate into a machine. Samples of these include:

- air-powered press;
- screwdriver;
- hot glue melt;
- automatic fill heads for liquids;
- ultrasonic welding.

Rarely does one want to start the design process by replicating these. Because there is usually more than one vendor able to supply these often-used devices, competition normally keeps the pricing reasonable. It is foolish to think that one could design a hot glue melt applicator cheaper than the systems that have been around for 30 years or more. However, that never stops some engineers, who want to design the entire machine, even at the risk of higher costs and greater delays!

A custom workstation would be the actual detailed design of the device to tie bows on packages, as discussed in Chapter 2. During the required actions described in Chapter 2, a concept on how to do the process was established, but none of the real details as fleshed out. A gripper was seen as a desired tool in the process steps, but we did not determine if the gripper was to be placed on a robot or a custom designed pick and place device. Also, the gripping surfaces need to firmly hold the ribbon without destroying its polished appearance.

The determination of whether or not to use an available workstation will often take the following steps:

1. Look for available devices.
2. Determine all of the combinations of workcell layouts and the needed workstations.
3. Play "what if" games if one is not limited to available devices and can make custom devices.

Workstations

4. Establish a trade-off matrix to compare all options.
5. Use the "best" answer, but always remember the other options. The best answer may be controlled by nontechnical issues such as economic conditions, safety, union contracts, and so on.

It could be that by using commercially available workstations, a designer is painted into a corner to create a solution similar to every other company's attempt at automation. A novel concept may drive the overall system layout to a better solution.

In the case of tying a bow, if one could find such a workstation, the task would be moot. It is the key problem to this project. As for determining all possible combinations, a quick hand sketch of what the different designs would look like would be very useful. As one might imagine, the work discussed in Chapters 2, 4, 5, and this chapter are almost always an iterative process. If after several hours or days of automation concept development, no possible gripper design to grab ribbon and pull it through the loop can be found commercially or custom designed, then perhaps the great idea in Chapter 2 is just a dream. One must bring all of these steps into focus, or else one needs to look at the second best design, or maybe even will have to say that it cannot be done!

So since this is an iterative process dealing with multiple chapters of this book, not all automation engineers will proceed with such linear thinking. They may jump to an answer based on earlier successes with other applications, or they may be well versed in one particular robot language, and see that solution path as being the simplest and less risky path to take. Is it the best path overall? Sometimes the answer is "yes" and sometimes the answer is "no".

6.1. WHEN IS IT A WORKSTATION?

So with the above introduction, we could use these checkpoints to assist with defining the tasks of a workstation:

- something that does some operation;
- does it repetitively;
- uses some type of power.

Other issues for workstation designs to consider are:

- Do you perform the operation online or remove it from the conveyor?
- Do you consider a few complex motions or many simpler motions?
- Do you integrate structural members with the actuators?
- What are the roles of sensors?
- What kind of accuracy is needed?

- What kind of adjustments will be needed, and how will they be done?
- What kind of safety standards will you have to adhere to?

With these definitions and raised concerns, one would likely assume that a conveyor to move a stream of products from point A to point B would not be a workstation. And in most cases this is true. But every now and again, some external device will be integrated with the same type of conveyor, and just like the confusion over what is and is not a robot, a workstation is born. Again, if the customer wants to call a conveyor system a workstation, and they are happy with your price, why not go along?

6.2. WORKSTATION BASICS

The building blocks of most workstations will include:

- structural members;
- bearing devices;
- drive mechanisms;
- actuators;
- sensors.

One might ask about the distinction between a drive mechanism and an actuator. A drive mechanism could be a gearbox or chain and sprocket system that takes the motion and power generated by an actuator (an electric motor for example) and produces the appropriate torques and speeds. Actuators will be covered in Chapter 8, so they will be generally left out of the discussion here. Likewise, sensors will be discussed in Chapter 9.

Many automation builders do not always divide the total machine into workstations; they just build the machine in segments of the overall machine, but the trade-offs and issues of this chapter still remain. A designer needs to know what components can be purchased and what needs to be built from scratch, and how to effectively integrate and interface them. And the laws of physics cannot be ignored.

6.2.1. Structural Members

If you work alongside automation builders long enough, someone is bound to make the statement, "well, it doesn't have to fly." This implies that the strict weight restriction that aircraft designers need to adhere to is not a concern in the automation world. Since the machine is probably shipped by truck only once in its life, the total weight is usually allowed to accumulate faster than one could ever believe, particularly the machine frame. No-one wants the frame to bend or buckle. And since the frame is often designed before the entire machine is laid out, one is not sure what the static or dynamic loads will be (Norton, 1999). This is not the best practice, however.

The Frame

Many machine frames are constructed from steel — lots of steel. If it is to be located in a food processing plant, it will need to get washed down once each day and therefore is likely to be stainless steel. Rectangular hollow steel tubing, perhaps 2 in. × 2 in., that could be welded into something that might support a military tank is a natural reaction from experienced machine builders.

Sometimes the thought of bolting steel members together sounds a little more flexible in the design phase when there is still some doubt of exactly where some facet of the machine needs to be located, but since most automation machines will be humming along for years to come, the thought of bolts coming loose from the vibrations gives experienced designers concern.

Another concern for the frame (as well as the entire machine) would be the effects of vibration and any resonance modes. These resonance modes can be one of the biggest surprises after a piece of automation has been built and is undergoing test runs. Sometimes the resonance occurs at the normal operating speed, but more likely it can happen when the machine is accelerated from rest up to the operating speed. If the resonance transition time is short, and there is little that can be done to stiffen the frame and moving members, it may be deemed to be acceptable. Or it may shake the entire machine apart in six months. This is something that some computer analysis can assist with, but the level of modeling required still may produce simulation results that are shifted significantly from the real-life occurrences.

The support frame could be extruded aluminum components, since they are available from more than a dozen suppliers. They are easy to order, and some brands have a fantastic number of stock items for joining lengths of aluminum extrusions to form a structure or to have some limited motion (perhaps an access door). Various clear plastic sheets can be cut to fit within the frame, giving a very professional look in a very short amount of time. However, these members, being aluminum, are not as strong or rigid as steel. And they are not welded together, they are bolted together. They will deflect more than one might want, and are often only used for external guarding or very limited load-bearing situations.

Whatever materials and construction are used, the frame will be custom designed and built for your machine. One cannot purchase a meaningful frame from a catalog.

Moving Members

The mass of the moving members does have an impact on motor selection, overall speed, and the response of the machine. So a greater effort is taken to try to minimize the weight and inertia effects. Most automation builders will use a very detailed CAD drawing to determine the size and assembly characteristics of the moving members. They will use the CAD program to calculate weight and inertia properties for them. Only some automation builders will actually run a finite element analysis (FEA) on the moving members to see what kind of deflections will occur. For some machines, deflections may not be critical due to relatively slow speeds, but for other machines, one should do the FEA so as to limit any surprises. Not enough machine design houses feel the need or are comfortable performing FEA.

Most moving members are custom designed and built, either in-house or at a subcontractor to a machining builder.

6.2.2. Bearing Devices

The frame and moving members are usually obvious to even the novice automation designer, but when it comes to attaching the moving members to the frame, some people might not think enough about friction, wear, and the need for bearings. And one cannot just look at standard roller bearings similar to what is found in one's automobile wheels. It all depends on the speeds and loading situation. Something will become a consumable, either lubrication that will be added periodically, or parts that will need to be replaced on a specified timetable.

Much of the concern is based on friction between two surfaces. Figure 6.1 shows two scenarios, where the magnified large peaks on the left will potentially have less frictional effect than the lesser peaks on the right due to the reduced contact area. This is why engineers sometimes guess incorrectly on what will improved a high frictional situation. However, when it comes to bearings and bushings, the profiles will hopefully look significantly smoother than either scenario.

Loads are generically found being applied to rotating shafts. This is because there are often a great number of rotating shafts involved with the major operations of the automation. The loading can be radial (Fig. 6.2), thrust (Fig. 6.3), or a combination (Fig. 6.4). The types of bearings will differ signifi-

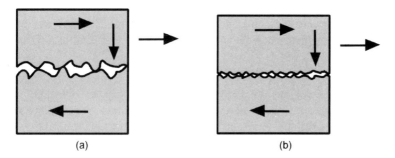

FIG. 6.1 Magnified frictional surfaces; (a) very rough; (b) moderately rough

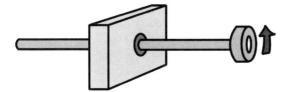

FIG. 6.2 Radial loading of rotating shaft

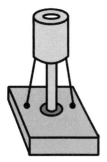

FIG. 6.3 Thrust loading of rotating shaft

cantly for these three cases, so one cannot simply guess and select a bearing that happens to be sitting on the shelf nearby.

The form of the bearing can take several forms for each of these loading situations: thrust bearings (Fig. 6.5) with washers (Fig. 6.6); sleeve bearings (Fig. 6.7); and flanged bearings (Fig. 6.8). And then there is the more well-known roller bearings (Fig. 6.9), with their several internal components (Fig. 6.10).

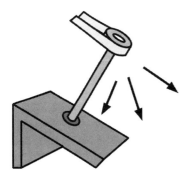

FIG. 6.4 Combined loading of rotating shaft

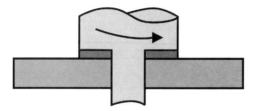

Fig. 6.5 Thrust bearing

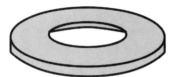

Fig. 6.6 Bearing washer

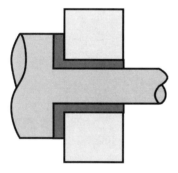

Fig. 6.7 Sleeve bearing

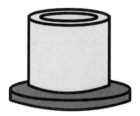

Fig. 6.8 Flanged bearing

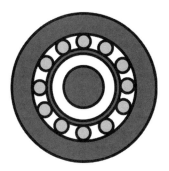

FIG. 6.9 Roller bearing

If the size of the bearing needs to be very compact, sometimes designers will opt for needle bearings. Figure 6.11 shows a cutaway view of such a bearing. If the loading is of greater forces, or there is combined loading, then the options are cylindrical (Fig. 6.12), tapered (Fig. 6.13), or spherical (Fig. 6.14). Spherical offers the ability for the shaft and bearing to be self-aligning, something that may be well worth the extra costs. This allows for accumulated errors in machining and assembly to be accommodated when the automation machine might be 95% completed.

Sometimes an older more traditional bearing surface will work just as well. A sleeve bushing, made of material that will either create a hydraulic film of lubrication, or will run in a dry condition, will be useful. Figure 6.15 shows a standard sleeve bearing. These can be press fit into the frame or other moving member, and their compact attachment properties can be attractive (Fig. 6.16).

Other times, sleeve bearings are placed into pillow blocks (Fig. 6.17). Another option is to replace the pillow blocks with a split sleeve bearing made in

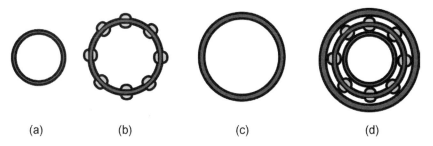

FIG. 6.10 Roller bearing: (a) inner ring; (b) race with embedded roller balls; (c) outer ring; (d) assembled unit

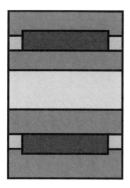

FIG. 6.11 Needle bearing: cutaway view

two halves (Fig. 6.18) to assist with assembly. No one wants to force a six-foot long shaft through a tight sleeve bearing, or worse yet, several tight sleeve bearings.

If the shaft or rod is to reciprocate in the bearing block, then there is a linear bearing available that is commonly used. There are several versions out there, but the classic version uses ball bearings that recirculate through channels (Fig. 6.19) to allow proper load reactions between the balls and the reciprocating motion of the rod, and the bearing block. These bearing are more expensive than roller ball bearings. But when the application calls for them, they are the right solution for the task.

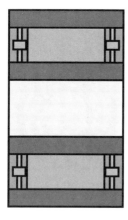

FIG. 6.12 Cylindrical bearing: cutaway view (cylinders with end pins spin within race)

Workstations 135

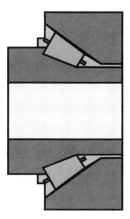

Fig. 6.13 Tapered roller bearing: cutaway view

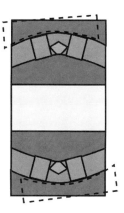

Fig. 6.14 Spherical bearing: cutaway view (dashed lines show alignment possibilities)

Fig. 6.15 Sleeve bearing

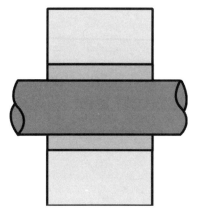

FIG. 6.16 Sleeve bearing press fit into support frame

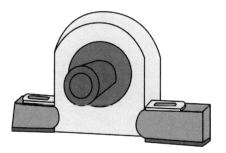

FIG. 6.17 Pillow block

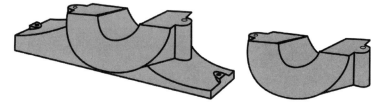

FIG. 6.18 Split sleeve bearing

6.3. DRIVE MECHANISMS

The drive mechanism takes the actuator output motion and transforms it to another location, power ratio, or manipulates it to some other characteristic. It could be a standard drive belt and pulley system (Fig. 6.20), either with similar sized pulleys or different sizes to change the speed and torque ratio. Belt sizes and types are stan-

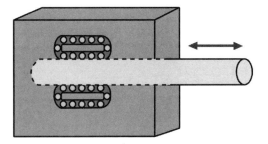

FIG. 6.19 Linear bearing

FIG. 6.20 Drive belt and pulley system

dardized, with the most common being the V belt (Fig. 6.21). There can be an idler pulley to take up the slack and reduce wear and noise (Fig. 6.22).

There are chain drives (Fig. 6.23) with which most people with two wheel bicycles can claim familiarity. As with a bike, most chains need lubrication. So their use is often coupled with a maintenance schedule to check on lubrication, and the need to clean up excess grease, but chains do have the added benefit over belts in that there is no slipping between the two shafts. The chain can stretch and wear out, but the relative timing can be assured.

A more up-to-date version of a chain, which is much quieter and can have fewer maintenance issues, is a timing belt (Fig. 6.24). The belt has teeth molded

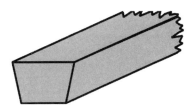

FIG. 6.21 Standard V belt

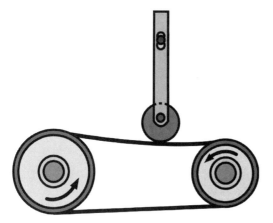

Fig. 6.22 Belt drive with idler pulley

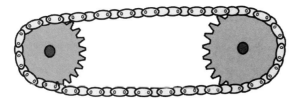

Fig. 6.23 Chain drive

into the inside surface, and the pulleys have matching surfaces that keep the belt registered. So as long as the tension is maintained (and timing belts can stretch from use) the belt should not slip. The allowable amount of force transmittable is given by each supplier.

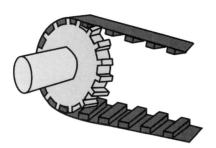

Fig. 6.24 Timing belt

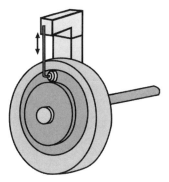

Fig. 6.25 Cam system

A traditional method used by automation designers of old is the cam (Fig. 6.25). It is still very valid today, but it lacks the flexibility or reprogrammability that is available through today's controllers and actuators. There is a significant cost in milling out the cam groove accurately, and the groove surface is costly to repair if it gets worn or damaged. However, the resulting motion can be very deterministic for many operational cycles.

Other drive mechanism issues consist of brakes, such as a band brake (Fig. 6.26) and a combined clutch and brake (Fig. 6.27). These items are needed when the machine's large rotation inertia needs to come to rest quicker than just letting it slow down on its own. Safety reasons alone may require a braking system. If it uses the right kind of bearings, it may take quite a while to stop without some assistance.

The last area to consider is the joining of two shafts. If one does not think too hard about this problem, they are likely to use a simple rigid shaft coupling as in

Fig. 6.26 Brake band to stop rotating shaft

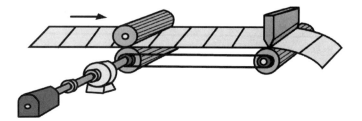

Fig. 6.27 Clutch brake system

Fig. 6.28. This is fine if the shafts are to rotate once per hour, or something similarly eternally slow, but for any rotational motion greater than this, the stress on each of the two shafts, and their bearings and support frame, will lead to trouble.

It is better to use either a universal joint (Fig. 6.29), when one knows that there will be significant shaft misalignment, or some other type of helical cut shaft coupling (Fig. 6.30) if the misalignment is much less. The helical joints can torsionally wind up when significant loads are applied to a coupling that is perhaps under-rated, but the benefits outweigh the new problems created.

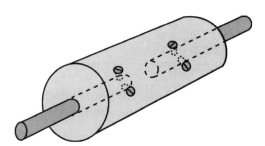

Fig. 6.28 Rigid shaft coupling

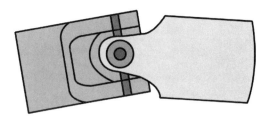

Fig. 6.29 Universal joint

Workstations

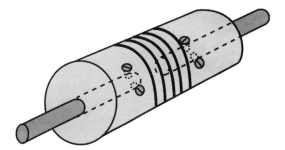

FIG. 6.30 Helical coupling joint

The misalignment of two shafts with high stiffness bearings can result in huge loads, which equal a loss of energy and a reduction of life.

6.4. CASE STUDY NUMBER 1: TBBL WORKSTATION DESIGN

If we now consider some of the details of the books on tape case study in Sec. 4.7.4, there is a list of questions to be first developed and then hopefully answered. Figure 4.23, repeated here (Fig. 6.31), shows some required motions to align either of the two different thicknesses of cases and then firmly hold the case in place for the latch unfastening action to occur. This was determined to be easier than adjusting the latch unfastening device vertically to accommodate the two different thicknesses.

So how is the case lifted up? And when does this happen? The case is probably moving forward by a conveyor system, as per the linear asynchronous design in Chapter 4. A hard stop, either fixed or sliding into the path of the tape case, will align the case by its leading edge (Fig. 6.32). Then the next step can occur if the automation system knows the hard stop was actually hit. A light beam (Fig. 6.33)

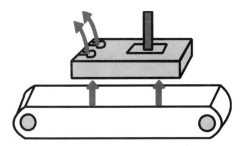

FIG. 6.31 Motions to lift and unlatch a case

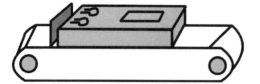

FIG. 6.32 Movable hard stop aligning leading edge of a case

or limit switch (Fig. 6.34) can be used as sensors to tell the controller this has happened, among other options. These two options are most likely to be cost effective compared to say a fully outfitted computer vision system, which would work fine but likely be overkill. Each of these sensors has its pluses and minuses (left for discussion in Chapter 9 on sensors), but they are fairly interchangeable for our planning here.

So how can the case now be lifted up? The motion would logically seem to be pure translation, so a series of rotations would be possible but lacking in efficiency. We could lift by:

- moving the case by vertical translation or sliding;
- using a scissors jack approach;
- inflating an air bladder;
- using magnetic power;
- using a four-bar linkage.

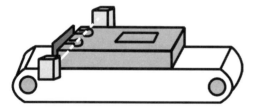

FIG. 6.33 Use of an electric eye beam to detect a case

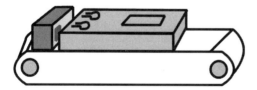

FIG. 6.34 Limit switch on hard stop to detect case presence

For some of these options, the form of an actuator (or power) is stated, as with the air bladder, but for others, like the scissors jack, the discussion on types of actuators needs to continue.

6.4.1. Moving the Case by Translation or Sliding

Because the lifting motion is not very high, and the case is fairly light, perhaps the easiest thing to do is to eliminate some possibilities first. We can therefore rule out the need for hydraulics, using its pressurized oil like an excavating backhoe, since this type of actuator would be used for lifting tons of mass. And since the lifting needs to occur only until the top edge has reached a target height, we do not need to consider a servo feedback driven system, which is more complex and would cost more.

We can consider compressed air, as both a motion strategy and a power source. The compressed air (or pneumatic) cylinder can lift (Fig. 6.35) until the reaction force of the lifting against a hard stop balances the internal air pressure (Fig. 6.36). It should not damage the case unless the supplied air power is excessive. It could easily handle an infinite number of case thicknesses without any significant sensory feedback or controller requirements. If compressed air is available, it is a potentially ideal solution.

A side issue would be the structural requirements for the lifting. Many air cylinders are designed to only handle loads along the axis of motion. Any side loading will quickly wear out the cylinder. And the concept in Figs 6.35 and 6.36 are idealized, assuming that the center of mass of the load lifted is directly about the cylinder. However, there may be an odd number of tapes in the case, so this assumption may not be true.

It would be better to have a pair of linear bearing rods and supports integrated into the design. Figure 6.37 shows one possibility. This looks easy enough, but for those who have tried it, getting the air cylinder and the two linear rods aligned for relatively frictionless motion through the entire range of motion is

FIG. 6.35 Air cylinder for lifting

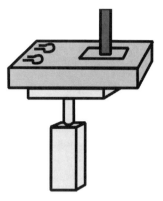

FIG. 6.36 Lifting until reaching a hard stop

not trivial. So, many engineers select a linear slide that is an air cylinder integrated with the two linear bearing rods as a single component (Fig. 6.38). A plate to distribute the support under all of the case would be attached. It is then almost impossible to apply the component incorrectly (although the author has seen it done).

A second option to the air cylinder would be a screw drive powered by a DC electric motor. This motor need not have any complex feedback associated with it. There would need to be a limit switch or electric eye beam similar to the case registering to the hard stop in Figs. 6.3 and 6.4. For the screw drive, there is a need for at least a single load-bearing rod, so as to restrict the case lifter to not spin (Fig. 6.39). A second rod is usually used for balancing forces (Fig. 6.40). Most screw drives units are not designed to see significant side loading or off-axis torques, since these will create early product failures.

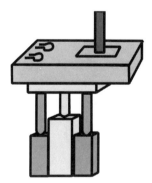

FIG. 6.37 Two load-bearing rods located in parallel to the air cylinder

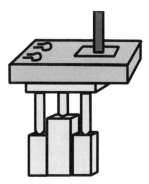

Fig. 6.38 Integrated air cylinder and load-bearing rods

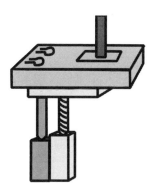

Fig. 6.39 Screw drive and the resistant to twist single bearing rod

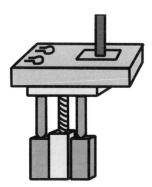

Fig. 6.40 Screw drive with a pair of bearing rods

FIG. 6.41 Scissors jack lifting a car for tire change

This option will probably cost a few more dollars than the air cylinder approach, but the cost differential is not much. If there is no compressed air available, or if compressed air causes a problem for other reasons, the electric motor screw drive could be the best choice.

6.4.2. Use a Scissors Jack Approach

Some cars use a scissors type jack to lift a car when changing a tire. Figure 6.41 shows one in use. The horizontal threaded screw shaft and nut are designed such that the jack is self-locking; that is, the frictional forces do not allow the weight of the car to back drive the screw and handle to spin freely. This is almost always a good thing when applying a scissor jack.

Using a scissors jack to lift the case in our workstation (Fig. 6.42) has some benefits. The top of the jack is held quite parallel to the reference surface of the jack's bottom frame member. The case would be effectively supported under the latch unfastening device. The actuator drive would most likely be a DC motor to spin the screw shaft. There are rotational air-powered motors, but they are less commonly found, and can be noisy.

The scissors jack has more moving parts, and might be more expensive that the linear slide units, but the scissors jack might be more compact vertically than

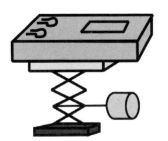

FIG. 6.42 Scissors jack lifting the case

Workstations

the linear slides. This dimension may drive the automation's base frame to a lower reference surface, assuming that the vertical height of the incoming case has been set as a requirement (and most likely would be). Yes, a conveyor system could be used to lift the cases up so as to be over a taller lifting device, but this adds to the cost that is possibly not needed.

One will not know which type of lifting device to implement until one finds candidate commercially available units and investigates their dimensions, but there are still other options to explore.

6.4.3. Inflate an Air Bladder

Many trucks on the road today use a series of air-filled bladders to act as both a leveling device and a shock absorber. The lifting function is not as precise as a feedback-driven actuator, but does compensate for uneven loading within the truck's body. These bladders can be a bulging cylindrical or a bellows design (Fig. 6.43). Some packaging machines (Davis, 1997) use similar bladders to cushion the multiple products being manipulated. Imagine the need to absorb the shock when a glass container filled with liquid is inserted into a box by controlling its dropping motion only in the slightest way (Fig. 6.44).

To use an air bladder of either design to lift the books on tape case would require some additional moving structure. The relative motion of the top surface of the bladder, and therefore the case itself, is not well defined (Fig. 6.45). The case most likely would not be aligned properly under the latch unfastening device. Several linear bearing rods would be a viable solution, but then the function of the air bladder has been reduced to that of an air cylinder, which will in general be a fraction of the costs of the bladder. And the air bladder's rubber makeup is not as industrially hardened as an air cylinder. So we will not select this option.

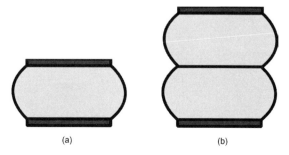

Fig. 6.43 Air bladders: (a) bulging cylindrical; (b) bellows design

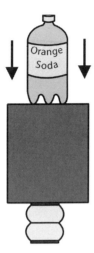

FIG. 6.44 Air bladder cushioning bottle being dropped into case

6.4.4. Using Magnetic Power

One has several choices when it comes to magnetic actuation, but they are in real life limited to several specific niche operations. The main choices are:

- solenoids;
- magnetic levitation.

Solenoids are used in many products found in industry and around the home. The door that holds in the soap until the correct timing of the cleaning cycle of your dishwasher is actuated by a solenoid. They are cheap, small, and can work a reasonable number of cycles. They are made from a wound coil of wire that creates an electromagnet. The center core is a material that responds

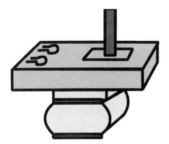

FIG. 6.45 Air bladder lifting a case: not well positioned horizontally

Workstations

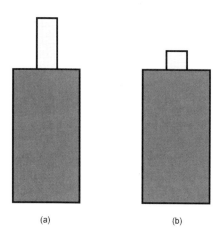

FIG. 6.46 Solenoid: (a) core normally out; (b) core pulled in by electromagnet

to the magnetic pull or push created (Fig. 6.46). Many solenoids have internal springs to return the core to the home position. They can buzz or hum when actuated.

They do, however, have a small range of motion, and they cannot handle any significant side loads. They do wear out after so many 10,000s of cycles, so they many not be sufficient for high-speed operation without significant preventative maintenance scheduling. It is not clear that a solenoid would have the strength and range of travel required for this application. Their motion limit is usually around 0.5 in.

Magnetic levitation may seem like a high-tech dream being applied to a down-to-earth problem. Maglev, as it is referred to for short, is used in Japan for high-speed trains. And would it require that the cases, now made solely from plastic, be modified to be magnetically actuated materials? Or at least a pallet or part of a conveyor is moved under the case each time a new case comes in to be unlatched?

But this perhaps silly discussion will help to raise a red flag not just for Maglev, but for the solenoid! The cases contain magnetically recorded audio tapes. Will Maglev, or even a simple solenoid, cause damage to the tapes? This concern is big enough to remove any use of magnetic manipulation from our considerations (Fig. 6.47). It even helps us ask the questions, "If we use an electric DC motor, where will it be placed, and will it be too close to the cases and cause tape degradation?"

So, adding what might be dismissed as a silly or even stupid suggestion to the list on possible workstations sometimes does bring out the red flags early on, when it does not cost any money to change a potential problem. Imagine if after

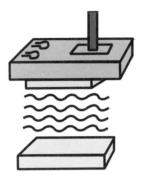

FIG. 6.47 Maglev lifting of the case: tapes would be damaged?

building the machine one finds that the DC motor is directly next to the case at one point in time, and by activating the motor one causes multiple gaps in the audio tapes. One might not even get word of such a problem for weeks or months after the machine is supposedly installed and paid for, and then this bug would tend to drive one crazy, and be very costly to debug and repair!

6.4.5. Using a Four-Bar Linkage

The use of a four-bar linkage would seem initially to be a poor choice. As seen previously in Fig. 4.19, the general motion of any coupler path is a cross between circular motion and a deformed banana. It does not normally lend itself to straight line motions. There are some special mechanisms known to designers for generations (Norton, 1999; Chironis, 1991) that do generate straight line motion by using only rotational joints, but these do not apply to this problem for several reasons:

- The required space for such a linkage is quite large, and makes little sense.
- The costs of the many rotational joints far surpass the costs of two linear bearing rods.
- Repair and maintenance would be more costly.

However, this design can work, and would be useful if this was part of a synchronous machine with a common drive shaft, in particular if we move away from a four-bar linkage toward a slider crank mechanism. This is how pistons in your car's engine work. These are also well documented (Norton, 1999; Chironis, 1991). With a slider crank mechanism, the input rotation of the crank is transformed into linear motion of the slider. The crank rotation would then need to be controlled to stop when the case had reached its upper limit sensing, and then most likely reversed so as to lower the case after latch unfastening. This reversing

Workstations

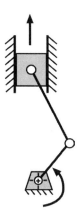

FIG. 6.48 A slider crank mechanism

might seem like more work than it is worth, but remember the DC motor also needs to reverse in the previous designs for linear lifting and the scissors jack.

A slider crank mechanism (Fig. 6.48) would produce the same effect as an air cylinder, yet the slider output most likely has already been designed using a single linear slide. This slide might be beefed up to have two slides or a double slide, so when the case is lifted, all side loading would be handled within its specified limits. A slider crank does have higher cost than the linear sliding device, since there are additional members and bearing supports (Fig. 6.49).

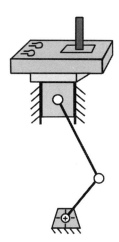

FIG. 6.49 Lifting using a slider crank mechanism

6.4.6. Conclusions

The deck of cards stating possible workstation concepts was stacked in this case study to present the two better solutions first, but we have seen that the consideration of the other concepts, if even for only a few minutes, has provided some additional insight and red flags. There are still many details for this case study (as well as each specific project) such as the CAD layout of all parts and a complete bill of materials, which we will leave for Chapter 12 on Specifications.

6.5. CASE STUDY NUMBER 2: AUTOMATED SCREWDRIVER WORKSTATION DESIGN

Depending on one's environment, one can be exposed to screws and screwdrivers at an early age. Predominately, the world seems to be assembled with slotted head screws (using a flat edge screwdriver) and Philips head screws (that look like a plus " + " sign). Usually one finds a Philips head screw and screwdriver the easier to use since the screwdriver is less likely to slip off of the screw's head, but with either type of screw, a manually operated screwdriver is something that most people can use.

Now imagine you are to make a workstation to insert and tighten screws. From your own experiences, you think that this would be a relatively easy task. Why even a child can use a screwdriver, or so you think, but let us look at the steps required:

- Pick up a screw from a table top or from a package of multiple screws.
- Align the screw with its head on the top.
- Insert the screw into the previously drilled hole.
- Align the screwdriver blade to the screw head.
- Twist the screwdriver and apply a downward motion.
- Stop when the screw is snug, but do not overtighten.
- Remove the screwdriver from the screw head.

This list probably seems longer than what you would normally think about while you were performing these steps. Many of these steps you would take for granted. It is these "take for granted" steps that are usually the more difficult for automation to replace the human effort.

If this task was to be performed by a human operator for a significant part of one's work shift, then the person would probably use a powered screwdriver rather than the simple manual one (Fig. 6.50). The screwdriver could be powered by an internal electric motor and battery, or it could have an electric power cord attached so as to not be dependant on batteries, whether rechargeable or of standard type. The electric screwdriver could also be configured with an adjustable

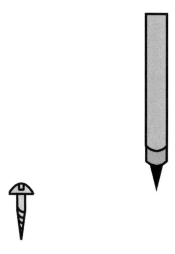

FIG. 6.50   Traditional slotted screw and powered screwdriver

slip release clutch to have a repeatable torque release when the screw has been completely inserted and is at the desired amount of snugness.

The screwdriver could also be powered by pneumatics (compressed air). These types of powered screwdrivers are not usually found in one's home toolbox, but are common in industry. Similar to the model using an electric power cord, there is no downtime as batteries get weak and replaced, and compressed air might be safer in the assembly of some items than electricity.

There are clips found on some home-type screwdrivers that will hold the screw head onto the blade while the screw is being twisted into the hole. The clips slip away from the head (most times) as the screw head comes in contact with the surface. The benefit of holding the screw firmly during insertion, particularly if the process is inside a complex environment, is worth the effort of sometimes having to pull the clip away from the screw head before giving the screw the final twist.

However, Fig. 6.51 shows the initial environment for automation. Many a student in an automation laboratory will solve the problem of obtaining a screw and orienting it correctly by placing a series of screws into a rack that has holes in it for just such a purpose. We have just created another task to be either humanly supported, or need another smaller automation workstation, that of filling the rack with screws.

The issues are now how to grip the screw head and align the slotted blade with the slotted head. One might think that the problem would be simpler with Philips head screws (and it would), but some applications wanting to automate

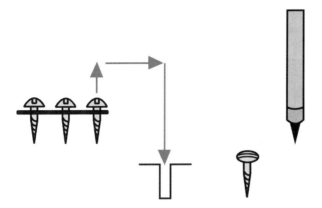

FIG. 6.51 Screw feeding issues and driver head alignment

still select fasteners that are not automation friendly. Grips similar to the hand held model can be used, and even the Philips head needs to get aligned to a degree. So some type of spring-loaded device is often used that allows for the blade to spin in contact with the head for so many revolutions until the blade and head must be aligned.

Students (and other researchers) have investigated magnetic forces to hold screws onto the screwdriver. But these types of forces are not very robust if the automation system slightly bumps the screw into the side of the hole while the insertion is being processed.

6.5.1. Automatic Screwdriver Workstations

Most commercially available powered screwdriver workstations have evolved using a different strategy than the human operation. They use a vibratory bowl feeder (see Chapter 7) to create a stream of screws, all pointing in the same direction (usually point first). The stream of screws is then conveyed to the workstation area using a hose and compressed air. Figure 6.52 shows the basic concept.

The screws are blown into place when needed. There might be a movable gate to keep the next screw in line from falling into place too early, which would prevent the screwdriver from properly engaging. A set of grips similar to the hand-held model stop the screw from being blown out of the system entirely. The screwdriver blade then travels forward, spinning to align the blade and the screw head. The motion device (robot or other transfer arm) carrying this automatic screw machine would then move into location above the desired hole. The screw would be rotated until the screwdriver clutch would slip. The

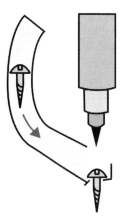

FIG. 6.52   Automatic screw feeding and screwdriver clutch

grips can be retracted if needed to avoid being stuck under the screw head. This means the correct amount of torque has been applied. Then the motion device would move to the next hole while the next in line screw advances into place.

6.5.2.   Conclusions

The workstation in this case study has two major components. First is the bowl feeder to supply the screws in the correct orientation. The second component is the screwdriver head that fits at the moving end of a robot or transfer arm. This case study is significantly different from the earlier one, since there are two components only joined by the air hose carrying the screws, and that it is a situation where one should not design their own workstation.

6.6.   MACHINE DESIGN AND SAFETY

You are responsible in the real world for the safe operation of any machine that you design and/or build. Whether you know about applicable safety standards or not, if there is an accident involving your machine, the courts may find you liable. Ignorance is no excuse. Machines can be designed with safety in mind, or they can be corrected later with external guarding, and so on, but this can be a life and death issue that can haunt you for decades to come. It has put companies out of business.

Early machines, like some of the lumber mills powered by waterwheels 200 years ago, are a great example of how NOT to do safety. Overhead common drive shafts were coupled to giant saw blades by wide leather belts. The belts had no cover or shield to prevent someone's clothing from getting caught between a pulley and a belt, so this could be a gruesome situation without even coming into contact with the saw blade. And then the saw blade had little or no guarding itself, in comparison to today's radial arm saws. Accidents were often seen as a way of life. Lawsuits did not rule the day as in the present. There were no advertisements on television for lawyers who promised you big bucks when you had an accident at the lumber mill. Power takeoffs on farm equipment are notorious for fatal and mangling accidents, too.

Now everyone in the chain of machine designer, builder, installation group, maintenance team, and so on can be a party to a lawsuit. It is unfortunate but unavoidable today. One needs to do homework to have a strong case so as to be able to walk away without a judgment against one. This starts with knowing what the safety standards are, and also applying common sense.

The concept of designing a safe machine can be found as early as 1914 in the Universal Safety Standards (Hansen, 1914), where the Workman's Compensation Service Bureau stated for steam engines that "All dangerously located moving parts, such as fly-wheels, cranks, eccentrics, cross heads, tail rods, fly balls of governor, governor sheaves, etc. to be guarded in approved manner." Now some of these devices, such as mechanical governors, are no longer used in today's machines, but a similar statement was made for power transmission (Hansen, 1914). So the message has been clear for decades. If it moves, it should be guarded.

An early textbook by Hyland and Kommers (1929) states that high-speed machines are a hazard, and need to be guarded. They document that in 1927 the American Engineering Standards Committee adopted a safety code for Mechanical Power-Transmission Apparatus. Their guarding statement was augmented from the 1914 version stating that guarding needs to go all the way to the floor.

6.6.1. Pinch Points

The classic problem for safe machine operation is defined by the term "Pinch Point." Figure 6.53 shows three different occurrences. The location where two rollers press something, called the nip, is a standard situation. This is where any loose clothing can get drawn into the rolling and drag the wearer into it too. The moving lever also can come down and do some harm. Often there is some gearing or other mechanical advantage being applied, so a human cannot arm wrestle the lever with much success. The third situation is often an oscillating press. Again, due to the processing requirements, the forces involved are life threatening.

Guarding these problems is a must. None of us can claim to be born and practicing engineering before 1914, so one cannot claim guarding is only a new fad no-one knows about. And to add more evidence that the standards for

Workstations

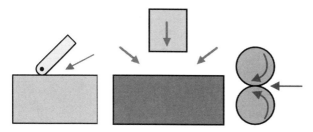

FIG. 6.53 Pinch points

safety do not stay fixed on 1914 definitions, the American Society of Mechanical Engineers (ASME) and American National Standards Institute (ANSI) have established and revised safety standards for Mechanical Power Transmission Apparatus (MPTA) (ASME, 1994). Because of the fact that most if not all automation machines have power transmission somewhere within the machine framework, the MPTA standards apply.

To make things even more clear for some automation machines, there are ANSI standards for Packaging and Packaging Related Converting Machinery (ANSI, 1994). Current versions can be inspected on their website (www.ansi.org). There are safety codes for robots (ANSI/RIA, 2003). There is a great deal to be concerned with depending on the specific machine and application, but guarding or shielding is a basic must.

The timeframe of when the machine was first designed and built is the most critical. A machine designer who incorporated all of the current standards of the day will most likely not be faulted if an accident occurs that deals with more rigorous standards 20 or 30 years later, but if the 30-year-old machine is still being produced for all of the 30 years, and more rigid safety standards are adopted, the issue of whether the manufacturer needed to issue a recall, or simply a warning advisory to all known customers, is a situation one should involve a lawyer to decide. In a court of law, only the standards of the day can be included for consideration of whether it was a good or negligent design.

Guarding can be in the form of a permanent wall, but access is often required to clean up leaking grease or clear out a jam. Figure 6.54 shows three ways to handle this situation. A door can open to one side. A cover can lift above, usually assisted by springs, sealed air cylinders or counter weights. Or a series of light beams can be used to create a curtain that will not allow a human hand to enter without shutting down the machine's operation. The physical moving doors usually incorporate a sensor interlock switch and control system.

There is always the need to get into these critical moving parts, if one asks a machine operator or maintenance person, and although it is against all safety

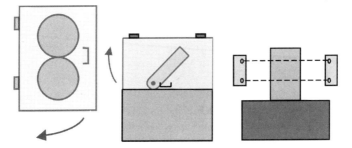

FIG. 6.54 Shielding pinch points and moving devices

standards, some operators get tired of proper shut-down procedures, particularly if the machine jams every 15 minutes. So sometimes safety interlocks are defeated or bypassed, and the consequences are troubling. Never allow for any safety bypass situation without proper backup procedures.

6.6.2. Lockout/Tagout

Before the controller's front cover of an operating machine be opened, one should investigate if the type of wiring and power usage requires a Lockout/Tagout arrangement. This means that the energy sources, usually AC electric current, must be turned off before the door can be opened. The lever that switched the power off may be coupled to a physical door locking device. Lockout/Tagout (ANSI, 1982) lists how and when the lever gets padlocked, and the labeling required to let everyone know what has been turned off, why it was turned off, and who has the key. A licensed electrician should know this process and assist you.

6.6.3. Warning Labels

It is very good to use warning labels (Fig. 6.55), but they are not sufficient. You still must guard a dangerous situation. However, warning can help. There are warning labels all around us now, so much so that we may be getting complacent about them. There are standards for the proper label type and location from ANSI and the National Electrical Manufacturers Association (NEMA) (1991).

6.6.4. Risk Assessment

The Packaging Machinery Manufacturers Institute (PMMI) in conjunction with the ANSI 2000 version of the B155.1 safety standards has embraced the concept of risk assessment (PMMI, 2000). Any safety standard sets the minimum level of

WARNING MOVING PARTS

FIG. 6.55 Warning labels

safety required. Machine designers and manufacturers need to do even more. They need to look at:

- What different types of machine occurrences can happen.
- What is the anticipated misuse of this machine.
- What is the level of possible injury.
- What is the probability of occurrence.
- What can be done to rectify the problem.
- Document the process and results.

If this process is done with due diligence, it should greatly minimize the number and magnitude of safety-related issues. Done well, it will also speak well of the machine designer's efforts if and when an accident comes to court. Accidents are sometimes assigned by juries a certain percentage due to faulty machine safety design, careless people near the machine, and/or a freak occurrence of nature. The determination in the courts today may not always be logical, but doing one's homework will be better than not!

6.6.5. Safety Responsibility After Delivery

Very often, after one delivers an automatic machine to a customer, there will be several if not many opportunities to review the machine in operation over time. Then is when the designer, builder, and even the salesperson need to keep a sharp eye towards how the customer is using the device. One should inspect to see if the customer has removed any guards, or defeated any safety interlocks. If so, the customer should be notified by some type of certified mail that this practice is not endorsed by your company. Mr. Fred Hayes summed this up well in two recent PMMI documents (Hayes, 2003; PMMI, 2003), when he stated "If you see it and condone it, you own it."

6.6.6. Safety Standards are Just a Starting Point

National safety standards are viewed by many to be a bare minimum. Many large companies have more stringent requirements than OSHA or ANSI. One needs to be aware of these before quoting a system. Other times in court cases for injury lawsuits, experts are questioned if the applicable safety standards are sufficient. If one explores the situation for any length of time, one may conclude that there are other seemingly common sense items that need to be guarded or accounted for. Energy can be stored in a spring or compressed air and can do damage after the machine is turned off. Do not simply rely on safety standards. Look at the risks involved! Is it obvious that someone might try to defeat a safety interlock because of the need to adjust the equipment while it is running?

6.6.7. Real-Life Accidents

The author has inspected more than 20 potential machine safety failures related to product liability lawsuits, both for the injured party and the machine builder. Some of these accidents seemed likely to occur someday, such as when an operator is daily to clean grease from a moving chain and sprocket near a series of knitting needles. Other accidents seem highly less predicable, not that the root cause was hard to identify, but the resulting accident (as witnessed by several people) was from a chain reaction similar to one only seen on Saturday morning cartoons. Not that the accident was in any way funny. However, the chance happening of one unguarded device hitting a second obstacle that in turn moves and injures someone 30 ft away, would seem impossible to anticipate or reproduce.

Safety needs to be incorporated into the design of automation in today's society. One needs to think of all possible ways for misuse or stupidity. To wait until it is all constructed, and then to guard it, knowing there will be operators reloading or cleaning it, is just foolhardy.

6.7. CONCLUSIONS

Workstations are when the concepts and dreams become reality. This process of detailed work is generally not as creative or as much fun as the earlier stages, and some student teams lose their excitement. However, without good thinking and investigating, you will find yourself using the equivalent of Maglev and within week have thousands of damaged tapes. The real engineering heroes earn their rewards at this point.

PROBLEMS

1. Create a workstation to mix the chocolate chips into a cookie batter for your home use. Determine what would happen if one used their home lower powered hand mixer, and the batter was fairly thick. Determine

the impact temperature has on the batter's viscosity. Would the need for this workstation be the same in both the Arctic and at the equator?
2. Investigate on the Web for available workstations to perform various tasks required by students or engineering employees.
3. Design a workstation to automatically butter your toast as it leaves your toaster. Determine the relative needs for:

- structural members;
- bearing surfaces;
- drive mechanisms;
- types and ranges of motions;
- general sensing needs;
- safety.

PROJECT ASSIGNMENT

Using one of the projects from the Appendix (or any other projects), perform the following:

1. Create some concept sketches for the needed workstations. List the requirements for the:

   - structural members;
   - bearing surfaces;
   - types and ranges of motions;
   - general sensing needs.

2. Compare the concepts and lists from step 1 above and perform a matrix comparison, stating the pros and cons for each concept. Create a list of the unknowns or the areas of additional needed research or investigation. Determine what parts of the process seem to have some risks associated with it.

3. With the associated risks and benefits of the workstation concepts in the steps above, determine the most likely set of workstations and machine configuration. If this is solely an academic exercise, you may wish to go ahead with a choice that has more risk but will be a richer learning experience. In real life, you may want to limit your exposure to risk.

4. Make an initial safety review of your leading design. Determine where and how shielding should be implemented. Walk through the normal operation process, and any re-supply or error correcting actions and list possible concerns. Try to predict when, how and why someone

might want to defeat your safety system, and see how you can anticipate and account for this.

REFERENCES

ANSI. (1982). New York: Lockout/Tagout requirements Z244.1.
ANSI. (1991). Washington: NEMA Product Safety Signs and Labels Z535.4.
ANSI. (1994). New York: ANSI B155.1-1994 Bulletin.
ANSI/Robotics Industry Association. (2003). Washington: R15.06 Robot Safety Standard, www. roboticsonline.com.
ASME. (1994). New York: MPTA ASME B15.1A Bulletin.
Chironis, N. (1991). *Mechanisms & Mechanical Devices Sourcebook*. New York: McGraw Hill.
Davis, G. (1997). *Introduction to Packaging Machines*. Arlington: PMMI.
Hansen, M. (1914). *Universal Safety Standards*. New York: Universal Safety Standards Publishing Company.
Hayes, F. (2003). Product Liability and the Machinery Manufacturer. *PMMI Reports*, 13(5):8.
Hyland, P., Kommers, J. (1929). *Machine Design*. New York: McGraw Hill.
Norton, R. (1999). *Design of Machinery*. New York: McGraw Hill.
PMMI. (2000). *Risk Assessment Basics*. Arlington: PMMI.
PMMI. (2003). *An Overview of Product Liability for the Packaging Machinery Manufacturer*. Arlington: PMMI. (www.pmmi.org).

7
Feeders and Conveyors

To the automation novice, it may seem that all of the difficult problems with which one must be concerned have been covered. Yes, one still needs to select the right actuator and add the right sensor to make things function correctly, but are we not done with the challenging stuff yet? Depending on what is being processed, there is one tough nut yet to crack. That nut is the feeding of the products into the workstations within the machine configuration. Sometimes this is as simple as selecting a feeder from the catalog, other times, your worst nightmare.

Feeders, as the name implies, feed parts into the workstation. A conveyor can also function in this capacity, but its role is sometimes a gray area. Does an asynchronous conveyor with a series of pallets constitute a parts feeder or a machine configuration (it can be both)? When is a conveyor a feeder and not part of the machine configuration could be debated by lawyers until the cows come home (perhaps this is a useful way to occupy some lawyers you know). Within this text we will not try to split hairs on these definitions of conveyor functions, we will simply look at them for possible solutions and move on.

To realize the role of a feeder, imagine the bow tying workstation discussed previously. There needs to be an influx of packages to have bows tied on them. And there needs to be a supply of ribbon. Ribbon traditionally comes on spools, so we will try to use this format as a means of keeping costs low. Requiring ribbon to come in a cartridge, internally jumbled like some older dot matrix printer ribbons cartridges were designed, would cause a supplier to create a new packaging machine. This is most likely a higher cost to you, unless you are the one to make the new machine for them too. So we will target ribbon coming on a spool.

However, what happens when the spool runs out of ribbon? What happens if the bow tying machine has 90% of the ribbon it needs for the current package, and runs out?

- Will sensors stop the process before wasting time trying to use a piece too short?
- Will the machine malfunction?
- Will it jam like some office copiers, requiring much effort to clear?
- Can a huge roll be used?

The last point seems to be very practical, making the supply roll huge. This makes the percentage of time one has this issue become smaller, but then the force required to pull off the correct amount of ribbon becomes much higher with the large rotational moment of inertia. And once the inertia is moving, the supply roll might keep going so as to create a giant loop of excess ribbon, perhaps even reaching the floor or becoming a tangled mess. You may find yourself needing to give a motor assist to the supply roll to help pull off the correct amount, and then you might need either a servomotor to reduce the rotation, or a braking system. With the changing diameter of the ribbon as the roll gets smaller, the amount the motor assist operates is not a fixed number of turns. It may need to have the ribbon travel through some pinch rollers to gage the length of ribbon removed.

So, our original idea of a simple feeder for ribbon has the potential to become a separate automation project. This happened without a great number of requirements, or so it seems. Another potential thorn is how an operator changes over the roll when it runs out. Does the roller system need to be threaded carefully, or can the ribbon be inserted into a starting point and is it self-threading, like some 16 mm movie projectors in the 1960s and 1970s. How long this takes to refurbish the ribbon supply will have an impact on many other machine configuration concerns, including:

- Does the system need an accumulator or can packages back up, perhaps on a slippery conveyor?
- Does the operator need special training to change over the roll, or can the floor operator quickly slap it into place?
- Will the changeover time mean that other critical operations will go unmonitored?
- Will there be the need for an additional operator?
- Does the operator have time for a break, or do they need a break replacement?

We will look at some of these issues in this chapter, but some of these questions cannot be answered easily for all situations.

7.1. FEEDERS

If an old-fashioned lemon juicer is used to make lemonade, one takes a lemon, slices it in half, and places it into the juicer to be squished. Manually this can be done by a 10-year-old to sell lemonade on the street in front of their home. Lemons can be purchased by the case or in bulk and then placed in a bag when shopping. But what if this process is to be automated? Then it is a challenge. No one lemon is the same size as the next. Each has its unique shape and characteristics. They are far more difficult to feed than most components of your cell phone.

As a human you simply take your hand and grab a lemon from the box or bag. You could train a monkey to do this (they might even do it without the training, but they get bored easily and will not work 24/7), but selecting a product from a bin and then determining its orientation is not easy for automation. It has been one of the ongoing research problems for robotics for over 30 years, and the results are not always cost effective. However, let us step back and see the big picture of what can be fed and how.

A simple list of how to present parts for a process is given here.

- Place parts in a row, with some way to affix their locations and orientations.
- Similarly, place parts in a grid pattern so as to have a bigger supply.
- Stack these grid patterns onto multiple trays and move the trays when needed.
- Have an egg carton and shake parts into mating depressions.
- Stack parts and take from top or bottom.
- Have a black box spit one part out at a time.

The first three options all seem to require the need for an operator to align the parts either in a single row, or on a tray at some point in time. Perhaps if this process takes little time compared to the automation process by which the parts are being consumed, then this option might be economically palatable. However, for a high-speed process, you are creating a human slave to feed a machine. There are tray-handling devices that make machine configuration sense, but a different feeder in another location fills these trays. It only moves the problem from one site to another!

Researchers have built and a few companies have sold machines with a plastic egg carton type bottom designed and sized to a specific part. With a combination of vibration and letting loose parts fall back and forth over these dedicated carton bottoms the cavities eventually get filled with parts in a predetermined orientation. For specific plastic parts that are too lightweight to use other feeder types, a few companies have used this method in practice, but the proper shaping of the customized egg carton bottom is an art that only a few people in the world understand, so the author will not in general encourage it.

The stacking of parts and taking of them from the top or bottom will remind the reader of a card dealer at a gambling casino. If dealt from the top, a robot or transfer device simply goes to a position above the feeder and lowers until it hits an object, the product to be handled. There is little engineering involved to do this, but there is a hitch. An operator trying to resupply the stack, either when it is empty or simply when it is convenient to the operator, needs to be concerned about being in the workspace of the robot. One does not want to get into a game of chicken with a robot. The top load/unload feeder could be swapped in and out if there were two of them and it did not restrict the cycle time.

7.1.1. Escapement Feeders

Unlike in gambling, dealing from the bottom is a good alternative. If a stack can be safely replenished from above, this would be more desirable. Figure 7.1 shows the first alternative. Here a rigid part with moderately tight tolerances is stacked. The upper finger with its tapered edge moves to the left to restrict the possible motion of the three parts above it in the stack. The lower finger then moves to

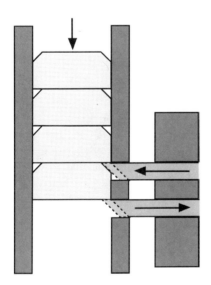

FIG. 7.1 Ratchet escapement feeder

Feeders and Conveyors

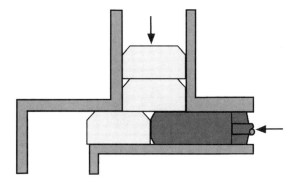

FIG. 7.2 Slide escapement feeder

the right to allow the lowest part in the stack to drop. The cycle is completed by the lower finger returning to the left, followed by the upper finger returning right, allowing the stack to drop by one part thickness. The part itself can allow for the fingers to ratchet along the part's surfaces, hopefully without damaging it. Figure 7.2 shows a second alternative.

These designs are often used, where appropriate, when there are a fixed number of parts used in a long production run. In some processes multiple parts need to be fed, so the idea jumps predictably to Fig. 7.3, but none of these three concepts would be good for our lemonade, since the lemons are not consistent in size. Any of these designs would potentially prematurely slice the lemon, and probably get gummed up from the juice and broken peels in a few hours.

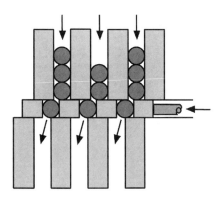

FIG. 7.3 Slide escapement feeder: multiple parts dispensed

FIG. 7.4 Bowl feeder

7.1.2. Vibratory Bowl Feeder

The list of feeder options stated a black box that spit out parts might be used. In many ways, this is what a vibratory bowl feeder is. A vibratory bowl feeder is commonly found in many assembly operations. The bowl sizes vary from 4 in. in diameter to 4 or 5 ft in width. The alternating vibrations produce a motion that will in practice cause the product to proceed up the spiral ramp on the inside curved surface of the bowl (Fig. 7.4).

The upward spiral path has custom fiducials either attached to it or machined into it, so as to only allow parts in the predetermined orientation to make it to the top (Fig. 7.5). Parts that are not correct are dumped back into the bowl. So any particular part may make several short-circuited trips before reaching the top properly aligned. No matter how many times the author has seen a vibratory bowl feeder work, they are always a joy to watch. The product's motion up the spiral seems to defy gravity. This is not actually true, since the vibrational energy is causing the part to go up the spiral each cycle more than it goes down. Figures 7.6 to 7.8 show the force balances at rest, upwards, and downwards. Parts do get shaken tremendously before they reach the top.

The key to the net upwards motion is a combination of the frictional coefficients, bowl incline angle, and angle of bowl vibratory motion, vibrational frequency and

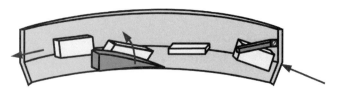

FIG. 7.5 Bowl feeder: customized fiducials along inclined slope

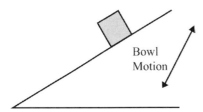

FIG. 7.6 Bowl feeder: part on incline

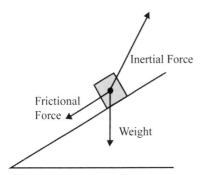

FIG. 7.7 Bowl feeder: forces at peak of upwards vibration

amplitude. Figure 7.7 shows how the inertial forces at the peak of the upwards cycle need to be higher than the frictional forces to move, and yet in Fig. 7.8, the inertial forces need to be smaller than the frictional forces to not slip backwards.

Bowl feeders can be coupled to a hopper to keep the product flowing. A bowl cannot hold too many minutes worth of product without getting clogged. And the clogging of parts can be a nightmare if the parts geometry allows for interlocking. Bowls can be "stirred" often manually or mechanically to keep in production.

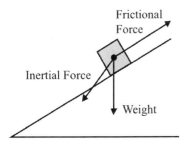

FIG. 7.8 Bowl feeder: forces at peak of downwards vibration

A mid-sized vibratory bowl feeder can cost around $3000 for the base bowl, but up to $12,000 for the custom work in creating the fiducials. There are a few companies who specialize in this work. You send them several hundred of the parts you want to be fed. Their efforts are perceived to be more of an art than science. Turnaround time to get the finished feeder can often be 12 to 15 weeks. The resulting bowls are for one product only, so any change in a part's geometry can create a small boat anchor and another 15-week wait.

A final note on these feeders is that some products will never be successfully vibrated, or that the process will damage them even if it does work. It is highly likely that a lemon should not or will not be vibrated up a spiral ramp, but the author has not had the opportunity to try this (on someone else's bowl, in case it is messy).

7.1.3. Centripetal Feeder

Some products, in particular plastic parts such as empty shampoo bottles, do not feed well with bowl feeders. Another method is to place a batch of products into a spinning bowl, and let the centripetal force move them to the outer edges (Fig. 7.9). Similar to the vibratory feeder, fiducials are placed near the outer edge to send improperly oriented products back to the center to try again. If a product has a center of gravity located offset to the dimensional center, this can be used for predictable dynamic responses.

Centripetal feeders can work in the 100–150 parts per minute range, and for some smaller parts, much faster. They can occasionally become clogged, but the types of parts handled usually do not present much trouble. Like bowl feeders, centripetal feeder companies need several hundred samples of your parts.

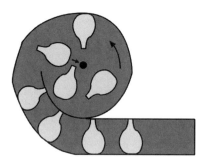

FIG. 7.9 Centripetal feeder

Feeders and Conveyors

7.1.4. Flexible Feeders

As one can see from all of the different styles of parts feeders discussed so far, there is a significant financial investment for a single part, and that investment may not be fruitful if the part changes in any significant way. One might wonder if there is a generic feeder available that would be cost effective for flexible assembly or other processes.

There are several variations commercially available, which all function similarly to the one shown in Fig. 7.10. In this figure, the parts being fed are shown as simple disks, but in practice they could be of any shape or size. The computer vision system camera takes an image of the parts location and orientation on the lower conveyor's left end. A robot arm (not shown) would get this information from the vision system, and would grab a part for processing. It two or more parts are entangled, or the vision system cannot determine the correct part geometry, or even if a part is upside down, the unprocessed parts travel to the left off the conveyor into a trough. When the trough is full, it goes through the motion shown to the upper position and dumps the parts onto the upper conveyor. This dumping process tends to untangle the parts, and flip them over if they were upside down. The upper conveyor is timed to dump parts onto the lower conveyor so as to attempt to spread them out. Sometimes simple optical sensors are used to smooth this process of determining a nice part distribution.

These feeders are a practical solution to difficult-shaped parts, often-changed production lines, and are economically long-term justifiable. They do imply a robot arm to transfer the part off of the conveyor belt, so this may be a significant expense compared to an escapement feeder or even a bowl feeder. Flexible feeders are also very quick to implement compared to waiting for a bowl feeder to come from the experts. There is no general right answer to

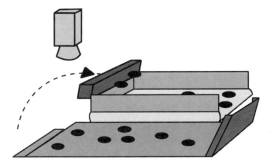

FIG. 7.10 Adept flexible feeder (from Carlisle, 1997)

which way one should go. It usually boils down to costs, timeframe, and the long-term economic outlook.

7.2. CONVEYORS

As was previously mentioned in Chapter 4 concerning machine configurations, there are almost an infinite number of types, brands, and sizes of conveyors. The simplest continuous belt conveyor (Fig. 7.11) as seen in the checkout line you have probably used without much thought. The addition of flights (Fig. 7.12) would seem to have merit, particularly when the conveyor is inclined. This would limit the distance a part could slip backwards.

However, it does not stop there. The belts themselves can be:

- plastic;
- woven;
- coated;
- curved;
- metallic mesh;
- food grade (washable);
- heat resistant;
- flexible;
- rigid and segmented.

7.2.1. Segmented Conveyors

The segmented belts (Fig. 7.13) can be designed something like a tank tread, where each piece is connected to the next by a pin (Fig. 7.14). The profile of

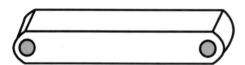

FIG. 7.11 Continuous belt conveyor

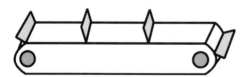

FIG. 7.12 Continuous belt conveyor with flights

Feeders and Conveyors

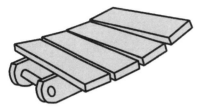

FIG. 7.13 Segmented conveyor belt: top view

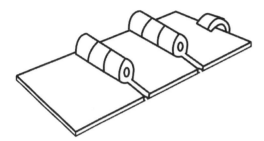

FIG. 7.14 Segmented conveyor belt: bottom view

the hinge portion can be very slim (Fig. 7.15) if this is needed. Flights can be attached to specific segments (Fig. 7.16) to properly space parts during transfer, and these flights can be custom molded when better performance is needed.

Some flexible segmented belts can move along conveyor frames configured to move upwards, downwards, and to the right and left. If one gets an opportunity to see a system like this, either at a trade show, a website animation, or best yet, in a factory setting, they are fun to watch. By properly squeezing a product from both sides with two such segmented conveyors properly aligned, products such

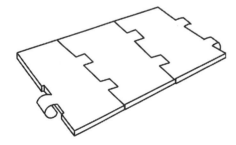

FIG. 7.15 Segmented conveyor belt: low profile segments

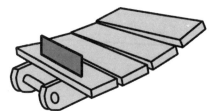

FIG. 7.16 Segmented conveyor with flights

as filled bottles or even paper towels can be moved across the room to the next packaging operation.

7.2.2. Other Conveyor Options

For some products, where the handling requirements may dictate very delicate motions, one could use a conveyor based on air jets. Designed something like an air hockey table found in many arcades, the product is manipulated by air jets having a specific directional orientation (Fig. 7.17). This directional quality is not seen in the arcade's air hockey table, however, or one player would have a distinct advantage.

This type of conveyor is not widely used. The amount of air needed to move products is not a trivial concern, both for the associated operating costs and the possible need to filter the air so as not to contaminate sensitive products.

Other conveyor material handling techniques include the use of a pusher connected to the conveyor's side (Fig. 7.18). When an optical sensor detects a part or box, a pusher plate is actuated to slide the part off the conveyor onto a target location. The target may be a counter top for manual intervention, another conveyor directing the product to the new desired spot, or a processing machine. With these kinds of side pushers, the part's orientation can be scrambled from any

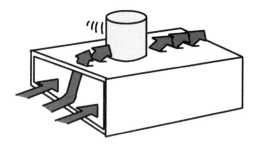

FIG. 7.17 Air conveyor: product floats and has motion

Feeders and Conveyors

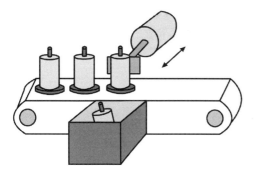

FIG. 7.18 Side pusher on a conveyor

previously alignment, so additional sensing or physical orienting will need to be performed. The side pusher is usually powered by an air cylinder and compressed air due to low initial costs and limited amount of compressed air usage.

An alternative to a conveyor belt is to use a series of powered rollers. These rollers need to be spaced so that parts will not fall into the spaces between rollers, so they are used more often for large parts, pallets, and cardboard boxes. Figure 7.19 shows how a common drive shaft is connected to each roller by an industrially hardened rubber band. The rollers and drive bands are often designed so that if a box is held by a movable stop, the restrained boxes will slip in an expected manner, and the nearby people will not hear any squealing noises. The movable stops can be spaced between the rollers.

With the space between the rollers available for useful devices, some manufacturers have added additional drive mechanisms to a second set of movable stops. A standard movable stop would restrict and align a box or pallet when directed, and then the second sets of moveable stops would lift the box and propel it to a perpendicular direction, onto another conveyor or whatever. There are usually small round belts on pulleys on these lifting stops as a means of propul-

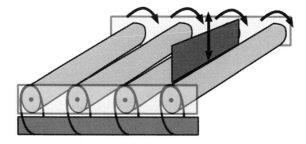

FIG. 7.19 Powered roller conveyor with moveable stop

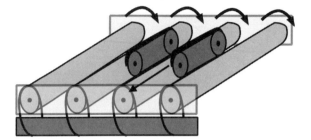

FIG. 7.20 Powered roller conveyor with right angle transfer device

sion. These somewhat complex conveyor systems can manipulate a box or pallet in many right angle paths, but they are not as good for moving long twisted parts such as a replacement exhaust pipe for your car (Fig. 7.20).

These commercially available conveyor systems can be highly programmed and a good means of redirecting parts and products for flexible assembly and processing. One would be hard pressed to construct a system made of many of these transfer devices in a cost-effective manner from scratch.

7.2.3. Timing Screws

If one has a steady stream of empty plastic bottles being constrained by guard rails while being transported along a relatively slippery conveyor belt, and there is a movable stop that makes them bunch up like bumper-to-bumper traffic on a holiday weekend, then it may be difficult to pull one bottle at a time from the pack to be properly spaced under a filling machine. Years ago someone very clever determined that if a variable pitch screw was custom made for a particular shaped product, then one could take the bumper to bumper group and define some order to them. Figure 7.21 shows such a timing screw used to perform this task.

The output of a timing screw can be directly to a flighted conveyor, to a pallet, or any other device that will keep the product properly spaced while being processed. Each timing screw is custom made from either your CAD drawing of the part, or from several part samples.

FIG. 7.21 Timing screw

Feeders and Conveyors

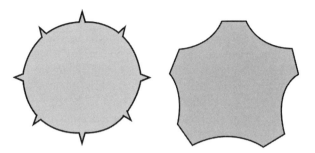

FIG. 7.22 Star wheels

7.2.4 Star Wheels

An alternative device to the timing screw is a star wheel. Instead of trying to pull bottles away in the direction of travel, a star wheel grabs or sweeps a bottle in a motion perpendicular to the conveyor motion. Figure 7.22 shows several different shape star wheels. Again, like a timing screw, most star wheels are custom designed for a specificly shaped part or container, but since the motion of the star wheel is in one plane (while the timing screw has a three-dimensional swept volume challenge to it), their design is more straightforward.

Star wheels can be used to directly feed a part into a process, or to place into a pallet system or flighted conveyor (Fig. 7.23). Many star wheels have contours to specifically orient a part for future processing. An example is a bottle to be filled that has an off-centered spout. If the bottle has a cross-section other than a circle, the shape can be used to its advantage. If the main cross-section is circular, as is, say, a standard liquid bleach bottle, then the star wheel might have two levels. The lower level would support the circular cross-section, and the upper level orient around the spout.

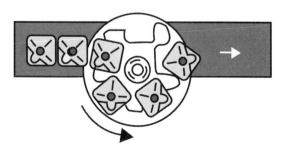

FIG. 7.23 Star wheel separating product

7.3. ACCUMULATORS

Sometimes one does not want to bunch up a stream of parts while they accumulate on a slippery conveyor. They may be fresh baked goods, and will need to have good airflow on all sides while they cool before packaging. There may be some high-speed production process upstream, and even a few minutes of a downstream process can mean hundreds of extra parts need to be accounted for. So for these situations, many different models of accumulators are available.

The serpentine model in Fig. 7.24 is one of the basic options. Here products will travel the entire available route whether they need to or not. Some models will move in the side curved rails to lessen the number in the accumulator when the upstream process is off line, and there is no need to have a gap within the accumulator's contents. Then if the amount of the contents needs to expand, the curve rails begin to expand to the side until they return to the position shown. This is similar to those endless lines at Disney theme parks, when the one hour wait tries to keep you moving, although you may see the same people traveling by you over and over again.

Another form of accumulator uses more of the available space in a factory. It spirals the product up one conveyor, and then spirals it down the other (Fig. 7.25). This is a prime model to be placed in a cooling tower to chill the baked goods before packaging. Depending on the model, however, it may not have the capability to contract its route and adjust its content. So if the timing gets mixed up, there could be a five-minute gap between the second to last coffee cake to the last coffee cake. And since there might not be more than one speed of the conveyor system, one must wait that extra five minutes to get the last one out.

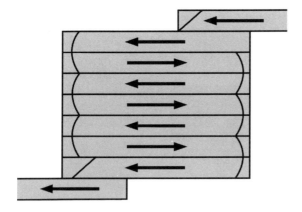

FIG. 7.24 Serpentine accumulator: top view

Feeders and Conveyors

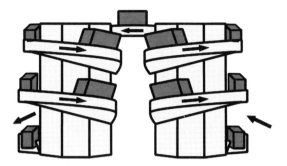

FIG. 7.25 Spiral accumulator

7.4. PICK AND PLACE FEEDERS

Exactly where a feeder ends and the processing machine begins is yet another gray area. Some people would state that the feeding stops where the bowl feeder presents the part. Others would say it ends after the pick and place device (or robot) places the part into the process. The argument is usually moot if one is designing the entire machine, but it is very critical when you are sending out feeders for subcontract.

One could use a six-jointed or SCARA robot to take parts from a feeder, for example a bowl feeder, but this may be overkill. If the presented part is simply moving up, over, and then down without tight tolerance requirements, then a pick and place device is probably the way to go. Some are electrically powered, but most are based on configuring several pneumatic air cylinders and bearing rods and frames. They are marketed as pick and place devices, and either come with a robotic gripper where you can select a model or design your own (Fig. 7.26).

Perhaps one of the biggest strengths of such a pick and place device is also one of its biggest weaknesses. It can only be adjusted by the manual manipulation of setscrews. There is no controller or sensors capable of changing the position an actuator moves via means of software alone. The device is not a robot in any sense of anyone's definitions. The device is very useful, but limited in certain applications.

This means technicians must be knowledgeable on how to adjust the device, and most likely the device will need to be in operation on the factory floor to be properly set up. Sometimes once the setscrew positions are determined, and if there is no drift from temperature changes, product variation, or the like, the maintenance staff is eager to use a locking type of glue to guarantee that the setscrews do not shift from repetitive vibrational motion. This can be ben-

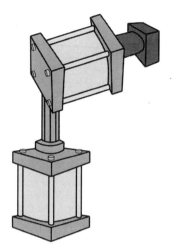

FIG. 7.26 Pick and place feeder

eficial if there are no changes, or frustrating if one has to scrape away the glue to make that compensation no-one ever imaged would ever occur.

7.5. CASE STUDY NUMBER 1: DROPPING COOKIES

A student project was to create a machine to place raw cookie dough blobs onto a special conveyor for baking. This was a conceptual project only, in the sense that no actual company was sponsoring it during the automation course at Rensselaer. Several automation consultants had expressed the general challenge as one that was rich (in calories and tasty too!) to work on in the classroom environment.

The cookie dough was similar in form to that shown previously in Fig. 2.14. The overall process looks like an extrusion of the dough, with a means to cut off a measured amount (a blob) each time. These blobs look like the commonly found slice and bake chocolate chip cookie dough found in your grocer's dairy case, either in the traditional tube form where one slices off the raw cookies, or now in the pre-molded squares or circles.

The automation process is to cut off a blob of dough on top of a conveyor, creating a series of rows. A single extrusion and cutoff workstation would transverse across the row and the conveyor would index when the row was complete. This creates a very intermittent motion requirement for the conveyor.

Feeders and Conveyors

This process is not particularly challenging, except that the conveyor system through the baking tunnel has continuous motion. Trying to match the intermittent motion of the dough depositing across the row directly onto this moving conveyor would produce diagonal rows. And the timing of the dropping process is not dependable enough to drop cookies diagonally on a moving conveyor. While this might not seem like a significant problem to a home baker, having a row where the product comes into the oven on a diagonal row can upset the professional baker's apple cart. If one remembers from Chapter 2 on the baking cups discussion how an extra paper baking cup can impact product baking quality, one can see that things need to be very rigidly adhered to in the industrial baking world. The master baker in charge does not eat the slightly burned ones like you or I would do at home.

Since a blob of raw cookie dough is relatively sticky, it does not bode well for accumulation or any of the slippery conveyor surfaces. Once dropped onto a conveyor, they can either be moved with something like a spatula flipping a pancake, or it can be dropped again onto another surface. And since the conveyor belt through the continuous baking oven is a metallic heat resistant, somewhat open grid, it would be difficult to attempt any tricky maneuvers.

7.5.1. Bull Nose Conveyor Solution

There are devices available that butt the cylindrical curved end of one conveyor to the cylindrical end of the next conveyor, acting like a bridge. Some have small motorized rollers on them. Others simply let a product get temporarily stuck on them until the next product pushes it off. This does let the "last" product get stuck until the next shift of production, and is not recommended for raw cookie dough. These devices allow for the synchronization of the belt with the dough having been deposited onto the continuous belt for baking, but again the cookies are somewhat sticky and gooey.

Another solution is to deposit the extruded cookies onto a bull nose conveyor. These types of conveyors seem to be only made by some custom automation houses. Perhaps it is that each application of these devices is so specific in its demands that creating a generally available device is not profitable. The overall concept is that one uses a fixed length of conveyor belt and four rollers rather than the normal two rollers, but one changes the location of two of the rollers to achieve a specific motion.

Figure 7.27 shows the bull nose conveyor with the "nose" extended and contracted. A lower surface or conveyor would receive the results of the upper bull nose one. This would allow the product on the upper conveyor to fall onto the lower surface an entire row at a time. The thickness of the bull nose needs

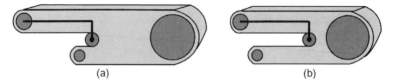

FIG. 7.27 Bull nose conveyor: (a) extended; (b) contracted

to be small, so that the falling of the product is such that the landing orientation is predictable and repeatable (Fig. 7.28).

7.5.2. Experimental Testing

The student team conducting this project had several concerns. One was, how does a blob of cookie dough fall, and the second was, how would it fall from a bull nose conveyor. It was easy enough to test blobs of slice and bake dough to see what height they could be dropped from and not cause a flipping rotation. Under a controlled static environment, something around 1.5 in. was found to be satisfactory, but they needed to test how it would perform when the bull nose was retracted, and the entire row would be effectively propelled over the cliff representing the bull nose curvature.

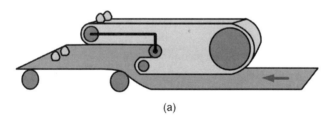

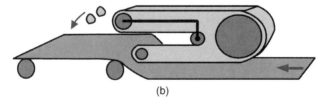

FIG. 7.28 Cookies: (a) extruded onto bull nose conveyor; (b) falling onto baking conveyor

Feeders and Conveyors

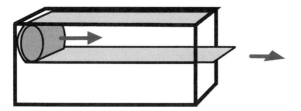

FIG. 7.29 Student team implementation of the bull nose conveyor

So a test model was built. Using 1 in. × 1 in. wooden structural members, a frame was constructed (Fig. 7.29). Since the testing was only to model how the cookies would fall when the retraction occurred, the motion of the conveyor advancing on the bull nose system was ignored. The team selected to use a vinyl window shade, since the material was close to what they desired and the cost was minimal. The window shade was not even made into a continuous belt. They did not use the internal torsional spring that normally raises the shade when installed in a common window frame. They pulled the shade to the right by tying it to a string and connecting it to a weight hung over a pulley to get a repeatable power source to actuate the motion. They purchased several other wooden dowels and made loose fitting bushings out of plastic scraps. They moved a lower piece of cardboard by hand to represent the continuous baking conveyor.

7.5.3. Project Assessment

If the right weight was chosen, it could be actuated and then the raw cookies would be dropped to the surface below. This process worked well for dropping onto either a moving continuous baking conveyor or a stationary surface below. It was anticipated that if the conveyor belt itself could actually move, that would give an additional variable to the degrees of freedom of the cookies motion, and that the right combination of belt speed and bull nose retraction speed would be found, but their experiment showed that there would be less risk to implementing this on a real machine.

The choice of a vinyl window shade was reasonable for modeling, but probably not food grade. Any machine in food processing needs to be washed down every 24 hours. There was a general accumulation of some cookie dough after a few rounds of testing, so in practice, the bull nose conveyor would need to be continuously cleaned or wiped down. As for a solution to the challenge, the students' design was a good one.

7.6. CASE STUDY NUMBER 2: FEEDING OF TBBL CASES*

The green books on tape cases detailed in Chapter 4 did present a feeding challenge, one that has yet to be implemented at the New York State TBBL facility. The case study about the BMC unloader mentioned how the contents were deposited onto a belt conveyor whose height was only a foot from the facility floor, but as stated before, the green cases need to be oriented so that the latches are on the top surface facing forwards, and there are more than just green cases in the BMC. So how does one sort and feed these products?

There were no commercial feeders found during the investigation period. The only possible option was the bin picking robot scenario, and this was dismissed for possible high costs and unknown risks. So the R&D team at RPI conducted a series of experiments to develop a crude working operational model (Simon, 2000). This effort had many trials and errors until a series of steps gave somewhat repeatable results, but it should be noted that this module of the automation was always considered to be the highest risk, and has yet to be implemented because of this.

The process was broken down to the following steps:

1. Remove all of the BMC contents that are not green cases.
2. Rotate the cases to have the latches leading.
3. Flip the cases over, if needed, to have the latches on the top.
4. Spin the case, if need be, to have the latches leading.

There were some additional smaller steps generated as the best step processes were developed.

To filter the contents to determine a green case from the rest of the possible contents, all of the cases needed to be separated from each other. The term often used for this separation process is to singulate. Since the conveyor stream from the BMC was almost always a pile of cases two to four pieces deep (Fig. 7.30),

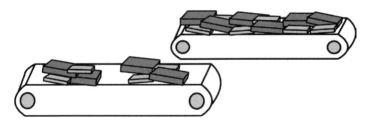

FIG. 7.30 Step 1: pulse BMC conveyor to make piles of mixed cases

---

*This case study supplied by Mr. Matthew Simon.

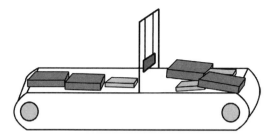

FIG. 7.31 Step 2: knock down piles to a single layer

a weighted flap was used to spread them out (Fig. 7.31). This would work well if the piles were intermittently situated. So before that step, the original BMC unloader conveyor was pulsed to deposit piles of the mixed cases onto a second continuously moving conveyor (Fig. 7.30).

Now there would be a single layer of mixed product. The research team then conducted experiments to distinguish the characteristics that were present in the green cases but different in the other expected cases. For the most part, the green cases were smaller in their two major overall dimensions than the other expected cases. It was assumed that if somehow the U.S. Postal System delivered an unexpected case that was near in size and shape to the green cases, a sensor would trigger a notification alarm to get an operator to clear out the problem. The goal was that it was not cost effective to identify any and all possible trouble occurrences specifically.

The best results are shown in Fig. 7.32, in a mechanical filter referred to as the "waterfall." The cases would be slid off the second conveyor shown in Fig. 7.31, looking like a dry discrete item waterfall. The green cases would be small enough to fall into a slot fashioned above a third conveyor, running perpendicular to the second. The larger cases would not be able to fall into the slot in time, and would simply continue their fall into a bin.

Now the cases were in a slot on their edge being transported by the third conveyor. The latch side could be either on the right or the left, and the case

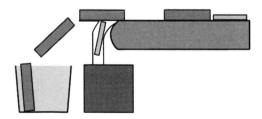

FIG. 7.32 Step 3: waterfall sorting of cases

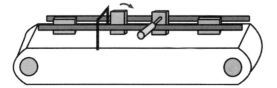

FIG. 7.33   Step 4: case pinching and limbo bar rotation

might be on its end rather than on its side. Experiments showed that a simple rod referred to as the limbo bar would easily rotate the cases on their end to their sides, with the conveyor below the case providing all of the needed motion. However, if two cases happened to fall from the waterfall immediately after each other, they were too close to allow the limbo bar to work. The second case prevented the rotation. So, as Fig. 7.33 documents, there is a need for a small air cylinder or electrical solenoid to just pinch the second case enough to retard its motion for the duration of the first case's possible rotation. Several electric eye sensors were needed to be located so that the proper timing could be accomplished. If there were not two cases immediately following each other, the pincher did not need to be activated.

Now all of the cases would have the same orientation, and it was just a matter of the four possible configurations of the latches. There would be the need to possibly flip and spin the cases, so it was determined that these operations were easier to constrain if the green case was now in a more stable position. The next step was to flip the case 90° along the axis of the conveyor motion. This was done by constructing a metal guide or plow, as it is often called (the term plow does have its roots in the plow used by farmers to turn over the soil). Then limit switches with rollers could be used when the case was temporarily held by a movable stop. This would determine if the case was ready to be opened, or if it needed one or both of the next steps. If the case was on its front and needed to be flipped onto its back, an air cylinder would rotate the case. This is similar to trying to flip a hamburger with a spatula (Figs. 7.34 and 7.35).

The latches should now be on the top, but they may not be leading. If required, another movable stop is activated, and a suction cup comes down

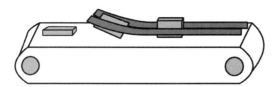

FIG. 7.34   Step 5: rotate case onto its back or front

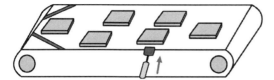

FIG. 7.35 Step 6: flip case if needed

from above. The case is lifted a fraction of an inch, and the case is rotated 180°. Then all cases are presented to the latch unfastening module.

7.6.1. Conclusions

This process was demonstrated to be feasible at the RPI facilities, but the TBBL administration was limited is obtaining funding in major chunks over many years of time. It was determined that this feeding process did not have the highest risk for repetitive motion injuries, and would not have as big "bang for the buck" as other modules would. So this effort is waiting for the right economical and political timing to be approved.

7.7. CASE STUDY NUMBER 3: DONUT LOADER MACHINE*

Sometimes within an automation challenge one will find the need for a feeder that when discovered, impacts the entire machine. Such is the situation with this case study. Creating a high-speed transfer or loading machine has been one of the many challenges for machine designers for decades. It is a classic problem used to help define the area of rigid body guidance in linkage kinematics (Norton, 1999). It is usually viewed as the transferring of a product (such as a bar of soap) from a horizontal conveyor and rotating and transferring it into an awaiting package (such as the bar of soap's box). And developing a parts feeder to keep this high-speed machine running is an equal challenge.

However, when the product rate increases to a certain point, and the product is loaded in multiples into a single package or box, other challenges come into play. When designing an automation machine, one often needs to account for the incoming product stream to temporarily stop for an unknown amount of time. The machine builder has no control over the upstream production and its reliability. If a machine was to load 10 items side by side into the box, and production has halted before filling a box, the number of items already placed into the box must not fall over. Products that did fall over would create a second

---

*This case study supplied by Mr. Clay Cooper.

process halt, requiring human intervention to remove the product in question. Timing issues would be very difficult.

In this case study, the example is to load donuts side by side, six to a box. The donuts are individually wrapped and sealed in clear cellophane. The rate of loading was set by the customer to be 300 per minute. The incoming stream of donuts must be inspected by a simple visual sensor, for removal of incomplete donuts. The sensor scanned the width of the donut and calculated the sum of the scan signals to find the product's area.

The resulting donut loader needs to have the donuts enter the system averaging 300 per minute, but if a donut was rejected from the input conveyor, the entire donut loader needs to pause for the 0.2 seconds while the empty conveyor spot passes by. Thus the timing of the feeder conveyor and the machine itself needs to be highly integrated. It also means that the loader methodology must be able to handle pauses without dropping donuts.

7.7.1. Design Approaches

Many designers have tried to use high-speed four-bar mechanisms to load the item (donut, bar of soap) into a box. The challenges of using such a linkage come from two different sources. First is the dynamics problem. As the speeds get higher and higher (due to product loading cycle times), the linkage can see perturbations due to play in the joints, link members bending, and the effects of wear over time. A second problem is the percentage of time that the mechanism's motion is performing useful work. As the input crank moves in one 360° cycle, the amount of useful work may occur only during a small portion, say 90 to 160°, for example. The remaining time, percentage wise, is wasted. So the mechanisms must move even faster, to make up for this lost time.

Other Transfer Mechanisms Considered

Besides the classical linkages mentioned above, several other standard industry solutions were investigated. These solutions come from the packaging industry, a group of machine builders based more on practical experience rather than much of any use of theoretical developments.

Bomb Bay Doors

As the name implies, product is accumulated over a pair of doors, acting as a bottom surface. When actuated, these doors pivot to allow the product to fall into the package or box. While this method is popular in the packaging industry for rigid items, freshly baked donuts create a different problem. The donuts that one buys in the market have cooled, and they shrink 10 to 20% in their width. If a box was designed to use Bomb Bay doors to load fresh donuts, by the time one purchased the donuts, the shrunken donuts would look like one was being

Feeders and Conveyors 189

cheated. The extra box material and volume for shipping costs is also an issue. So the use of Bomb Bay doors was ruled out for this product.

Loading Box While on its Side

A second alternative was to load the donuts while they were stacked on their sides. They would be transferred as a stack into the open box. Since the donuts were being produced in this orientation, this approach seemed to have some merits. After loading, the rigid and easier to move box could be rotated. However, the stacking of six cellophane wrapped donuts is not trivial. Again due to the donuts being warm (and fatter), the stack would be higher than the box height. The stack would need to be compressed. Forming a series of stacks of donuts (at 300 donuts per minute) created continuity problems. The amount of machinery, levers, sensors and the like were great. So this alternative was eliminated.

Paddle Wheel Transfer

A simple paddle wheel design, where the angle between the fingers does not change, was also considered. This concept could be implemented by simply replicating a paddle wheel found on an old steamboat. Circular arcs beneath the paddles could help hold the donuts in place during part of the transfer motion. However, they could not be used over the open box. Donuts would be falling into the box, without enough control. So this concept was abandoned.

7.7.2. Dual Cam Track Mechanisms

Very few mechanisms textbooks demonstrate the usefulness of dual cam tracks. Cams and their tracks are usually to create a vertical displacement, such as lifting a valve in an internal combustion engine. The nominal mode of operation is that the cam plate or barrel rotates, forcing a spring-loaded follower to displace at a predetermined motion that corresponds to the cam profile. A single degree of freedom motion vs. time is created.

However, by using a dual track cam, both the displacement and orientation can be controlled. A chain provides the relative motion of the follower to the cam surface. In this application, there are multiple followers, located every chain length (a little more than an inch). Each follower has two rollers, constrained in two cam slots. The follower's orientation at any position along the path is therefore constrained. Slight changes in the distance between the two slots are transformed into small angular changes, sufficient to grip or release a product.

While creating such a track is not generally considered Modular Automation, the resulting machine component can be modular. Any changes of performance require an adjustment of the cam slots, either as a pair or relative to each other. These changes need to be cleverly engineered, with adjustable cam slot locations, or a great amount of remachining would need to be done.

Cams are at the opposite end of the spectrum compared to Mechatronics and computer control, but in this application, the use of these mechanisms consistently fixed the position and orientation no matter what the speed, acceleration, or any other dynamic issues. The donuts can be firmly held at all times and conditions. Donuts falling into the box if only partially filled are not a concern.

7.7.3. High-Speed Donut Loader Details

This concept required a true dual cam track to implement (Cooper and Derby, 1998). The overall concept (American Dixie Group, 1997) is shown in Fig. 7.36. The donut enters in a horizontal position from the left. The left-most conveyor is where the scanning sensor would have determined whether a donut is good or bad. An air jet would blow off the bad ones. The second shorter conveyor above between the first conveyor and the donut loaded is the actual feeder to the machine. It shoots donuts between the open fingers. The moving fingers, controlled by the dual cam track, are spread open to easily accept the donuts. A circular arc plate assists the holding of the donuts while rotating the majority of the required 90°.

After the donuts reach the travel bottom, the dual cam track acts to compress the donuts between the fingers. Then the track path, in essentially a triangular path, lowers the donuts into the open box. The second half of the triangle removes the fingers from the filled box. By use of the continuous motion

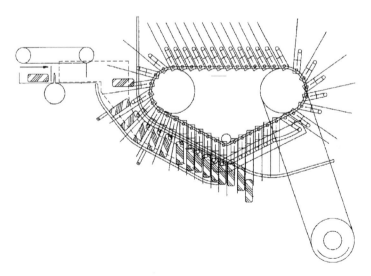

FIG. 7.36 Donut loader and feeder (from American Dixie Group, 1997)

Feeders and Conveyors

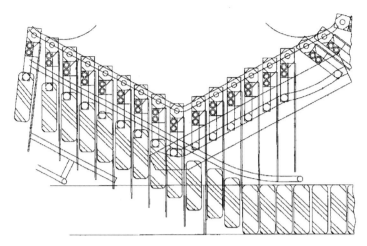

FIG. 7.37 Donut loader closeup (from American Dixie Group, 1997)

available from the dual cam track, boxes can be loaded, one after another. The open empty boxes are lined up end to end.

Figure 7.37 details the region where the high-speed loading occurs. The fingers (Fig. 7.38) have slots to accommodate the cylindrical rails that assist

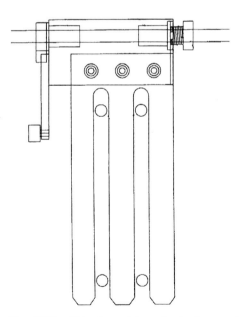

FIG. 7.38 Side view of donut loader fingers (from American Dixie Group, 1997)

the transferring donuts. The bottom rails guarantee that the donuts will not fall out prematurely, while the upper rails force the donuts to fall into the box.

7.7.4. Conclusions

The use of a dual cam track to manipulate industrial products is a very viable option compared to standard linkage mechanisms. The resulting feeder/machine will not lose product orientation if shut down at any point in the loading cycle, and does not pose any dynamic problems or concerns when working at 300 cycles per minute. The machine is more robust than any attempt to use high-speed controls or Mechatronics solutions to solve the same problem.

7.8. CONCLUSIONS

Some products are easy to feed and one will have many options. Others, like lemons, will be difficult. Does one pre-sort lemons to select only the correctly sized ones that can be automated? Do you humanly stack them in a tray when there is a lemonade production lull? Could one make lemonade in a lights out facility? Is it cost effective to do so?

It may seem silly to spend so much time here discussing lemons, but many of the automation challenges still left to address have parts more difficult than lemons to handle.

PROBLEMS

1. Go to your local supermarket or department store, and observe the checkout area. Note how the conveyor system and barcode scanner are functioning, and how the clerk bags the products. Make conceptual design recommendations on how this system can be approved, and how much automation could be accomplished.
2. Design a pair of feeder systems to handle the standard No. 2 pencil before the eraser is crimped on, and the eraser itself. Determine what kind of feeder would seem to be best for unsharpened pencils due to their length, and how the high friction of multiple erasers would cause a challenge.
3. You are an automation design engineer, and are tasked to fill a standard desk stapler with a row of staples before the stapler is package in its box for retail sale. Determine how you would transport the empty stapler, the rows of staples, and how you would insert them without breaking them apart.
4. Investigate on the Web competitive vendors of segmented conveyors that are "flexible." These conveyors can be routed via their modular

framework to transport products both up and down, as well as turning to the right or left. Their paths can be snaked through existing walls, beams, and so on. Determine any concerns you might have if the product was either a can of soup, or a roll of paper towels.
5. Brainstorm on some future possibilities to levitate products into a box for packaging, eliminating the need for robots or transfer devices. Use all physical properties known to humans.

PROJECT ASSIGNMENT

Using one of the projects from the Appendix (or any other projects), perform the following:

1. Determine the viable options for feeders for your product's components. Find as many commercially available options, and conceptualize custom options. Determine what if any sensory feedback is needed to assure component placement accuracy when required.
2. Determine the viable options for conveyors for your product's components or final assembly. Find the type of conveying surface (smooth, slippery, high friction) to gain the proper advantage. Compare with the need for product accumulators vs. the relative costs of each type of conveyor.
3. Review the entire automation process to see if there is a need for any product component accumulators. Estimate the pros and cons of using accumulators, and the relative costs for either approach. Establish the trade-offs of using a set of accumulators where needed.

REFERENCES

American Dixie Group. (1997). Packaging Machine. U.S. Patent No. 5,664,407. September 9.

Carlisle, B. (1997). Adept's Flex Feeder. U.S. Patent no. 5,687,831. November 18.

Cooper, C., Derby, S. (1998). Using dual track cams for high speed rigid body motion. ASME Design Technical Conference, Conference CD-ROM Paper DETC98/MECH-5841, Sept.

Norton, R. (1999). *Design of Machinery*. New York: McGraw Hill.

Simon, M. (2000). *A Modular Automated Handling System for the New York State Library: Investigation, Design, and Implementation*. Ms Thesis, Rensselaer Polytechnic Institute.

8
Actuators

At this point in the automation machine design process, one might think that the remaining steps would simply be filling in the details. Find the right size of motor and be done with it. Let someone else do the dirty work and move on to the next challenge, but this could not be further from the truth. Finding the right type and properly sized actuator (a device that produces motion with some force or torque) will often return a designer to an earlier step in the design process, sometimes all the way back to the original concept.

The human muscle is one of the most compact and efficient actuators ever built. Trying to emulate this wonderful design will cause one to be frustrated. Perhaps this can best understood by noting that there are currently no prosthetic devices available to replace a missing arm or leg that have self-contained power and actuation. In such an application, if one could achieve this result, it would seem to be more cost justifiable than its application in a generic automation device.

8.1. TYPES OF ACTUATORS

If one looks past the human muscle and focuses on the types of currently available actuators, one will generally see a list something like this:

- compressed air (pneumatics);
- hydraulics;
- electric motors.

Actuators

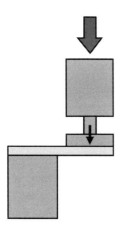

FIG. 8.1 Machine using linear output

There are subcategories to each of these actuator types, and there are some newer developments that might not seem to fit, but these types of devices where power of some sort is used to generate a motion with a force or torque are the building blocks for automation.

As an example, let us look at a simple application. In Fig. 8.1, there is a metal plate that is originally flat. Ultimately, in Fig. 8.2, the plate is bent. The questions needing to be addressed include:

- How thick is the plate?
- What is the plate's material composition?

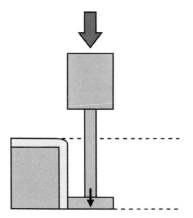

FIG. 8.2 Machine completing operation

- What is the bending force?
- Is there lubrication used to assist the bending process?
- How fast does the bending need to occur?
- Will the plate have springback?
- How many plates will be bent on this machine?
- What are the currently available power sources?
- Where will this bending machine be located?

Each of these questions will help us to understand the challenges of actuator selection.

The plate's thickness, composition, and bending force are all related. For most materials in use today there are handbooks that will assist the designer to determine these relationships. For a given thickness and material type, the information to calculate the bending force will be provided. However, if it is a newer composite material, it is possible that the designer will have to perform some tests to determine these relationships. So a known process is less risky than an unknown process, to no-one's surprise.

Other answers to these first three questions might produce other, more obscure, concerns. If the plate material had carbon fibers internally, and if the plate could be bent successfully, would the process release some carbon fiber particles in the working environment? Would these carbon fibers mean that the actuator needs to be mechanically shielded from the surroundings? This is not a big deal for hydraulics or pneumatics, since the shielding option is readily available, but what if the electric motor uses air flow for cooling? The surrounding air, if it contains carbon fibers, could create havoc with the electrical contacts. Perhaps the contacts become dirty very quickly, or, on the other end of the spectrum, the contacts get shorted out.

Does the bending process need lubrication? Will the lubrication cause the actuator to fail in a few weeks or months? Will the machine need to be cleaned periodically? Or will the actuator leak hydraulic fluid onto the plate, and thus the work area become contaminated?

The length of time allowed to bend the plate will greatly affect the sizing of the actuator. If a gearbox can be used, and assuming that gearbox friction does not become excessive (and in real life this cannot be ignored), one could power this plate bending press with a squirrel in a cage. Most likely the squirrel would have to run for hours to do this task, similar to the length of time an analog wind-up clock takes to move the hour hand. It can be forceful, but not practical in industry.

Springback occurs with most metal plates. Even though the metal chosen is not technically to be used as a spring, it will recover from the pure $90°$ bend as shown in Fig. 8.2 to perhaps $89°$ or at least $89.99°$. This springback

Actuators

issue is a concern for the manufacturing engineer who wants to use this machine, but it also can create issues to the automation engineer. Will there be a frictional force to return the moving bending fixture to the original home position? What will the force level be, and can the process be done quickly so as not to waste time?

Imagine if the squirrel cage method worked (Fig. 8.3), and imaging that the springback forces were minimal. But the squirrel was made to turn around and rotate the cage in the opposite direction. Now it would take about the same amount of time for the moving bending fixture to be raised to the home position. The gearbox ratio would possibly be excessive for the returning motion, assuming that the bending fixture was not extremely heavy. Could one shift gears like one does for a 10-speed bicycle or a manual transmission automobile to make the return motion happen quicker?

So in real life (without the squirrel), the actuator requirements might be such that the return to home motion be a different mode than the bending motion. It should return faster and hopefully with lower loading requirements, but maybe not. Changing gear ratios is not cheap, so other machine design options might need to be explored for the optimal costing and performance cycle time.

The question concerning how many plates are to be bent addresses the life expectancy of the automation machine. This is parallel to you as a home owner purchasing an electric drill at a lower price than that of a commercial grade drill. As a homeowner, you might use the drill 10 minutes to a few hours per year, depending on your skills and your (or your spouse's) ambitions. You expect the drill to last for a while, but if during a large construction project it

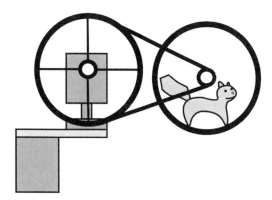

Fig. 8.3 Low power source geared to generate a large but slow force

up and dies, you purchase another one without much disappointment. It was not too expensive in the first place. However, if you are a contractor drilling 8 to 12 hours per day, you spend double or triple the amount to have a drill that you expect to last for four or five years of heavy use. Your drill cannot let you down when there are deadlines or you have storms approaching and the roof is not attached yet.

Similarly, for the bending of plates, is this a homeowner's level of use, or 24/7 operations? There are both actuators that can handle the high level of use, and those from manufacturers who seem to specialize on inexpensive (we will not use the word cheap here, though perhaps appropriate) actuators for a machine that is not destined for longevity. Most actuators come with some type of duty cycle data, which is invaluable in making this judgment. Some customers will specify the machine life expectancy, which will dictate the performance level of the actuators.

The question about available power sources can be huge. Many large electric motors used in larger machines are 240 V AC three-phase power. For many industrial manufacturing buildings this is already available, but if the building is in a rural locale, or if the area is solely residential, three-phase power can be very costly to install. And as a machine builder, one has a limited potential in the horsepower output of 240 V single-phase motors.

Pneumatics (compressed air) can be seen as a limitless supply of energy if a factory is already outfitted. But the issue of capacity can sneak up and surprise you. The current number of machines might be maxing out the compressed air system. Your new machine might be the sole reason an upgrade or improvement is required, and your machine might need to bear the entire costs. The next machine may get a free ride on your machine's back.

Hydraulic power supplies are usually separate for each machine they power. Rarely will there be a plantwide hydraulic power supply available to tap into. These supplies are often located near the equipment they power. However, the accumulated noise of many such powered machines might become excessive upon the implementation of your machine. Although chances are operators would be using ear protection before your machine gets installed, it could be the straw that breaks the camel's back for noise pollution.

Related to this is the final question on our list, the machine's location. If the machine is handling unpackaged food products, it usually must get washed down once each day. This can be a big issue in deciding to have waterproof electric motors. Or it can limit the hydraulic actuators to not having any drips of fluid that can get into the food. Even pneumatics, which will often vent out the air from one end of the cylinder during motion, must be filtered to keep out oil droplets since many compressed air systems deliberately add oil particles to keep the cylinder's internal systems from rusting out.

Actuators

8.2. APPLICATION CONCERNS

The number of different actuator applications is numerous, but there are a few issues that bear mentioning in general:

- Amount of work to be done;
- Frictional losses;
- Dispensing viscous materials;
- Gearing and inertial;
- Conveyors;
- Intermittent motion.

These each represent a larger list of possible concerns that make actuator selection more difficult than one would expect.

8.2.1. Amount of Work to be Done: Power = Force × Velocity + Friction

The basic amount of power a machine needs to provide can be calculated idealistically. If one returns to the machining processes, one can look at the anticipated force and multiply it by the velocity and have a determination for power. If it is a rotational situation (Fig. 8.4) where metal is being removed, one needs to determine the anticipated torque and the rotational velocity to calculate the power.

If the automation concerned is not a machining process, then one needs to look at the motions and weights or inertias to determine the idealistic power requirements, but just like the problems in high school physics, it supposedly takes no work for one to walk on a level floor. In reality, there are always losses to account for, and therefore one needs actuators larger than the physics world states.

8.2.2. Frictional Losses

Friction is normally the machine designer's enemy. Except when one wants friction to assist with bringing a machine to rest quickly, friction adds to actuator size

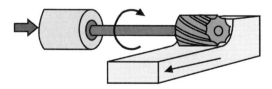

FIG. 8.4 Rotary output power

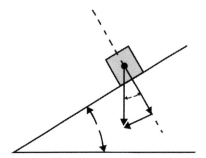

FIG. 8.5 Static friction

and power consumptions. Many machine components need to overcome both static friction (Fig. 8.5) and kinetic friction (Fig. 8.6). The proper use of bushings and bearings will hopefully keep fiction minimized as much as possible, and prevent a situation where there is significantly larger static friction compared to kinetic friction occurring. It is difficult to anticipate the resulting motion when high static friction finally gives way to the lower kinetic friction, particularly when the static friction may change over time, temperature, and number of cycles. If this motion is to happen in a predetermined time sequence, it can be troublesome.

There are also process situations where an actuator needs to overcome a viscous frictional situation. This can appear in many configurations, but two common forms are linear (Fig. 8.7) and rotary (Fig. 8.8). The rotary form can also be seen in journal or sleeve type bearings, but is hopefully less significant than the friction one sees in a process such as Fig. 8.8.

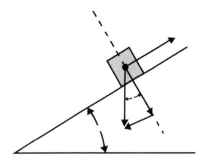

FIG. 8.6 Kinetic friction

Actuators

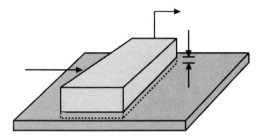

Fig. 8.7 Viscous friction

8.2.3. Dispensing Frictional Materials

Some frictional situations cannot be avoided. And in some cases, one does not want to avoid it, but wishes to have it under control in a known environmental condition. The dispensing of a viscous material (Fig. 8.9), such as an adhesive or a silicon product, cannot significantly modify the liquid's frictional properties. If the material was too runny, it might drip from the dispensing head. Or if it freezes up when the plant's interior is no longer heated to 65° during the winter months as a cost-saving effort, the entire automated process may have changed so as to threaten quality or production rates.

8.2.4. Gearing and Inertia

There are many textbooks available to assist one in calculating the rotation moments of inertia. And if one has infinite time to conduct a project design, one can calculate the effect of every tapped hole, screw body and head, and length of sensor wire. As stated in Chapter 6, these machines are most like "not going to

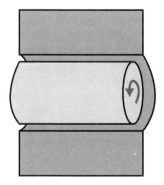

Fig. 8.8 Rotary viscous friction

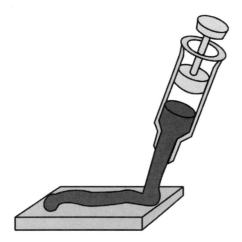

Fig. 8.9 Dispensing viscous materials

fly," and the prudent thing to do is to calculate the significant contributions and add in a slight factor of safety to cover the minor details. Now if a CAD model will produce the entire results from one mouse click, then do not ignore the results, but student teams have gone through the erroneous effort of determining inertia contributions down to the 0.1% level, only to add or subtract some component two weeks later in the design.

There are two common approaches to inertia calculations when one designs a machine. When there is a gearbox involved (Fig. 8.10), one will want to know

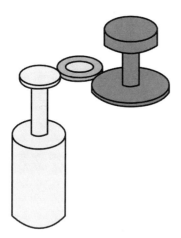

Fig. 8.10 Gear drive application

Actuators

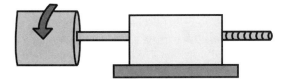

FIG. 8.11 Rotary to linear drive application

what the impact the load shown on the right will have on the motor on the left. The gear ratio is used to calculate the reflected load seen by the motor for sizing considerations. This is analogous to calculating the torque one produces by pressing the foot downward on a bicycle petal compared to what the torque is at the rear wheel. In order to climb a hill at a desired speed, it depends on what gear has been engaged for the required effort.

For many motion applications in automation, a rotational motor will use a threaded shaft to move the load in a linear fashion (Fig. 8.11). So for a target linear speed and force, the pitch of the shaft and the block will determine the torque and speed required of the motor. If one has all the time in the world to linearly move the load, the shaft pitch can be very fine, and one can simply use a rechargeable screwdriver to power it, but most automation applications will never be satisfied with this approach.

8.2.5. Conveyors

Conveyors at first glance do not look to be anything special for actuator sizing purposes. There is a motor usually connected to one roller (Fig. 8.12) in one of several arrangements. More commonly, the motor is connected via a right angle gearbox, so as to match a reasonably sized motor to the needed belt speed, and to take up less floor space, but a conveyor may see a significantly varying loading depending on the application. If the conveyor is to be used to load luggage into an airplane cargo space, it may be started in motion with no suitcases on it. Then the loading will increase as baggage handlers toss on your hopefully nonfragile suitcase. However, if there is trouble unloading it at the top, the conveyor may be stopped abruptly for safety reasons, and then abruptly restarted.

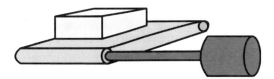

FIG. 8.12 Conveyor drive application

Whether or not a conveyor in its application is allowed to gracefully accelerate the belt and load up to speed (and gracefully decelerate to a stop) has great motor sizing impact compared to needing the belt to immediately resume top speed or to come to a stop in a fraction of a second. Nothing is more frustrating than a piece of automation built around a slower accelerating conveyor, and then having the customer update their requirements to having it instantly start at 60 miles per hour.

8.2.6. Intermittent Motion

Some requirements in automation have dependable intermittent motion. Dependable and repeatable (hopefully like your heartbeat) so that having an actuator go on and off all of the time makes a machine designer look to older proven technology. Dial Machines (or Rotary Tables) can be used to index 24 hours a day, and therefore have been using a constant speed motor to drive the input (right) side of the Geneva mechanism in Fig. 8.13. The pin on the disk goes into the slot and rotates the six part form on the left in 60° increments. The normal geometric motions cause some relatively smooth accelerations and decelerations, but there is still a variable loading occurring to the drive motor. The cyclical application of the increased and then decreased inertia almost begs for a flywheel to even out the required loads. This can be beneficial to a limit, until the flywheel adds so much moving mass that the system almost must continue to the next motion increment rather than stop on command.

8.3. PNEUMATICS

Pneumatics is the use of compressed air through pipes and hoses used to move something. Most times the compressed air goes into a rigid device that is capable

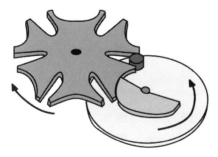

FIG. 8.13 Rotary table with Geneva mechanism

Actuators

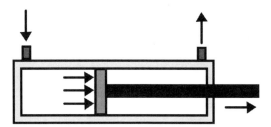

FIG. 8.14 Pneumatic air cylinder: force = pressure × area

of a defined motion. The great majority of these devices can be modeled by Fig. 8.14. Here, air enters though the hose fitting on the upper left of the cylinder and fills the left chamber. The increased volume of air moves the piston (and rod) to the right. The air in the right chamber is normally allowed to exit through the hose fitting on the cylinder's upper right. If the air in the right chamber is not allowed to exhaust, it will become compressed. Then the resulting motion will be limited to some proportion related to the difference in pressures on each side. Rarely is this used in practice.

Since air is quite compressible, the most easily achieved repeatable motion is to allow enough air to go into one side (or the other) so as to drive the piston to one end. Then the speed and motion of how it got to the end is hopefully not critical. If the desired motion has very specific velocities and/or accelerations, it is probably not best to use pneumatics.

There are hundreds of types, models, and variations to the generic air cylinder shown. A few of them, represented by their stick figure models, are shown in Fig. 8.15. Almost all of these are designed to work at standard shop air pressure of 80–90 psi. Air cylinders do wear out from frictional issues and leaky seals around the piston rod. Contaminated air can induce particles that can cause havoc to the cylinders' moving surfaces and the control valves, so filters on the air line are a must.

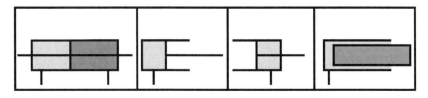

FIG. 8.15 Pneumatic cylinder types: (a) rod on each end; (b) single direction rod external support; (c) single direction rod internal support; (d) piston serving as the rod

Major types include:

- Single action direction (Fig. 8.15b and c);
- Single action direction with spring return;
- Dual action (Fig. 8.14);
- Pancake style: fat but very short with limited stroke;
- Telescoping: multistage, similar in design to multiple part TV antenna;
- Rotary;
- Multiple position: other small air cylinders stop piston from traveling to end

Other options include:

- Location of hose fittings;
- Cylinder mounting hardware;
- Integrated sensors to confirm travel is complete;
- Internal cushions to dampen piston deceleration;
- External shock absorbers to dampen piston deceleration.

To make the cylinder in Fig. 8.14 reverse direction, the hose connections need to be changed. Manual valves can be used to switch which cylinder end is connected to the compressed air source, but using manual valves is not practical for automation. To have a controller operate a pneumatic cylinder, electrically signaled valves are used. Most of these use an electromagnetic solenoid to move a small piston opening or closing one or more ports. The machine builder is responsible to configure the hoses to the correct ports to get the desired performance. In most applications, pneumatics is more convenient than hydraulics since there is no effort to recover the exhausted air, so the hose and valve configuration does not have to be as complex.

The size and shaping of all of the hoses and fittings will have a great impact on performance (Lentz, 1985). Figure 8.16 shows the general trends of most pneumatic equipment. As one moves away from the origin, each curve represents a larger valve diameter. The impact of both air lines and valves of insufficient diameter is sometimes hard to predict. It is not uncommon in developing totally new automation approaches that the entire pneumatic system needs to be upgraded to a larger diameter to get satisfactory performance.

Pneumatics components are available in English (inches) and metric (mm), and various commonly used sizes are so close to one another (and sometimes not properly labeled) that a pile of leftover parts is difficult to distinguish. A cross-threaded metric part into an English fitting can cause a world of problems. Another issue is that the working area of the piston is not the same on both sides due to the rod's connection to the piston, so the motion may not be the same in both directions for the same air supply.

Actuators

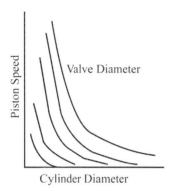

FIG. 8.16 Cylinder diameter vs. piston speed with respect to valve diameter

8.3.1. Advanced Pneumatics Devices

One design issue is the fact that the generic cylinder shown has the rod sticking out. The overall dimension in the direction of motion is a little more than twice the piston motion. Sometimes a designer cannot fit the cylinder and the rod into the machine envelope, so an alternative design is the rodless air cylinder. One common method is to remove the rod and attach a strong permanent magnet to the piston, and add to the cylinder an external sliding magnetic piece that follows the inside piston magnet. As long as the applied loading is less than the magnetic attraction, it will stay coupled. There are more than a dozen other rodless cylinder designs available, many using tricks other than magnetism.

Another recent pneumatic device is the air cylinder with servo control. For many years this was but a dream, since air is highly compressible compared to a liquid. If one was to use a manual valve and try to keep a cylinder piston at some point between its limits by flipping the switch back and forth, the results were very crude. Even electronically, the results were within an inch or two at best. However, recent research has helped industry to produce commercially available air cylinders with great servo control. A dedicated microprocessor monitors a continuous position sensor to determine the current piston position. Then the microprocessor switches the input source back and forth at a tremendous rate, also understanding the compressibility effects of air and the time delays in the motion response, and the result is highly accurate. This product is a true outcome of Mechatronics research.

8.3.2. Pneumatics Vacuum Generators

A valuable subset of the pneumatics toolbox is the method of using compressed air to generate a vacuum. Normal vacuum pumps driven by electric motors are available, but they are in general quite substantial in size. If the vacuum pump is not closely located to the suction cup, there is also a time delay while the connecting hose draws down to the vacuum level. If one needs a vacuum cup to grab something in a tight environment, or if payload restrictions eliminate the possibility of a vacuum pump, a Venturi vacuum generator (Fig. 8.17) can create an unbelievable amount of suction in a very short amount of time. There are even dual and triple stage units that can produce low pressures and reasonably high flow rates.

The Venturi device has compressed at flowing through from the left-hand side. The restriction in flow near the middle of the device creates a pressure drop (this is perhaps the only useful item found in a fluids course textbook that automation engineers can use to their advantage). There are only two real limitations to this device. The first is that they take a constant supply of air pressure flowing through them to hold an item. This is different to an air cylinder, which does require continued pressure to keep something in place, but not the continued flow rate of a Venturi device. Secondly, if the environment of use is dirty, the device will pick up particles. There is nothing as frustrating as cleaning out instant cocoa drink powder from a Venturi device when some of the transported pouches break apart.

If an automation application requires multiple suction cups, and there is even the slightest chance that one of the cups might not get a good seal, so therefore would not achieve a vacuum, it is common to see every one or pair of suction cups having their own Venturi generator. The incremental cost is not unreasonable, particularly compared to losing the grip on a costly transferred product.

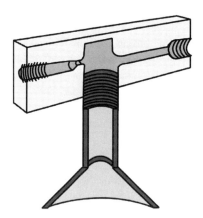

FIG. 8.17 Vacuum generator and suction cup: cross-section

Pneumatics systems are relatively cheap to purchase, cheap to install, and cheap to run. Some machine builders swear by them, others, because of nasty operating conditions, swear at them. Some builders will never use them if given a choice, but by having the power source (the air compressor) at some other location, there can be a great deal of action happening in a compact working volume.

8.4. HYDRAULICS

Hydraulics is like the big brother to pneumatics, a much stronger, noisier, more dangerous, and potentially messier big brother. The basic concept of the air cylinder in Fig. 8.14 stays the same, but some of the operating conditions change tremendously. Hydraulics use a fluid similar to your car's power steering system rather than air, so compressibility is orders of magnitude less. Many hydraulic systems operate in the 2000–3000 psi range, so the cylinder, hoses, valves, and power sources all need to be upgraded. One could never use a pneumatic device for a hydraulic application. A leak in a hydraulic hose or fitting can produce a high-pressure stream that acts like an abrasive waterjet cutter, able to slice off one's fingers and the like.

Figure 8.18 shows the traditional components in a minimal hydraulic system. The fluid starts out in the reservoir. The pump brings the fluid up to the 2000–3000 psi pressure and it fills an accumulator. The accumulator's role is to have a large supply of high-pressure fluid available for short bursts that might normally outstrip the pump's capacity. It acts analogously to a capacitor in an electrical circuit. However, the pump runs continuously (a real noise source), and if the accumulator and the hydraulic cylinders are not in need of the fluid, a valve between the pump and accumulator dumps the excess fluid back into the reservoir. So the energy consumption of the system is always on full throttle as soon as the pump is turned on.

Any thoughts of turning the pump on only when one needs the pressurized fluid are dashed by the requirements of the systems. The fluid needs to be warmed up and any bubbles removed. Bubbles of air are compressible, and would compromise the overall performance. This has been one of the limitations of hydraulics systems.

A system of valves is used to control a hydraulic piston as shown in Fig. 8.18. Since the supply pressure is constant, the larger the cylinder, the larger the force. In this simple representation, the valve will only allow the piston to move to the left. The fluid in the cylinder's left chamber will be pushed back into the reservoir. This, in practice, is quite useless. There needs to be a more complex valve setup that allows the cylinder to move in both directions. This is achieved by directing the pressurized fluid into either side, and allowing the returning fluid to route to the reservoir.

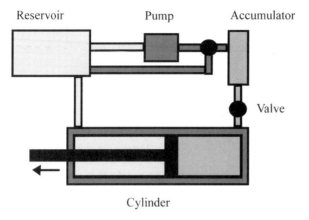

FIG. 8.18 Hydraulic actuator system: force = pressure × area

Besides being noisy, hydraulics is often messy. When there is an oily fluid being forced at 3000 psi through fittings and held in by flexible seals, one can imaging that there would be leaks, either visible drips or a smelly film in the air. One only has to look at the grease (and the dirt it attracts) found on most backhoes or other construction equipment to see the potential concerns. One would not like to see a backhoe used in the baking of cookies, for example, but in high-force applications like metal working presses, there are no logical alternatives.

The early development of industrial robot arms was totally based upon hydraulics. This was chiefly because the electric motor industry was nowhere near where it is today. Early robots, having hydraulic pumps and reservoirs, were used to lift heavy metal parts into such applications as foundries and the welding of car bodies. In many situations, this was seen as a great advance at the time, and the leaks were seen as something one had to live with.

An interesting automation development by IBM in the late 1970s and early 1980s was their RS/1 robot. It was a gantry style robot that used a novel method of cycling multiple rollers over a machined metal repeating sine wave. The multiple rollers were actuated by four small hydraulic cylinders, similar in size and configuration to the valves found on a trumpet. By energizing the four valves in a definite pattern, motion in one direction or the reverse could be achieved. It was a technological marvel of the time. However, it was a robot designed to assemble electronics that was hydraulically driven. It gave off an invisible yet detectable (by the human nose and fingers) film that would coat everything in the working environment. Experiments at RPI by the author showed that it took only several hours of operation for the friction coefficient between the robot's gripper fingers

Actuators

and a block of aluminum to change from 0.4 to 0.1. The great technology advancement that it represented was limited by its type of power source. IBM then converted the next generation RS robot to use electric motors.

8.4.1. Hydraulics in Automation Today

Except where hydraulics is still needed to generate a great force with a remote power source (the reservoir, pump, and accumulator do not need to be directly next to the actuator), the number of automation applications using hydraulics has dropped. There are almost no remaining robot designs using hydraulics, and many older packaging machines have been upgraded. Electric motors have evolved to become far more efficient and compact than they were 20–30 years ago, and they have made a huge inroad into automation.

8.5. ELECTRIC MOTORS

The basic DC electric motor has been around for a very long time. It is still used for some applications, but has often been replaced by newer technology. The basic DC motor as perhaps you wound in a high school technology class uses a stator (Fig. 8.19) and has windings of lacquer-coated insulated wire (Fig. 8.20).

The motion is generated by the timed changing of the stator field, as shown in Fig. 8.21. Depending on different motor parameters, speeds could be changed by varying the applied voltage.

This kind of motor is useful if one is designing a printing press for making the morning paper. Motors and the attached hardware can come up to speed gradually or quickly, always producing a newspaper per cycle, no matter what

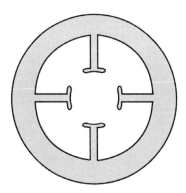

FIG. 8.19 Electric motor stator laminator

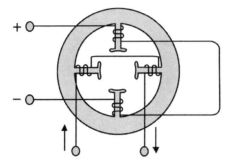

FIG. 8.20 Four-pole stator

the speed, but for something like a robot arm, where one wants the robot to go to a programmed position and stay there, a standard motor is not too useful.

8.5.1. Electric Servomotor

So the servomotor was developed, where the motor was designed to rotate to a fixed angular position and remain there. This is something one cannot achieve with a normal electric motor found in a power tool such as an electric drill. This process relies on a sensor located on the motor output shaft to report what the current angular position is. A comparator in the motor controller (Fig. 8.22) looks at the current position and the target position, and makes an appropriate signal to move the motor. Experienced motor controller manufacturers will use some moderate level strategies to use the feedback intelligently to achieve optimal performance. This is a rich topic for a good book on motion control theory.

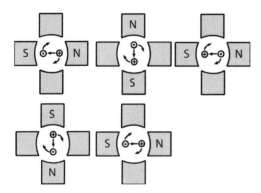

FIG. 8.21 Rotating stator field

Actuators

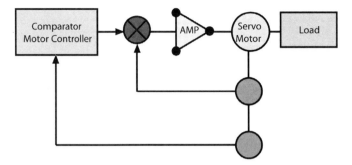

FIG. 8.22 Servomotor drive block diagram

So if one dictates a desired final position, speed, and acceleration, one can get a reasonably close output trajectory that will usually suffice for most applications. Servomotors and their supporting electronics are not outlandish in costs, but are not cheap either.

8.5.2. Electric Stepper Motor

Many applications of motors, not limited to automation, want to control an electric motor and hold it at a fixed position. Products like the dot matrix printer helped to drive the electric stepper motor technology so that very low-cost control could be achieved. Advanced stepper motor technology created significantly larger motors, many capable of moderate tasks within automation. The basic stepper motor shown in Fig. 8.23 has two sets of windings, and a permanent

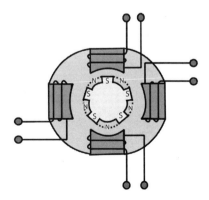

FIG. 8.23 Permanent magnet stepper motor

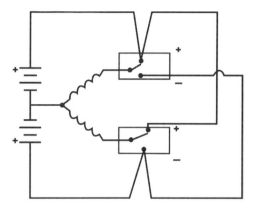

FIG. 8.24 Stepper motor circuit

set of magnets connected to what looks like teeth labeled "N" and "S" for North and South on the rotating center core or rotor.

By systematic switching of the windings (Fig. 8.24), the N and S teeth of the rotor are propelled or repelled to create a repeatable rotational motion. The motion can be in either direction, and can be controlled (Fig. 8.25) for speed and acceleration. The original single step motion process was first augmented to a half step process, and now uses micro stepping. The jerky motion of original dot matrix printers can now be as smooth as a finely tuned servomotor.

Stepper motors can work either with open loop control (no feedback) or closed loop (feedback from a rotary position sensor). The largest application base is in the open loop mode, since products like current ink jet printers are highly cost competitive (and sold for a loss to get you to buy their ink!). Perhaps the only real limitation to address in open loop control is the need for the motor to "home" itself

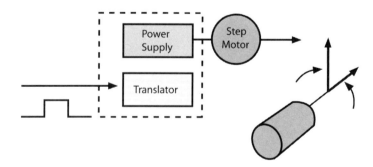

FIG. 8.25 Stepper motor with translator control

Actuators

by creating a contact with a limit switch or optical sensor. Without this information, the motor is only moving relative to where it currently is. So it needs a reference point to have an accurate position. Stepper motors, if over loaded, will not cycle their rotor's teeth properly, and behave in a manner referred to as "slipping."

8.5.3. Motor Sizing

Determining the correct motor size can be straightforward in many cases, and other times a challenge. The key concept with most electrical motors, whether DC continuous speed, electric servo, or electric stepper, is that there is some trade-off between torque and speed. Figure 8.26 shows a very generic diagram that has a linear relationship between the two. For most motors it is far from linear. One should always investigate the actual torque speed curve supplied by a motor manufacturer before one ever starts the purchasing process. These curves are usually quite accurate, and are the primary information used for sizing.

Generally, an application's torque and speed want to be under and to the left of the curve. Where one is in this region is open to engineering debate. In comparing similar types of motors but of different sizes and power, one would look at:

- What is the correct size (torque–speed) motor available?
- What is the impact of the larger motor's size and weight?
- What sizes are the other motors on the machine?
- Will the motor stall out?

Let us look at each issue.

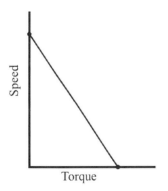

FIG. 8.26 Motor torque vs. speed

8.5.4. Motor Selection

A stepper motor could have a torque–speed curve that looks something like Fig. 8.27. This curve is showing a continuous duty cycle, where the motor is being used a high percentage of the time 24/7. If the motor was being operated only minutes per hour (called intermittent duty) it could be run at higher speeds or torques. This would usually be designated by a second curve on the same diagram, or an entirely separate diagram. This is analogous to a human muscle being able to do a greater task for a short duration than what can be done for an eight-hour shift.

An "X" has been placed at what could be a good operating point. There is nothing special about this X, other than it is not very close to the speed or torque limit. Since most automation is being designed from scratch, the actual output of the motor is not always crystal clear. So it is possible one selects a motor with a fair factor of safety and the cost impact is not significant compared to changing the motor months later when debugging. If this motor application was to be duplicated by the thousands, then one would want to optimize as much as possible to get the most out of the cheapest motor available.

Motors are being optimized to new levels by some of the European robot manufacturers. Since most robot makers do not manufacture their own motors, they buy the same motors as their competition. So if one producer wants to get higher performance, they will run a motor on the brink of trouble. They will run the motor so it gets hot and is seemingly ready to fail, but it does not. This emboldened strategy takes some guts to pull off, something the author does not recommend easily!

Motor's Weight and Inertia

Robotics is a good example of why not to select oversized motors. The basic industrial robot arm design has the motor which drives any particular link situated

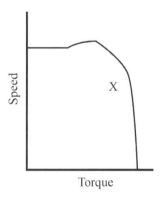

FIG. 8.27 Stepper motor torque vs. speed curve

Actuators

adjoining the link. So the motor closest to the robot's base must rotate all of the robot links, the payload, and the other links' motors. And then if the loads get bigger, the links must get larger to avoid deflections. You now have a vicious cycle of a robot weighting 2000 lb that can only carry 5 lb. In recent practice, a 2000 lb robot might carry 100 lb accurately, again not as efficient as a human, percentage wise.

In smaller applications, the rotating inertia of the motor's core is significant with respect to the rest of the mass to be moved. A quick analytical check is good to see if this will have impact, or is just a few percent. All of these comments assume that the motor connects directly to the output shaft. There is often a gearbox of some sorts, so the inertia the motor sees needs to be calculated on a level playing field.

Sizes of Other Motors on the Machine

Sometimes when all of the multiple motors on a machine have about the same torque–speed requirements, it is logical to select a single motor size for them all. The downside is that for one or two motors you have power overkill and extra costs, but for spare parts and support reasons the customer might be happier knowing that if they keep one extra motor on hand, their technicians can replace it in a matter of hours rather than wait for a service call. This can be an excellent reason to keep with one motor size, but it can also be an excuse for a lazy engineer to not bother with the sizing at all, and to just continue to use a favorite motor, that maybe happens to be sitting in inventory on the builder's shelf.

Stall Torque

This is a term that may be foreign to many, so one might be asking:

- What is it?
- Do you want to use it?

The best example of stall torque is using a battery operated power screwdriver. As one tries to twist a screw into a board that has a knot hole in it, the screwdriver will likely grind to a halt before the screw is completely inserted. The motor has stalled out, and the output torque has reached its peak, although not enough to do the complete job. The amount of this torque at the screwdriver's last gasp of life is the stall torque. It is that lower rightmost point on the curve in Figs 8.26 and 8.27.

Does a machine builder want to approach the stall torque? Most likely they do not. Similar to the battery screwdriver, one is creating the largest energy drain and mechanical strain on the motor system. This will produce a result that will likely be unpredictable. Imagine judging the results of how far a screw twisted into one knot hole, and assuming that at the next knot hole there will be similar results. Anyone who has worked with wood will know that this is improbable at best.

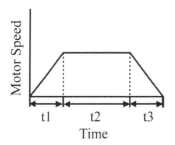

FIG. 8.28 Motor speed vs. time profile

8.5.5. Motion Profiles

A very large body of research has been performed to better understand motors and the effects different motion profiles have on actual performance. Figure 8.28 shows a common motion profile. The motor has an initial time segment where it is ramped up to a desired speed. The derivative of this curve implies a constant acceleration. Then the motor keeps a constant speed for a majority of the motion, followed by a deceleration or ramp down. This ideally should be easy for either a servo or stepper motor, but if it is an open loop control stepper motor and the ramp up or down is excessive, the motor can slip and the performance becomes unsatisfactory.

Assuming a gearbox of some type is used, the motion profile of the motor will match the profile of the application. Figure 8.29 shows a table that is to move, and its speed vs. time. A good point to note here is the reverse problem, which is usually what one sees when sizing a motor, and which is where the table speed has requirements for seemingly instantaneous acceleration and deceleration to and from a constant speed. This can be done successfully, but implies a much larger motor to reduce the ramping time. This will also drive up the costs. So the automation builder is wise to double check that the application really

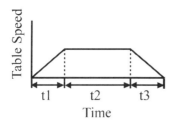

FIG. 8.29 Table speed vs. time profile

needs such fast acceleration (and larger motor) before designing the machine. Perhaps the customer does not know what their requirements are implying.

Advances in Electric Motors

There are many advances in electric motors and their control. With many markets using motors far beyond the limited realm of automation, the competition to develop better products is present. This includes:

- Stepper motors packaged with threaded shaft drives to produce linear motion;
- Linear stepper motors;
- Rare earth motors and other improved materials.

The first two choices give more off-the-shelf options to automation builders. The first saves considerable time and effort of designing and building a unit that has often been built from scratch in the past. The rotational stepper motor is a known quantity, and its motion is coupled via a timing belt to a threaded shaft. The shaft can then extend or contract in proportion to the stepper's rotation.

The second choice is even more compact. It takes the North and South teeth of the stepper motor rotor, and in a sense, unrolls them into a linear bar. The energizing coils are also unrolled, and thus the linear toothed bar can step to the left or right. The bar can be connected as a motion output rod.

Strengths and Weaknesses

The strengths and weaknesses of the various arrangements are shown in Table 8.1.

8.6. AMPLIFIERS, DRIVERS, AND TUNING

The available signals out of most controllers are either digital or analog, but at a 5 V level. The 5 V amperage is at the logic level that the processor board runs at, and at best can light up an LED on a circuit board. It will not be strong enough to run a 5 V

TABLE 8.1 Strengths and Weaknesses of Various Arrangements

	Pneumatics	Hydraulics	Servomotor	Stepper motor
Performance	Good	Good	Great	Great
Speeds	Fast	Fast	Fast	Moderate
Loads	Moderate	Heavy	Moderate–heavy	Moderate–light
Cleanliness	Needs filters	Messy	Good	Good
Noise	Banging noise	Pumping noise	High pitched	Moderate
Costs	Low	High	Moderate	Low

DC electrical motor. So the signal needs to be amplified, just like a portable CD player needs to be amplified before driving one' surround sound speakers.

There are several choices available to an automation designer, depending on the brand and type of electric motor to power. One can find that:

- Controller's analog voltage gets amplified to power the motor.
- Controller's digital signal gets converted by an amplifier to power the motor.
- Controller's digital signal instructs a smart amplifier (driver) to power the motor.

The first two are logical expectations, taking a signal and boosting it to make a large actuator move. There is usually an integral power supply or else an external power supply is needed. The third option allows for the controller to instruct a dedicated microprocessor (driver chip) to execute a trajectory sequence that the programmer can select beforehand. Any sensors to supply feedback send their signals to the driver chip, and the controller is only told by the driver that either the sequence was properly executed, or that there was an error.

Amplifiers have various levels that need to be set, the most standard being the gain. This is the multiplication factor by which the input is magnified. Some amplifiers have their settings adjusted by small set screws, so human setting of the gain with a small screwdriver is sometimes a guess since there is no feedback of the value being chosen. Trying to get multiple amplifiers to have the same setting is not trivial with this feature. Other amplifiers use computer software to download gain values, and this method is much more repeatable.

As will be shown in Case Study No. 2, there are also amplifiers that are self tuning; that is, they determine the gain and other parameter setting by exercising the actual motor connected to the actual load. The internal program may run anywhere from a few seconds to several minutes, putting the system through its paces. This can be a useful tool for creating new electromechanical systems, but it is not foolproof. The author had a system that was never tuned successfully, even though the product representative ran the self-tuning program 25–30 times.

A new approach performed in the laboratory has been developed by Dr. John Rogers (Rogers and Craig, 2003). This work uses open loop stepper motor control, and uses both computer modeling and actual hardware testing to determine the optimal motor stepping sequence. This sequence can then be repeated in multiple manufactured products like ink jet printers to achieve optimal performance without sensory feedback. As long as the products are similar in mass, friction, and motor strength, they will perform well. If a printer has excessive friction due to a manufacturing flaw, it will not, however, compensate for this deviation.

For all types of amplifiers and power supplies, it is common to place all of these and other electrical devices in a metal control panel. This panel can reside on the side of an automatic machine, or it can be mounted on the nearby wall. It

Actuators

also usually contains the circuit breakers or fuses for the line voltage for the machine, and has a primary switch that can be locked out for service and repair.

8.7. CASE STUDY NUMBER 1: STEPPER MOTOR SIZING*

In Chapter 11 there are detailed discussions about several new automation machines, one of them called the Trackbot. The overall application market and design goals will be left for that chapter. An integral subsection of the Trackbot will be presented here as an example for stepper motor sizing. This sizing effort was a challenging one that will not be totally finished until the next generation Trackbot is complete, but the lessons learned to date are worth noting.

The mobile part of the Trackbot is called a "Bot." Multiple Bots (2 to 12) travel around a closed loop track, come to a complete stop, and pick up a pouch from a conveyor. The Bot's pickup motion is the telescoping downward of a cylinder that has a suction cup on the end. The cylinder is contained within a larger tapered cylinder. This section of a Bot without its electronics, pneumatics, and cover is shown in Fig. 8.30. At the top is an electric stepper motor, which is connected to a mechanical housing, itself connected to the tapered external

FIG. 8.30 Bot with stepper motor, external body, and suction cup

---

*This case study supplied by Mr. David Brown.

FIG. 8.31 Bot without external body or motor

cylinder. A small section of the telescoping cylinder with the suction cup attached is visible on the bottom. The telescoping cylinder and suction sup can be lowered 6 in. from this home position.

The telescoping cylinder and suction cup are shown as a separate unit in Fig. 8.31. There is a band around the lower section of the cylinder made of a low-friction plastic. This is the sliding bearing surface that facilitates smooth motion with the external tapered cylinder. The hole in the top of the round cylinder is for a threaded shaft.

Figure 8.32 shows the round cylinder with the threaded shaft, although the threads are modeled as a rod in this figure. At the top of the shaft is an attachment plate that connects the shaft to the small cylindrical helical coupling. This small coupling (which was the weakness of the mechanical design) is connected to the stepper motor, and allows for slight misalignments to be forgiven. The supporting framework has been removed from the figure for clarity. Figure 8.33 focuses on the threaded shaft and the flat round nut that connects the cylinder (not shown) to the shaft.

This mechanical and electrical hardware setup represents many similar applications, but its implementation on the Bot has some unique requirements:

Actuators

FIG. 8.32 Stepper motor connected to telescoping cylinder

- The Bots are moving on a track that has two electrical rails. The Bot has roller contacts, similar to many commuter trains, to power all electronics and motors. There are no batteries on the Bot itself.
- The Bot is to lift a pouch with its suction cup that weighs up to 1 lb.
- The cycle time is 0.25 sec to travel either downward or upward, so the speeds are demanding.
- There is a significant size limitation for the stepper motor amplifier/driver.
- There is no position sensor, so the stepper will work in open loop control.
- Motion profile will be generated by a single board PC on the Bot.

The demanding cycle time implies that a normal threaded shaft would have to spin at extremely high speeds in order to move fast enough. Imaging how fast one would have to turn a nut to get it to move 6 in. along a standard bolt in 0.25 sec. There were no stepper motors that could turn this fast available. So a very aggressive threaded shaft and nut were used. Its pitch was for 1 in. of travel

FIG. 8.33 Stepper motor connected to threaded shaft via the helical coupling

for each revolution. You would not usually even refer to this as a threaded shaft compared to the standard bolt.

The track's electrical rails were earlier determined to be 12 V DC, due to some other constraints. The best stepper motor choices were in the 24 V operational range. So a step-up transformer was placed on each Bot. There were no step-up transformers for much higher voltages available in the year 2000 that would fit on the Bot.

The system was built and debugged for months. The stepper motor performance was erratic. The telescoping motion would appear to freeze up on random downward and upward motion. The stepper motor was slipping from its being pulsed too aggressively. The randomness came from environmental conditions, such as room temperature and the relative friction of the tapered cylinder on the round cylinder's low friction material band. Also, the nut on the very aggressive threaded shaft would sometimes stick rather than move smoothly, even though the nut was made from low-friction plastic.

There were other performance issues that clouded the debugging picture. One was that the Bot's internal connectors were not of high enough quality for the stopping and starting motion. The connectors used on this first prototype were standard electrical connectors use in PCs and similar stationary electronics.

Actuators

Loose connections happened way too often. One was never entirely sure which problem was causing a stepper motor failure when it occurred. Automotive type connectors are being used on the next generation Trackbot.

The time and motion constraints meant that the motion profile (Fig. 8.28) was being taxed. If the ramping up and down from the constant speed was too quick (causing the stepper to slip under load) then a longer ramp time meant that the constant speed portion needed to be at a faster rate in order to get to the end goal in time. It also did matter if the motion was up or down due to gravitational effects, and it mattered if the suction cup got a 1 lb pouch or came up empty.

Different ramping curves were tried, including a cycloidal motion (Downie, 2000). This motion was generated within the single board PC on the Bot, and did help to improve things, but not enough to get 99.999% performance. With the motor and amplifier system used, there was no self-tuning process that could have been used either.

As of 2004, there are several recommendations to try to improve the situation:

- Upgrade the rail power to 24 V, allowing for stronger stepper motors. If 24 V alone is not sufficient, use the existing step-up transformer to get 48 V. Or if 48 V is not enough, combine the 48 V with the 24 V to get 70 V.
- Smaller amplifiers/drivers that are more powerful are now available compared to 2000.
- Better amplifiers/drivers with built-in speed ramping options are now available.
- Investigate present-day electric servomotor systems.
- Use pneumatics if sliding positional accuracy not critical.

8.8. CASE STUDY NUMBER 2: SERVOMOTOR SIZING*

A United States based company designs and makes custom safety covers for swimming pools. While the company makes regular stock safety covers that fit standard-sized pools, their main product is the custom safety cover. Some homeowners build oddly shaped pools. Since these pools are not any standard size or shape, they cannot be covered with a stock product. Instead, these pools must have covers tailored specifically for them.

A typical cover is made of several pieces of material. Since covering a swimming pool with one large piece of cloth is infeasible, the cover must be designed from several narrow (3–6 ft in width) panels of cloth stitched together and reinforced by 1 in. wide strips of cloth webbing. Each panel has a unique

---

*This case study supplied by Mr. Peter Caratzas.

shape that, when stitched together, creates a cover that approximates an oversized outline of the swimming pool. In addition to support of the panel seams, the webbing also provides a means of mounting the cover to the decking around the pool. Buckles and hooks attached to the reinforcement webbing connect to anchors sunk into the decking surrounding the pool.

The process of making a custom safety cover begins with the measurement of the customer's pool. Once all critical dimensions are obtained, the information is faxed to the design department. Using AutoCAD, along with several functions developed especially to automate the design process, the draftsperson generates a plot of the finished cover, which states the amount of each type of material needed.

The completed design is released to the manufacturing floor. Here, according to the plots, workers called cutters draw out the general outline of each panel, the location of each reinforcement strap and other features necessary to manufacture the cover. General outlines of the cover are marked in yellow crayon and attachment information such as the locations of reinforcement straps and buckles are marked in white crayon. The transposition of information from the plot to the cloth is the most error-prone step in the entire manufacturing process. To attain a 100% quality of the output, this stage requires the cutter to maintain a very high level of concentration during the entire day — an almost impossible demand. In addition, this step requires a great deal of knowledge and experience because other factors that will determine the quality of the finished cover are set at this step.

The material used to manufacture the cloth is an open-weave polypropylene cloth. There is a good deal of variance in the quality of this cloth. In fact, the manufacturing process is so prone to flaws that it is rare to find a roll of material that is longer than 150 yards that does not contain at least one flaw. It is common for the loom to skip a thread, leaving a gap in the material equal to one thread length in thickness. This is a problem called a line flaw. This is a noncritical flaw that does not weaken the strength of the cover, but does affect its aesthetics. Knowing where the flaw is, the cutters can plan to locate a reinforcement strap over it, hiding the unsightly line flaw from the customer while reducing the amount of wasted material. A second flaw type is an area flaw, which is a defect in the material than needs to be avoided.

The manufacturer of the cloth inspects each roll of its product for the various types of flaws. Inspectors mark the area flaws with strips of yellow adhesive tape and line flaws with strips of red adhesive tape.

8.8.1. Proposed Improvements

The ideal solution would incorporate a system that would maintain the in-process quality checks that guarantee a quality product coupled with a manufacturing system that ensures predictable throughput. After studying the process, it was

Actuators

determined that the greatest area for the introduction of flaws was through human error. These errors were most likely to occur in the marking and cutting operation and had the greatest impact on the process throughput.

The RPI Center for Automation Technology (CAT) developed a machine that takes the human decision making out of the marking process (Caratzas, 2002). By eliminating the operator from interpretation of the design plan, the number of errors introduced in the marking process is reduced significantly. Instead, the operator runs a machine that interprets the design data and generates the design.

8.8.2. Hardware Requirements

Since the cloth contains flaws that cannot appear in the finished cover, and these flaws are marked by colored tape while the cloth is being manufactured, some sort of sensor had to be developed that could find these pieces of tape as the cloth was being pulled out. This sensor would have to signal the motion controller to stop the material outfeed and give the operator options to eliminate the flaw in the material.

The manufacturing facility is a light-industrial site. It does not have pneumatic or hydraulic lines installed, and only a very few 220 V circuits to run the large "Union-Special" style sewing machines. It was hoped that the CAT could supply a machine that could run at 110 V with minimal pneumatics or hydraulics so that minimal site preparation would be required.

The most important requirement to the company was the one that would justify their investment in the project. The completed machine would have to mark the cloth panels more accurately and quickly than the manual cutting method did. If the designers released designs that were correct, the machine should definitely make fewer mistakes, if any at all. Throughput was the second part of the requirement that would have to be met if the project was going to be a success. The company estimated that their cutters needed 15–25 minutes to lay out all the yellow and white lines on all the panels that make up one cover. The ten-minute spread in time was dependent on the complexity of the cover and number of panels in the completed cover. The company wanted to minimize this time. In the CAT's discussions with them, it was decided that they would be satisfied if the throughput time could be reduced to a range of 8–15 minutes per cover. After measuring the length of each line and estimating the speed at which the machine could mark, the machine's time to mark a cover was determined. This was added to the time required to pull the raw material out onto the table and remove the finished cloth from the table. The time required to mark the sample cover was estimated to be 9 minutes. To achieve this goal, the machine would have to mark at a speed of 24 in. per second and pull out material at a "rapid traverse speed" of 48 in. per second. In addition, when the machine was moving but not marking, as it might when moving from the end point of one line to the beginning of the next, it would operate at the "rapid traverse speed."

8.8.3. Basic Overall Concept

The company wanted this machine to be as similar in function to the manual method of marking as was possible for two reasons. First, space in their facility is limited. If the machine ever broke down for an extended period of time, they wanted to be able to use the space that the machine occupied for manual marking and cutting of covers. Secondly, they wanted the cutters to make transition from manual marking to automated marking with as few problems as possible. It was felt that if the machine mimicked the tasks that the markers performed, the workers could easily identify which task was being performed at each stage and monitor that the machine was operating properly. By reducing the learning curve associated with operating the machine, the company could realize benefits of the device quickly.

The basic design is analogous to a large flatbed x–y plotter. A photograph of this is shown in Fig. 8.34. In this case, the bed is a 64 ft long and 7 ft wide table. The bed can accommodate the 6 ft wide cloth that is used to make the covers. A gantry straddles the 7 ft width and is designed to move back and forth along the length of the table. A linear slide mounted to the cross member

FIG. 8.34 Basic machine overview

of the gantry spans the width of the table. A carriage on the slide holds two marking heads (pens) that spray the yellow and white ink onto the material. This configuration gives the carriage two degrees-of-freedom and allows it to move anywhere within the 64 ft by 7 ft work envelope. Each marking head is mounted to the carriage via a pneumatic cylinder. The cylinders allow each head to extend to the proper standoff distance when they are marking and retract when not in use. The carriage also holds two ink reservoirs, each of which holds eight ounces of ink, yellow and white respectively. There is a cable chain attached between the table and the gantry and one attached between the gantry and the carriage. Any cabling or pneumatic hosing passes through this cable chain to prevent their tangling and snagging as the carriage moves through its full range of motion. Solenoid valves are mounted to the gantry to control the actuation of the cylinders that raise and lower the heads and to activate the spray heads. There is also a flashing yellow safety beacon and warning beeper attached to the carriage that are activated whenever the gantry is in operation to warn anyone in the vicinity that the machine could move at any time.

The Drive System for the X/Y Axes

A variety of methods were investigated to drive both gantry and the carriage, and a variety of factors involved in selecting the drive scheme for each axis. Most important was accuracy. The company stated that the machine must be accurate to the nearest $\frac{1}{8}$ in. over the entire length of the table. Given the length that the gantry had to move, a chain drive or cable drive looked the most promising initially. Cable drive was almost immediately eliminated because there was concern that stretch in the cable would lead to inaccuracies. In addition, the cable would have to be checked periodically for fraying. These maintenance issues quickly eliminated a cable drive system.

The second method investigated was a chain drive. Once again this showed initial promise because of the ease of implementation. It would be relatively simple to purchase 128 ft of chain, join it in a loop and drive it from one end to get the gantry to move back and forth. Concern quickly arose about the accuracy of such a drive system. Given the mass of the gantry (95 kg), the design team questioned if the chain would not stretch as the cable would. In addition, the chain would add mass to the system being moved, meaning that larger motors would be required to drive the gantry. Finally, since the chain is flexible, there was concern that this lack of stiffness would adversely affect the speed of response of the system. Given the chain's tendency to bunch when pushed and stretch when pulled, it was felt that it would be difficult to predict the dynamics of the chain drive during acceleration and deceleration.

For the Y-axis (carriage), a lead screw drive was also investigated. Since the carriage only had to move over a length of 7 ft, the design team thought that some sort of lead screw or ball screw would be an alternative. With this setup, there would be no reason to worry about the dynamics of the system. The lead screw is stiff along the axis of motion, and its behavior over the motion profile is consistent. This system was eliminated for another reason. The rotational velocity was calculated to find how fast the Y-axis motor would have to spin to get the carriage to move at the specified linear velocity. For a typical helical lead screw, the rotational velocity was fast, but attainable by most servomotors. The concern was that given the spinning a 7 ft long lead screw at the angular velocities that were required might cause the lead screw to start whipping. Whipping could lead to premature failure of the lead screw's nut and inaccuracies in positioning.

The final method investigated was the rack and pinion drive scheme. The design called for gear racks to run up each side of the entire length of the table. Meshing pinions would be attached to shafts at each end of the gantry. These shafts would be connected to the X-axis drive motor by a gearbox that would split the input torque of the motor shaft into output torque at two separate shafts. By driving both ends of the 7 ft long gantry, the moments driving the gantry would be balanced and it would track straight along the table. Since this scheme is a direct-drive method, it was felt that positioning would be most accurate with this method. The only uncertainty in position would be the result of backlash in the gear train. The backlash could be controlled in two ways: first, by purchasing gear heads with negligible backlash; secondly, by designing adjustment into the shaft hangers that held the pinion gears in mesh with the rack. Adjusting the distance between the rack's pitch diameter and the pinion's pitch diameter could reduce backlash between the rack and pinion. The rack and pinion system provided the greatest benefits in our configuration and was chosen for both X- and Y-axes.

Unspooling Schemes

Developing the drive system completed one major portion of the conceptual design — accurately controlling the gantry as it traversed the workspace. There were other functions that the final design had to perform. The machine had to be able to autonomously spool raw material off a roll and pull it to an optimal length for the given panel it was making. The material would have to be clamped in position during the marking stage and then released when marking was completed. Finally, the finished cover had to be cut from the roll of raw material to which it was attached.

In deciding how to pull out the material, two schemes were explored. Each was developed around a bar that ran the width of the table and had clamps attached at points along it. The bar would be supported on wheels that would allow it to roll up and down the table. In the first concept, this bar would not

Actuators

be powered. This bar would be driven by the moving gantry. In order to pull material out, this bar would dock to the underside of the gantry. Once the bar's clamps were fed material and they had tightened on it, the gantry would pull the clamp bar and its payload to the desired length. When this bar had reached it destination, it would release itself from the gantry and, via brakes designed to attach the bar to the racks running up each side of the table, lock itself to the table. When marking was complete and the finished panel was removed from the table, the sequence would be performed in reverse, and the moving clamp bar would be brought to a home position to wait the next feeding of raw material.

There were disadvantages to this system. First the gantry had to make at least one extra trip up the table to either drop off or pick up the moving clamp bar. This trip would add to the cycle time estimates to complete the cover. In addition, since the moving clamp bar itself would add mass to the system, it was not clear how the tuned motion controller would behave when the added payload was attached to the gantry.

In the competing design, the moving clamp bar gantry would be self-propelled, driven by a third axis on the motion controller. There were several advantages to this system. A lightweight moving clamp bar could traverse the length of the table more quickly than could the gantry. Also, the gantry would not have to make the extra trip down the table that was required by the unpowered clamp bar. Pullout speeds would be faster and machine cycle times would be lower. The disadvantage to this setup was that a third servo axis on the controller would be required to implement this design. The added axis, while being used infrequently, would add considerable cost to the control system. In addition, the added motor drive would add more backlash adjustments and overall maintenance requirements to the system. After looking further into the amount of time that would be added by the gantry's extra trip down the table, it was decided that the benefits of the powered clamp bar did not outweigh the costs, and the unpowered clamp bar was chosen.

8.8.4. Actuator Sizing

Table 8.2 is a summary of the system as installed at the company. It gives the machine dimensions and specifications. Major hardware components of the drive system for both axes are also given.

The design of a motion control system (Fig. 8.35) is an iterative process. The system will include a motor coupled to a load. The coupling may take one of two forms: direct-drive or speed reduction. In direct-drive, the system is coupled directly to the motor. This may reduce the efficiency of the system, but increases it's accuracy, since the load will be directly coupled to the encoder, as well as the motor.

TABLE 8.2 General Machine Specifications

Overall dimensions (l × w)	64 ft × 7 ft
Maximum width of cloth	63 in.
Usable length of table	61 ft
Weight of gantry	205 lbs
Marking speed	50 in./sec
Pullout speed of material	50 in./sec
Number of inputs	6
Number of outputs	12
Controller	Galil DMC-720
X-axis information:	
Motor	Galil Nema 34
Gear reduction	Bayside 5:1 Planetary
Right angle gear drives	Zero-Max 5/8″ Shaft
Pinion	Browning 3/4″ face, 16 pitch, 24 tooth
Rack	Browning 3/4″ face, 16 pitch
Couplings	Lovejoy Spider 5/8″ i.d.
Amplifier	AMC 25A12
Power supply	AMC 80 V
Y-axis information:	
Motor	Galil Nema 23
Gear reduction	Bayside 5:1 Planetary
Right angle gear drives	N/A
Pinion	Browning 1/2″ face, 24 pitch, 24 tooth
Rack	Browning 1/2″ face, 24 pitch
Couplings	N/A
Amplifier	Galil Amp 71
Power supply	Integral to amplifier

On the other hand, to increase system efficiency, the load can be connected to the motor through a speed reducer. Speed reduction can be accomplished by gear reduction or pulleys. Speed reduction will reduce the size of the motor needed to drive the system, but may introduce inaccuracies into the system through backlash in the gearing components. The designer must determine the critical criterion.

The first step performed in sizing the motion control system of the machine was the determination of the system load. This was determined by modeling the mass of the gantry. Masses for each component were estimated and the total gantry system was determined. The equipment was proposed to run at a certain speed, and had to accelerate to speed from a stop. The maximum

FIG. 8.35 Control cabinet including all motion control components

acceleration was determined so that the machine could reach the proposed average velocity. With the system mass and acceleration known, the force required for the system was calculated. Once the force was known, the torque could be determined. Since the load was driven by a rack and pinion drive, the torque was calculated using:

$$\tau = r \times F \tag{8.1}$$

where τ is required system torque, r is the radius of the pinion gear, and F is the force required by the system.

At this point the process became iterative. The investigation was performed to determine the need for gear reduction. As previously stated, gear reduction allows the designer to select the most efficient motor for the load being driven. In other words, the designer can select the smallest and, most likely, cheapest motor for the application. Since torque is increased as a multiple of the gear reduction at the output shaft of the gear reducer, speed must be reduced by the same factor. Care must be exercised to ensure that the motor will be able to meet the speed requirements of the system if gear reduction is used.

To begin the analysis of whether or not gear reduction is needed, a motor was selected that met the necessary torque requirements. The rotor on the motor has an inherent moment of inertia, J_M. An optimal system will have J_M equal to the moment of inertia of the system, J_{LO}. If not, gear reduction must be used. For the system, it was determined that gear reduction was necessary.

If gear reduction is used, the inertia matching must be performed by comparing J_M to the reflected load of the system at the input shaft of the gearbox, J_L. In this case, the comparison takes the following form:

$$J_L = \frac{J_{LO}}{N^2} = J_M \tag{8.2}$$

where N is the gear ratio of the gearbox. Rearranging terms, it was found that the optimal gear ratio was:

$$N = \left(\frac{J_{LO}}{J_M}\right)^{1/2} \tag{8.3}$$

As a rule of thumb, J_{LO} should be less than ten times J_M.

8.8.5. Sizing the Amplifier

The amplifier drives the motor by providing power to coils in the rotor. Since $P = VI$, the two factors that affect motor performance will be current and voltage. The peak torque that can be generated by the motor is a result of the peak current provided by the amplifier. Motor manufacturers will publish a torque constant (K_T), which relates torque to amps provided. To reach maximum designed acceleration, the motor must generate its maximum torque (T_P) and the amplifier must supply its maximum current. To determine the peak current required (I_{max}), the following equation was used:

$$I_{max} = \frac{T_P}{K_T} \tag{8.4}$$

Similarly, for the system to reach the maximum designed velocity ω_{max}, the amplifier must generate its maximum voltage (V_{max}). To determine the maximum voltage, the following equation is used:

$$V_{max} = K_T \omega_{max} + rI_{max} \tag{8.5}$$

where r is the resistance of the motor. The first term is the back EMF of the motor, and the second is the voltage required by the circuit created by the motor's coil. The importance of these numbers cannot be understated.

8.8.6. System Troubleshooting

As the gantry was installed on the table for the first time, it was subjected to manual testing. The gantry was supported by rubber wheels, and interconnected to the table by the gearing system. Each side of the gantry had a spur gear that interfaced with a gear rack that ran the length of the table. To ensure that the marking would be as accurate as possible, it was necessary to minimize the backlash in the drive gearing. Backlash for the gearing was set by minimizing the gap between the spur gears on the gantry and gear rack on the table. This adjustment was performed as the gantry remained stationary at one end of the table.

When the backlash adjustment was complete, the gantry was rolled through its full range of motion by hand. As this test was taking place, it was found that the gantry was easier to move in certain areas, and more difficult to move in other areas. It was quickly determined that the gear racks on each side of the table were not parallel. The distance between the two gear racks did not remain constant along the length of the table. As a result, as the spur gears on the gantry rolled along these different areas, the backlash in the gearing would change. In some areas, where the distance between the gear racks was greater, backlash was increased, and the gantry moved more freely. In other areas, this distance narrowed and the minimized backlash in the gearing system would cause binding that would nearly prohibit movement in the gantry system.

The difference in backlash in different areas of the table would cause two problems. First, repeatability throughout the system would not be equal, since there would be more play in the system in different areas of the table. Secondly, since the torque required to move the system to different positions was dependent upon the amount of backlash at the current position of the gantry, concern arose about the controller's ability to compensate for the variation in the required torque. Both problems could potentially lead to inaccuracies in the location and repeatability of the gantry, and ultimately, the marked lines that the machine was creating. This would have been unacceptable.

Because each of these racks was 65 ft long, it was expected that keeping them straight would be difficult, and that some adjustment would be necessary to ensure proper operation. Because this machine was created at one location and installed in another, developing a repeatable alignment procedure was equally important. The alignment procedure designed was straightforward and was easy to implement in the field.

System Tuning

With the last screw tightened and the final wire installed, the first-time machine builder may think that the equipment is ready for delivery. This is not the case. Several steps remain to be completed in the process after the final dust cover is installed on the machine. Once the machine is assembled, the system is ready for power-up. Before the first pool covers could be made, however, the drive system for each axis have to be tuned. When motion control systems are involved, as was the case on this machine, the controller must be optimized to match the dynamics of mechanical system it is controlling. With multi-axis control systems, where a controller drives multiple motors, each of which drives a separate mechanical system, the controller must be tuned for each axis.

Traditional motion control systems use a Proportional Integral Derivative (PID) controller to drive the motor to its commanded position. The PID controller's name is derived from the three control methodologies used to generate the command signal: proportional, integral, and derivative. The proportional controller generates a command signal whose strength is based upon the difference between the actual position and the commanded position. As the mechanical system gets closer to its commanded position, the proportional controller will reduce the strength of the command signal. When the system is at its desired location, the command signal should equal zero.

The proportional controller is inherently flawed. As the mechanical system reaches its commanded position and the command signal decreases, mechanical losses remain fairly constant. As a result, this controller will always leave the mechanical system with some error between desired and actual location. This is called a deadband. If this error is great enough to cause the controller to generate a command signal that will overcome the system's mechanical losses, the mechanical system will approach its targeted position. Otherwise, it will remain where it is.

The addition of an integral controller eliminates the deadband. This controller sums the difference between commanded and actual position over time. As time progresses and these individual errors are summed, the integral of the error increases. Consequently, the strength of the corrective action of the integral controller increases over time. When this error becomes great enough to generate a signal to overcome the mechanical losses, the system will move toward the commanded position.

Like the proportional controller, the integral controller has inadequacies. An integral controller is more likely to cause the system to oscillate. The design of the controller causes this. As the error signal increases, the strength of the corrective action of the integral controller is increasing. As the system moves toward its commanded position, the magnitude of the error is decreasing. However, it is still error, and when added to the previous error, the control signal continues to increase. In fact, the integral controller will not send a negative control signal until the system overshoots its commanded position, thus generating a negative difference between commanded and actual position.

To remedy the oscillatory behavior of integral controller, a derivative controller is added. The derivative controller analyzes the rate of change in the difference between the commanded and actual position. This controller has a tendency to slow the system down if it is near the commanded position and approaching it quickly.

It is not necessary to always use the PID controller. Depending upon the system design and required accuracy to perform command-following, it is possible to

Actuators

implement a proportional-only controller or a proportional-integral controller. Selecting appropriate gains for each of the individual controllers can optimize performance of the system. Each gain acts as a scalar dictating the magnitude of each controller's action. For example, by setting a high gain on the proportional controller and low gains on the integral and derivative controllers, the proportional controller will have a greater effect on the commanded signal.

Gains are used as a means to adjust, or tune, the controller to complement the mechanical system. Gains are not chosen arbitrarily. The mechanical system will have certain frequency response based on the arrangement of different mechanical elements such as springs, dampers, masses, and forces. There are two strategies for determining controller gains. One strategy involves using modeling methods that can be used to determine the appropriate gains for the system. These methods determine gains by modeling the mechanical system and the controller.

Rather than model a system, a second strategy for tuning is to determine the frequency of the actual system and determine the gains based on real information. The advantage to this method is that it eliminates modeling error. The importance of this cannot be overstated. Modeling error can be significant, especially when trying to estimate such factors as frictional losses. Controller manufacturers generally provide tuning software with their motion controllers to aid the user in determining the appropriate controller gains.

Tuning was accomplished using software supplied by the manufacturer of the controller. The software determines the gains by applying step functions to the system. By measuring the response of the system to the step functions, a determination of optimal proportional, integral, and derivative controller gains can be performed.

When system tuning was completed, performance of the system could be verified. It was noticed that the covers were taking much longer to be completed than expected. Panels that were supposed to take two minutes to finish were taking eight minutes.

The motions that the gantry followed were analyzed to explain why the system was taking longer than expected to make the covers. It was determined that the acceleration for the system was set too low. Since the covers are comprised of many lines that are short in length, the gantry was never reaching its top speed. In this situation, the gantry's motion profile was triangular instead of trapezoidal — the desired profile.

To increase the throughput of the machine, the team attempted to increase the speed of the machine, both in marking and pullout. The first solution attempted was an increase in both the speed and acceleration of operation of the machine. This led to unstable behavior. The machine would overshoot wildly in the X-axis (the more massive of the two axes), and the position at which the lines were drawn would be inaccurate.

In order to make sense of the behavior, it was decided to make a second-order model of the system. The shafts that drive pinion gears were modeled as springs and the spring constant of the system was determined experimentally. Friction in the wheels and at the interface between pinion and rack was arbitrarily approximated at a value of 10%. The gantry was weighed on a scale, and from that, the system mass was determined. Voltage and current outputs for the amplifier were known.

The model was implemented on Matlab and the results showed that the system was stable for the gains that were being used. This meant, inexplicably, that the system should be stable, yet it was not. The only information the team had not measured and therefore were unsure of had been the current output of the amplifier. When the team measured the current output to the motor leads, the team found that the amplifier was current-limited. The motor was not supplying the torque required or capable of, because it was not receiving enough current from the amplifier.

When the specifications of the amplifier were investigated in depth, it was found that although this motor and amplifier were sold as a matched set, the amplifier was incapable of providing enough current to get the peak performance out of the motor. In addition, the motor required 80 V to run at top speed, but the amplifier was only capable of supplying 50 V of maximum potential. This meant that the motor could not reach its top speed.

Since the Matlab model was already created, it was decided to add current limiting to the input of the system. When this was added to the Matlab program, the model predicted the observed behavior. By having this simulation available, the team was able to input the specifications of larger amplifiers/power supply combinations and determine which components would meet our needs.

Once the new amplifier was purchased and installed, the system behaved as predicted, with accuracy and speed. There was no overshoot and the system acted remarkably "stiffer." With these modifications, the system was installed in the factory and it has been in daily use since April 1995.

Curved Lines

Soon after the new amplifier was installed, and the machine was operating at the designed speed, a new problem revealed itself. Straight lines that required coordinated moves by the X- and Y-axes were being drawn as curves. During the initial debug, straight lines were drawn as expected. As the run hours increased on the machine, the "curved line" problem worsened.

There were two possible contributors to this problem. The problem could have been caused by the slippage in the mechanical drive system or corrupted feedback information. Corrupted feedback information could have been caused

Actuators

by a faulty encoder, noise in the feedback signal, or an incorrect translation factor between encoder counts and distanced traveled. For the lines to appear as curved, the problem had to be nonexistent in one axis vs. the other. If the severity of the problem was equal in both axes, the lines would appear as straight, since both axes would be moving with equivalent amounts of error.

Two steps were taken immediately to try and determine the problem. First, all conversion factors between encoder counts and distance traveled were re-checked to ensure that the controller was using accurate information for its trajectory generation. This scenario was the least likely to occur, but it was the easiest to check.

Mechanical slippage in the drive system of one of the two axes was the most likely cause of the problem and easy to check. All couplings and gears were tightened to their shafts. Lines were painted on the shafts and couplings to indicate where these parts were aligned. The machine was run again and, over time, similar behavior showed that the lines had moved in relation to each other.

To permanently remedy this situation, it was decided to cut keyways into the shafts, so they would be rigidly fixed to their couplings. The machine was run for several weeks and this change seemed to fix the problem. Slowly, however, the problem began to worsen. The painted lines on the shafts still indicated that shafts were not moving with respect to their couplings and inspection showed that the keys in the shafts were intact.

The two right angle gearboxes were never inspected for slippage when the curved line problem was found. They were selected to match the specifications of the motor. When the new slippage problems were encountered, the two gearboxes were dissected to look for potential failures. This inspection revealed interesting results. The gearboxes are designed with two shafts intersecting at $90°$ to each other. Beveled gears are rigidly mounted to each shaft. The interaction between the two gears transmits motion from one shaft to the other.

To prevent the gears from slipping, a hole is drilled through the gear and its mating shaft. A pin is pressed through this hole. When these gearboxes were dismantled, it was found that the pins had broken, and slippage was occurring between the shafts and the beveled gears. The question arose as to why the pins had broken, when the components were sized to match the motor. The pins were modeled as beams, with forces applied at the interface between the shaft and gear. Maximum torque was resolved into forces at these points and it was determined that, indeed, the pin should not break. This calculation was followed by cyclic loading calculations of the fatigue life of the pins, based upon fully reversed loading. Failures were revealed with this calculation. The vendor was contacted, and verified that these gearboxes were designed for continuous motion applications, not the reverse loading of a servomotor system. The application exceeded the design limits of the system.

With the new data, two new right angle gearboxes were specified. The new gearboxes were designed specifically to meet the demands of a motion control system. The resulting replacement has been in service without fail since.

8.9. CONCLUSIONS

The correct sizing of actuators is needed if your new automation machine is to function properly. Simply using an oversized motor off your shelf can work sometimes, but other times come back to haunt you. To achieve correct sizing, one must calculate the loads and rely on the manufacturer-supplied diagrams and data sheets, but once in a while this information may be incorrect, so do not assume anything if performance is not what you expect.

PROBLEMS

1. Take a battery-operated screwdriver and create a stall torque condition twisting a screw into a very hard piece of wood. Devise a method to measure the stall torque with the aid of a spring scale or other device.
2. Using a battery-operated screwdriver, use your hands and the forward and reverse switches to attempt to create a very simple servo situation. Pick a direction corresponding to a map direction (South, Northwest) and try to servo the screwdriver blade until it comes to the correct direction. Mark the blade with a piece of tape as a pointer to help with this.
3. Borrow a battery-operated toy car and add extra loads to it, so as to create near stall torque condition (it can still move, but barely). Compare battery life to operation with the normal load.
4. Create a humanly operated vacuum cup system with a plastic straw and some other household items you can find (or go to the hardware store to be inspired). Determine how much you can lift by sucking on the straw if you have the open area of just the straw itself, and compare to some other larger opening by placing a larger opening on the end of the straw. (Be sure to use a stiff plastic straw and not paper.) Or use a larger straw in the form of a garden hose. From the weights lifted and the area of suction, calculate the negative pressure that you are able to achieve.

PROJECT ASSIGNMENT

1. Investigate the different actuator options for your project. Compare them and develop a trade-off matrix. Then select your first choice

and develop the various subsystems. Keep your second choice handy and develop that choice to a lesser degree.
2. Search on the Web for leading vendors for your actuator choices. Look for their specific sizing information, and find one vendor who has a Web-based sizing program to run.
3. Estimate the total power needs of your actuators. For pneumatics, find the flow rate needed. For hydraulics, find the working pressure and flow rate. For electric motors, find the voltages and amperages.
4. List the number and type of inputs and outputs to power and control your actuators. What trade-offs can be made for fewer actuators and greater control signals?

REFERENCES

Caratzas, P. (2002). Design and Manufacture of Automation to Mark CAD Data on Cloth. MS thesis, Rensselaer Polytechnic Institute.

Downie, J. (2000). Cycloids Take the Edge Off Sharp Motion Profiles. Machine Design, June, 110–111

Lentz, K. W. Jr. (1995). *Design of Automatic Machinery*. New York: Van Nostrand Reinhold Company.

Rogers, J., Craig, K. On-Hardware Optimization of Stepper-Motor System Dynamics. Mechatronics Journal, Elsevier Press (revision submitted 2004).

9
Sensors

As one designs an automatic machine for the first time, it is very easy to underestimate both the role of sensors and their sometimes limiting capabilities. Most humans with reasonable eyesight and tactile abilities take many tasks for granted, but it will surprise the novice how quickly one needs to add more sensors than could have been guessed at the start.

Imagine that your job is to grab boxes from a conveyor and to place them on a pallet. Now, imagine that you are having a really bad day. You have:

- Been up all night with a sick child or pet and have a headache.
- Your glasses are dirty.
- You have a splint on a broken finger.
- The rails that guide the boxes to you have moved out of adjustment.
- One of the two light bulbs in your workplace has just burned out.
- The water pipes in the ceiling above you keep dripping on your head
- The heat is off and it is cold in the building.

You could still do your job, even if there is no smile on your face. You would hope and pray for a better day tomorrow, but you can survive. The pallet will get stacked.

Automation sensors, on the other hand, will potentially fail from each and every setback listed, assuming that one looks at an analogous situation:

- Sensors would not get headaches, but they can get disconnected from the controller by loose connectors or worn wires and not be able to help the controller "think."

Sensors

- Dirty glasses refer to an optical sensor becoming dirty. Most sensors cannot deal with any significant amount of dirt or grime on their "eye."
- The broken finger can refer to a bent robotic gripper that is supposed to grab the box and therefore needs a new sensing and motion strategy and the existing sensor cannot help solve this problem. Or it can mean that the bracket holding the sensor gets bent and the sensing direction has been changed.
- Automation most times could not adapt to the rails moving out of adjustment.
- The burnt-out light bulb would not be acceptable if a machine vision system was being used. Consistent lighting is critical for vision success.
- The water dripping could be a problem. Many sensors are not rated for a high moisture environment.
- Temperature variations cause sensor trip levels to change.

So a worker having a really bad day can adapt (if they want to) and continue to work when automation would have ground to a halt. This means that it is not just the job of the automation engineer to determine what kind of sensors need to be located on the machine, but it is critical to anticipate how the sensor environment can become error prone.

9.1. SENSOR TYPES

Sensors are designed and built to address a very narrow market niche. Whereas human eyes and hands can seemingly do it all, most sensors are quite simplistic. There are machine or computer vision systems available that are more powerful than single function sensors, but they are not inexpensive to buy nor inexpensive to implement and maintain. So, many automation machines will use a host of simple sensors, and perhaps a single vision implementation at best. There are always a few exceptions of machines that use multiple cameras, but they are in the vast minority.

Sensors can determine some property, either as:

- a discrete binary (on/off) situation;
- a continuous value.

A discrete sensor would determine if a box had been propelled into a limit switch mounted on a movable stop. Either the box is there or it is not. It cannot tell if the box is getting close, or what speed or force the box has when it hits.

A continuous sensor can tell you some value that is likely to change quite quickly. A car's speedometer hopefully gives one accurate feedback on one's velocity while driving. If it is not adjusted correctly at the factory, then one will likely get caught in a speed trap.

A discrete sensor's output is most likely digital, actuating a switch that completes a circuit or disconnects it. An analog sensor can be interpreted for a digital application, such as the desire to sense a value of one's car stereo volume knob so as to not burn out the speakers (the author does not know if this has ever been implemented in real life, but he does desire it), but these applications are few.

Continuous sensor's output can be either digital or analog. It depends on the technology and circuitry being used. Many times it is an analog value that is converted to a digital signal. An example would be the capacitance sensor used for left turn lanes at traffic lights. Many times a sensor wire that connects with the sensor controller is embedded into the pavement. A car with its significant metal mass changes the capacitance field, and the sensor circuit takes the analog signal and determines that YES a car is there, or NO it is not. Then it tells the traffic light processor to take the option of giving a green left turn arrow or not. This sensor has some threshold value for a car, and will often not reach this threshold for a smaller motor cycle. And one will never trigger the sensor with a bicycle.

In the traffic light example, the sensor has its own dedicated circuitry or processor, so the traffic light control could be a simple set of relays, if it is an older model. One of the issues to investigate for one's automation machine is where the analog signal gets processed. And even if the main controller is capable of performing such processing, will it be a burden and drag down the overall machine performance. Sensors with built in processors have become quite popular for this reason, as well as having a system that best matches the sensor to the processor. Proper implementation is easier.

If a sensor has analog output, it will most likely need to be converted to digital for the controller to process the information. Almost all controllers work on digital signals internally for the decision process. Some controllers have built-in analog inputs where the signal is converted internally, but smaller and less expensive controllers often do not have this capability and need external conversion.

A final sensor type distinction is that of how the sensor determines its value. Does it make direct physical contact or does it do it from a distance. These are referred to as:

- contact;
- noncontact.

Contact sensors have two major concerns. The first is that contact means that there is usually something physically moving, and that moving closes a switch. Imagine how long a standard room light wall switch would last if some obnoxious kid stood there 12 hours a day and flipped the switch on and off several times each second. It most likely would not last the normal life of 20 to 50 years as in most homes. Something inside the wall switch, usually a metal

Sensors

spring or contact would break. Contact switches can fail in similar fashion, and since automation cycles many times over 24/7, breaking will happen and is just a matter of time (the author was stuck in an elevator in France due to a faulty limit switch). Looking at product life information is important in selecting brands and quality levels.

The second concern is that by contact sensing, the pressure or drag on the item being sensed may change the process. If one is winding a thin film of plastic, and a contact sensor can occasionally poke a hole in the plastic, the end customer will not be happy.

One might hear these arguments and declare that only noncontact sensors can and will be used, but in general, contact sensors can be significantly cheaper than the noncontact equivalent. Some noncontact sensors also require significant knowledge of how to implement and adjust them, while a contact sensor's mode of operation is usually obvious.

9.2. LIMIT SWITCHES

Somewhere either on or inside a limit switch is something that looks like a little plunger rod. This plunger is normally kept in its upper or outer position by a spring. At the bottom of the plunger rod's travels it makes something have an electrical connection. The spring makes this contact behave as what is referred to as "momentary contact." How the plunger rod is moved, how much it moves, and how much force it takes to move it are three of the many variables that range over the thousands of types and brands of limit switches.

The momentary contact part of the limit switch the plunger creates is usually referred to as "Normally Open" or "NO," but many limit switches have a second electrical connection where the plunger motion breaks a completed circuit, referred to as "Normally Closed" or "NC." One needs to know the function of the limit switch before selection.

Many compact limit switches simply have the plunger rod stick out of the switch body. This is great for limited lower speed cycles, like in your computer printer to sense if the cover is open. But when there is a greater force or relative sliding motion, rollers are placed on lever type springs to transfer the motion to the plunger rod of the switch. Figures 9.1 to 9.3 show three implementations. Figure 9.4 has a roller that is being viewed in cross-section for a fourth implementation. The plunger rods are protected within the switch body and are not shown.

Sometimes in lower loading applications the plunger motion needs to have a larger lever to catch the desired motion. Figure 9.5 shows a wobble stick that has a coil spring integrated into the levers to absorb any shocks or minor misuses of this switch.

FIG. 9.1 Limit switch: side rotary

FIG. 9.2 Limit switch: push roller

9.3. OPTICAL SWITCHES

Optical switches are gaining in popularity, both from easier implementation in recent years and from lower costs. Many contact limit switch applications can have a noncontact optical switch alternative. Figure 9.6 shows one of many

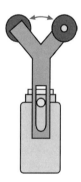

FIG. 9.3 Limit switch: fork lever

Sensors

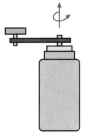

FIG. 9.4 Limit switch: top rotary

FIG. 9.5 Limit switch: wobble stick

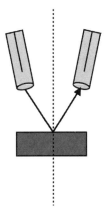

FIG. 9.6 Focal point of optical sensor

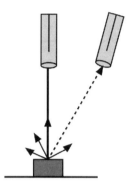

FIG. 9.7   Diffuse reflection

standard uses of optical devices. The device on the left is a light source that may have a primitive focusing lens, or it might be the end of a fiber optic cable. It transmits light in a narrow beam, and the light bounces off an object. This point on the object is the focal point. A detector device on the right senses the amount of light entering. If it is a certain value over the threshold value, it signals YES and the controller can act accordingly. If the object is not present, then there is no positive signal. The surface type of the object will have significant impact on the amount of light detected (Fig. 9.7). Both the transmitter and receiver can be in one physical sensor case.

The focal point is not really a point but a region. Figures 9.8 and 9.9 show the standard effective zone where a positive signal most likely will occur. Each optical sensor vendor should supply this type of information, since it not something about which one wants to guess.

FIG. 9.8   Focal point system

Sensors

FIG. 9.9 Focal point system

Many optical sensors do not use reflectance, but are line of sight. They are sometimes called through beam sensors. Figure 9.10 shows how this can be used to count packages on a conveyor. This method will work wonderfully if two adjoining packages do not touch each other. Through beam can also be used in many other applications (Figs. 9.11 to 9.16).

These line of sight sensors are used in most store checkout conveyors. There are issues, however, with the alignment of the light source and the detectors. If one had a row of sources and detectors as in Fig. 9.16, and the sources and detectors were hidden behind a shiny piece of steel that had small circular holes cut out for the light beams to traverse, the random reflections off the shiny steel might produce a false signal to one or more of the detectors. Or if the package being detected had shiny tape on it and the package was not totally square to the sensor system, a false signal can be sent by a neighboring sensor. This is a function of the sensor type and how close the detectors are located to each other. One solution was to alternate on each side the transmitters and recei-

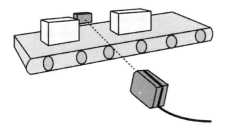

FIG. 9.10 Counting packages on a conveyor

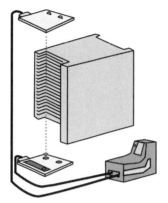

FIG. 9.11 Detecting presence of silicon wafer

FIG. 9.12 Bottle cap detection system

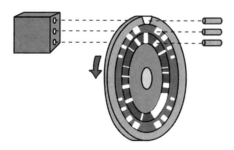

FIG. 9.13 Three-track encoder disk

Sensors

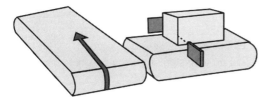

FIG. 9.14 Box detection for left conveyor motion activity

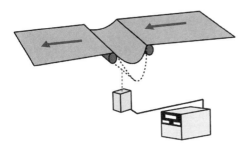

FIG. 9.15 Web handling excess material detection

vers to effectively double the spacing for false signal detection. The author has had to debug this system for quite a while with just such a problem!

9.4. OTHER SENSOR TYPES

There is a very large range of sensors commercially available beyond limit and optical switches. They can cost from a few dollars if they are made by the thousands, to thousands of dollars for one. It is recommended to the reader to search

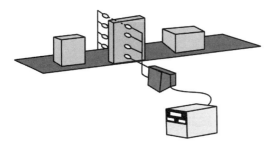

FIG. 9.16 Box height determination

the Web specifically for one's application to see what is available. The list given here is in no way meant as a limit.

- Pressure sensor: pneumatic line checking device. It is good to know that the compressed air is on and up to system requirements before assuming that air cylinders will function satisfactorily.
- Pressure sensor: hydraulic overload. Excessive pressure could blow out the lines and fittings and be potentially dangerous.
- Vacuum sensor: suction cup use. Detect whether a box has been successfully picked up by a robot suction cup gripper.
- Temperature: hot glue melt. Many packaging machines will not allow for operation if the hot melt glue is too cold, which would have produced poor box sealing.
- Weight: scale. A check to see if a package has received the proper amount of product.
- Force: strain gage. Used to detect excessive forces on key members during motion. Good for testing machines.
- Metal detecting: safety check. Capacitance sensors used to detect metal filings.
- Metal position: distance verification. Capacitance sensors used to obtain distance with noncontact. Also known as proximity sensor.
- Human presence: motion detection. Microwave transmission and reception can determine motion and/or distances.
- Human presence: thermal detection. Infrared detection of humans vs. machine signatures.
- Distances: laser range finding. Send a laser beam in the direction and interpret return signal to obtain very accurate measurement.
- Distances: ultrasonic detectors. Send out a sound wave and determine distance with reasonable accuracy at less cost than a laser.
- Package tracking: radio frequency. Use radio frequency (RF) tags on boxes or products. Can be queried even if on boxes stacked several deep.
- Color detection: optical sensor. Use of smarter detectors to interpret reflected light to obtain color information.

There are several issues that can bridge over many of these sensor types. They include:

- Drifting;
- Environment of use;
- Multiple implementations.

Drifting is when the circuitry of the analog device changes over time. This can happen from something as simple as a resister changing its resistance value as the device gets warm. The capacitance sensor circuitry used for the left turn

Sensors

lane example mentioned was proven to have significant drifting over time. The author was part of a research team at RPI that was using this system to sense humans around a moving robot as a means to create a safer workplace (Graham et al., 1986). The system's performance was frustrating when trying to decipher the difference between a single human and none as the system electronics warmed up.

The environment of a sensor can be nasty. Fine paper particles that accumulate in your computer printer can eventually destroy the required internal sensors. Image what cutting fluid, metal chips, or hot melt glue can do. Vacuum detection sensors do not like fine particles often accumulated inside when the environment is not clean. Some vacuum detectors become ineffective if they are turned on before a partial vacuum is achieved.

Multiple implementations of similar sensors are a problem that will probably not be found until final automation debugging. A single ultrasonic sensor can be simple to use. Development kits are available from suppliers at low cost, but if one uses two ultrasonic sensors at once, the signal from one can be detected by the other, and confusion will break out. System error codes may not even consider the possibility. One can scratch one's head for days trying to resolve this problem!

9.5. VISION SYSTEMS

Your standard VHS camcorder has some of the basics elements for a vision system. It has a camera that can take a series of images and store them electronically on magnetic tape. Newer models store them in digital format on tape or on computer hard drives, and one can view them later on a computer monitor or television screen. What other elements are missing for a complete vision system?

There is a huge amount of computer hardware and software beyond the camcorder. Volumes of books, courses, and entire college degrees have been written on them. Several decades ago one needed a degree in vision to properly use it, but no longer. Vision systems have become quite easy to use, with many issues transparent to the user.

There are three basic types of vision systems available:

- standard;
- line scan;
- stereo.

Standard takes the video image from a camcorder, or from a higher precision camera, and processes the information. Single items can be located and identified. Line scan uses a single row of elements to gain a visual slice of an image. There is more processing needed here but less cost for the camera. Stereo uses two cameras, similar to a human using two eyes, and can judge distances. Stereo vision is the least likely to be used in automation.

9.5.1. Standard Vision Systems

The most basic configuration is a camera that takes pictures in a 640 × 480 format. The numbers given are the picture elements (pixels) of the image, not necessarily the screen resolution of the PC monitor it is shown on. There are many other less common formats, some ranging to 1000 × 1000 pixels, or 1 Meg pixels total. Newer digital cameras for consumer digital still photographs are in the 3, 4, and 5 Meg range.

The next issue is to take the image from the camera CCD array and store it in the computer. This uses a frame grabber, something most PCs do not have internally when purchased. Then the data on the frame grabber needs to be processed. This processing can be done entirely in a Windows or Linux operating system, or it can be assisted with another PC board called an image processor. The dedicated processor board was the only logical option to select until recent PC operating speeds climbed so high. Now one can function well with either option. A newer option is that some machine vision cameras have internal processors that can be programmed as a stand-alone system.

Figure 9.17 shows an image of several pouches on a conveyor belt. If a human was watching this image on a monitor, they could easily determine the number of pouches, at least on the top layer. There may be some other pouches underneath, but the task would not stump them.

A computer vision algorithm could have a miserable time trying to sort out which pouch is on top of the pile. Since the pouches are mostly all white, and the edges are not rigid, the software will not be confident of which pouches are on top.

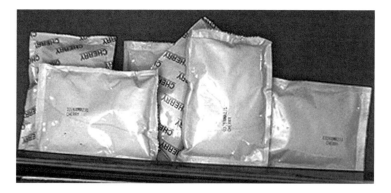

FIG. 9.17 Image of pouches on a conveyor belt

Sensors

Methods to improve the chances of algorithm success include the addition of different types of lighting. In vision, the three most important issues are:

- lighting;
- lighting;
- lighting.

This is not a typo. This is similar to the old business adage, "location, location, location." If the lighting is correct, then most algorithms can just process away. If the lighting is poor, one's chances are sunk.

Vision system algorithms can find any size object, or "blob," and then do any or all of these tasks:

- sort them by size;
- calculate area;
- find centroid;
- determine the number of holes.

These attributes can be used in the automation application to perform the desired steps.

9.5.2. Line Scan Vision Systems

When one mentions a vision system, one usually thinks of an image similar to what one views on a television or computer monitor. This is usually true, but not the only method of implementation. There are simpler devices called line scan cameras that can view 1 pixel by 1000 pixels. This information is normally processed as a series of scan lines, similar to most home and office document scanners.

This system is useful if the object to be imaged is moving on a conveyor belt. As long as the motion is constant, the line images taken over time can be integrated into a single two-Dimensional image. Another use is to scan long and thin items and place the scanning camera on a moving platform. This is like a flat bed scanner, but on a much larger scale.

Line scan cameras are not as easy to implement and as user friendly to a mechanical engineer, but they do have their uses.

9.5.3. Vision Approaches and Algorithms

Sometimes when selecting the items to be imaged by a vision system, there are software algorithms just sitting in a library ready to use. If one works on good lighting, then all is easy. Other times, there are no algorithms available. One

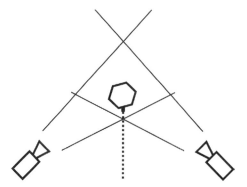

FIG. 9.18 Stereo vision viewing of projected laser line

needs to develop custom approaches to solve the problem. Two simple examples will be presented here.

A new method to achieve stereo vision was developed recently (Davis and Chen, 2001). The old method was to use two normal cameras in stereo vision mode. Here in Fig. 9.18, a laser line is projected onto an object that gets rotated while being inspected. The line is then processed from the imaged shapes (rarely a straight line if the object is not flat or a box) and the three-dimensional constructions can occur within the computer.

The newer implementation was achieved with a single camera and four mirrors (Fig. 9.19). The standard 640 × 480 camera viewing area was effectively

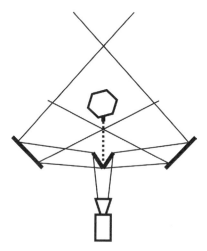

FIG. 9.19 Stereo vision using a single camera and multiple mirrors

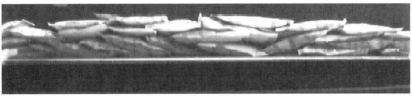

FIG. 9.20 Side image of pouches on conveyor: (a) raw image; (b) topology lines above pouches

split in half, the left half viewing the left side view, the right half the right side view. A laser line is still projected on the rotated object and the processing takes place. The processing time was similar to the stereo vision method, but the overall cost was reduced by the amount of one camera.

A second example is that of the conveyor full of mostly white pouches. The automation application was to lift the topmost pouches off the stack of pouches on the conveyor. Mr. David Brown located the camera off to the side of the conveyor, and tilted the conveyor slightly so they would slide against a transparent plastic side wall. Figure 9.20 shows a sample image. This image just shows part of a 4 ft long section of pouches.

Lighting of a 4 ft long section is not easy. Many techniques were attempted. The best results were obtained by using a section of light rope, which is a modern Christmas window decoration. The other issue was the algorithm to detect the highest pouches. Pouch profile could vary significantly depending on the amount the pouch had been bent while transported. Mr. Brown used a variation of a topological map algorithm, to effectively determine the highest mountain peak and added the specifics for pouches.

9.6. CASE STUDY NUMBER 1: USER INPUT MOTION DEVICE

A case study to place a single optical sensor would not be very informative, but redefining a current problem and solving it with a new implementation of existing sensors can be useful. That is the challenge in this case study concerning a powered lift device, something that would qualify as semi-automation.

In many factories, products need to be lifted from one location to another. They could be lifted from a conveyor and placed on a pallet. They could be car engines that will be placed into car frames. The lifted items weigh too much for a human to handle without some help. One option is an overhead gantry crane, where the operator manipulates the lifted item with a separate control box sometimes referred to as a teach pendant in robotic applications. However, differently from robotics, where the human teaches a trajectory once and the robot repeats it many times, the lifting equipment is dependant on the human to operate the device every time. This lack of a complex controller reduces the cost, plus the variability of where the lifted item is manipulated would make it hard for a robot to duplicate each cycle.

When an operator uses such a handheld teach pendant, they are not in direct contact with the lifted item. If the lifted item is a two-ton machining center, then the human should not be next to it anyway in case it falls, but if it is a car engine, one would want to be near it to carefully guide it into the car frame.

So several companies have built gantry style lifting devices where the human operator has their hand near the lifting hook (Fig. 9.21). The lifting cable is driven by three electric servomotors, one in each of the X, Y, and Z directions. By placing sensors into the hand control, the operator slightly tugs in the direction of desired travel. The sensors detect this tug, and send a signal to move in this direction. The sensors can be a series of limit switches (Fig. 9.22) or they can be linear potentiometers. Usually some springs are added to the hand control to help it return to the center (no motion) position.

Limit switches would produce a binary situation, either a GO in one direction or a STOP. If the tug is at 45° to the switch axes, two switches would be activated at the same time. When the lifted item it near to the end goal, the operator would stop tugging in that direction. For very slow motions and very light weights, this sensing information can be made to work with the feedback control

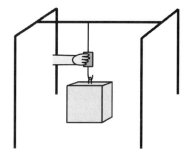

FIG. 9.21 Gantry lifting system

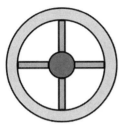

Fig. 9.22 Sensors in handle grip: top view close-up

circuitry, but when speeds increase and/or weights get larger, the lifting cable and item will start to have a pendulum effect. The operator must use the hand controller to stop the swaying, and this can be a time-consuming and frustrating task. A large lifted item swinging too much can become a wrecking ball to the other items already on a pallet.

The second method is to use linear potentiometers, or other sensors, that can give a proportional input. Depending on how strongly the operator moves the hand control in whatever direction will tell the controller how strong the motion in that direction should be. This way, as the lifted item gets closer to the target, the operator can reduce the applied force and hopefully get the item to slow down, but the lifted item still can sway like a pendulum, so fast motions and large loads are difficult to bring to a quick and graceful stop.

9.6.1. New Approaches to the Problem

Several different methods to this problem were explored by the author. These included:

- Additional cables from the corners of the lifting gantry system attached to the lifting hook for structural stiffness;
- Higher intelligence in the motion controller;
- A remote handheld tracking device.

The additional cables would require a significant number of additional servomotors and controllers, and the overall control would need to be quite coordinated. Also, the extra cables would act like a spider's web, unfortunately with all sorts of possible ways to get one's head nearly cut off. This seemed like a solution that could be made to work in the research laboratory, but not as a commercial unit.

The higher intelligence controller was the method several existing companies chose to distinguish their lifting assist product from others. Much research has been done on controlling loads on pendulums, if one adds enough motion detection sensors. In particular, if the angle of the cable is monitored where the cable is attached to the lifting motor, more information can be processed on the actual motions. This can improve the situation, but a better solution was desired.

A more significant variation would be to create a separate handheld remote control that would mark where the lifted device should be located next. It could be a cordless device that uses laser triangulation from the corners of the overall lifting gantry. Or it could use radio frequency (RF) signals or infrared (IR) light beams. Two concerns would be whether the lifted item or the operator would block the signals, and would there need to be a second transmitter located on the lift hook to state where the hook actually was with respect to the handheld transmitter.

This system was not implemented, although deemed to be able to work with enough development time, but this method was also seen to be too expensive. The existing gantry lift systems retail for around $20,000. Adding another $2500 to $3000 to the manufacturing costs for sensors and controllers did not seem wise.

9.6.2. The Leash Leading Method

So the author went back to the drawing board, or in this case, the brainstorming list. It did seem that by having an operator handheld device demark where the lifted item was to be located had great merit. The control strategy could then know what the approximate total trajectory should be, not just some incremental motor direction data as is presently done. And in most situations, this trajectory would not be over 5 ft high walls, and so on, so the path between the current location and the target location was unobstructed.

The author kept coming back to the amount of data to be sensed, and other less expensive methods to do this. The concept of a retractable dog leash came to mind. Except now the lifting device would be the "owner," and the operator the "dog." Another implementation of this concept is that of a tape measure, which pulls out of its housing with a slight resistive spring force. When the human-applied pulling force is removed, the leash or tape measure is rewound into it housing. The lifted item will follow the operator (Fig. 9.23).

The end of the leash would have a handle, similar to a coffee mug, originally located in the cable lifting the item (Fig. 9.24). A "dead man's switch," either an incorporated switch or squeezable grip part of the handle, would be desirable for safety considerations. As the human operator moves the handle away from the cable, the leash would be pulled out of its housing. The direction of the handle and leash relative to the cable would be sensed by a rotary device located at the leash housing (Fig. 9.25). The amount the leash is currently pulled out could be sensed by a rotary sensor detecting how many times the leash housing shaft has turned (Fig. 9.26). The vertical height differ-

Sensors

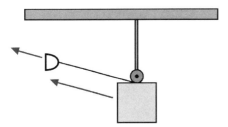

FIG. 9.23 Leash leading operation

FIG. 9.24 Handle in lifting cable assembly

ence of the leash handle with respect to the leash housing on the suspended chain or cable could be sensed by measuring the angle of the leash as it leaves the leash housing (Fig. 9.27). All three of these sensors can be economical rotary position devices.

The combined directional sensor signals would be combined with the amount of the leash pulled out of its housing to determine the direction and overall trajectory of the lifted item. The significant improvement of this

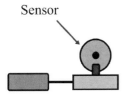

FIG. 9.25 Handle direction relative to cable sensor: top view

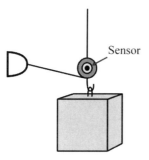

FIG. 9.26 Leash length sensor on housing shaft

method over current methods is the ability to better plan the deceleration of the generated motion. This will allow for less swinging of the lifted item on the cable, and reduce the number of collisions of the payload with other items in its work area.

9.6.3. Conclusions

It is very easy to jump to a conclusion of what kind of sensor one needs for an application, particularly if the industry has been doing it a standard way for a while, but if one is trying to differentiate one's product from the competition, it may be wise to step back, if even for a few hours, and review exactly what one is trying to sense. In this case study, by not accepting the industry standard, and by allowing the entire machine configuration to be modified around the sensing options, a novel approach was invented. If one was just willing to use new sensors without modifying the machine configuration, this appli-

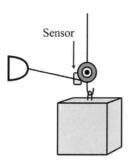

FIG. 9.27 Leash height angle sensor

Sensors 263

cation of a new strategy would have been frustrated at an early stage in discussions.

The greater the payload weight, the greater the benefit of the leash leading invention. This invention could be used as a device for human operators to dynamically move any powered device to a desired location. This includes industrial robot arms, fork trucks, and powered carts. The added cost to the existing system is minimal.

9.7. CASE STUDY NUMBER 2: PALLET LEVELING SENSOR SYSTEM*

This sensor case study also centers on a more complex problem looking for a solution. It deals with the trucking industry, where almost all products are placed on pallets. Many pallets are only stacked 3–4 ft high when the height of the truck is often 7–8 ft. It would be desirable to be able to stack one loaded pallet on top of another loaded pallet, assuming that the lower pallet can bear the weight, to maximize shipping efficiency.

Many loaded pallets that are not a full load (a full load target height is nominally 5–6 ft) are configured from distribution centers and contain a mixed load. Many supermarkets and discount department stores will restock their shelves by ordering only the amount needed to make the shelf look full. They do not want an entire pallet load of chocolate chip cookies, and then need to store 90% of the pallet in the back room. There is virtually no back room in these stores, just a receiving area for the deliveries each day.

These mixed load pallets are stacked similar to one stacking groceries into a brown paper bag. One has hopefully placed the heavier items (cans) on the bottom and filled the bag with lighter items (eggs) on the top. Now if a mixed pallet did have cases of eggs on top of it, and another mixed pallet is to be potentially placed on top of it, this will only be satisfactory if the egg cases can support the weight. If not, scrambled eggs for everyone.

So assuming that the mixed pallet can support the weight of another mixed pallet, the top layer of the lower pallet is most likely not a flat and level surface. It could be potentially made level if the mixed load palletizing process did a great deal of analysis, but currently no palletizing system can automatically achieve mixed loads with this criteria.

9.7.1. The Industry Need

There needs to be a way to stack one mixed load pallet on another. There could be giant shelves installed in these tractor trailers, but this is a big cost and does not work out well for integrating stacked pallets with full height pallets. It would be

---
*This case study supplied by Mr. Bernhard Bringmann.

desirable to be able to have the upper pallet adapt to the contour and level of the lower pallet.

This concept was developed by the author and proven in a laboratory demonstration by Mr. Bernhard Bringmann (2002). One of the key issues in the design phase was what kind of sensor system was needed to assess if the upper pallet was level, and did it belong on each pallet, where the pallet was stacked, or in the truck during transportation.

The location of the sensor was left open in the design stage, but the type of sensor was narrowed very quickly. Since the topography of the lower mixed load pallet was to be somewhat unknown, just filling in the lower locations with a filler material would not be very desirable. Sensing the topography was seen as expensive, and was actually meaningless. The upper pallet needed to be level within a few degrees, and the amount it would need to adapt to was unpredictable. This sensing could be done with triangulation methods coordinating with the room or truck walls, but this was expensive too. So a level sensor was seen to be the choice. An entire level sensor and hybrid signal conditioner was available for $100.

However, before explaining the sensor system in detail, more needs to be said about the novel pallet system with which it interfaces.

9.7.2. Self-Leveling Pallet Requirements

A device is needed that will actively change its height to establish an even level surface. This system should have the following properties:

- Be able to change height quickly;
- Take high forces;
- Align to uneven surfaces;
- Have relatively high static and dynamic stiffness;
- Exert relatively low pressures on cartons underneath;
- Be easily controllable;
- Be lightweight;
- Be inexpensive;
- Be suitable for everyday use (resilient).

Several alternatives were considered, including a system working with the same principle as a vacuum mattress (Porsche, 1988), pneumatic cylinders attached to the corners of the system or simply using flexible elements in different sizes that can be put on the pallet to fill empty spaces.

Eventually, it was decided to use an inflatable air cushion made of a resilient polymer. It meets many of the concerns listed. The laboratory prototype showed that it was feasible, and with a custom made model for series production, the first concern (speed) should not be a problem either.

The idea to use inflatable devices has a series of advantages when compared to the alternatives mentioned above:

- Great flexibility. The cushions are universally usable for many different loading states, for big or small stacking height differences. The volume utilization stays high. The stacking is not obstructed by the leveling devices.
- Simple and cost-efficient concept. Such devices could be produced and sold for very reasonable prices.
- Relatively easy integration in an automated environment. To use such a system in a fully automated process seems to be reasonable. The obstacles to overcome should be manageable.
- Potential alternative stacking method. The system would allow the stacking of both the upper and the lower pallets on the ground. After both loadings are completed the leveling system could be placed on the upper pallet, be properly inflated and then removed. This method would need less time if both pallets are loaded independently.

Air Cushions

Concerns when using an air cushion included:

- What basic geometry and dimensions of air cushions (cylindrical or spherical basic shape) should be chosen?
- Is a special material for the cushion necessary?
- How can the cushions be attached to the upper inflatable pallet (gluing, clamping, plastic welding)?
- Is the maximum allowable pressure sufficient to hold up large loads?
- Is the inflation time critical?
- Is it necessary that air is sucked out actively for deflation?

Upper Pallet

The upper inflatable pallet on which the cartons should be stacked should:

- Not use too much space;
- Be lightweight;
- Be stiff;
- Not easily slide on boxes;
- Be inexpensive.

For our prototype, a plastic pallet was used. Of course, it does not meet the space requirement. For commercial production, concerns may be:

- What material and manufacturing process should be used (deep drawn steel or aluminum or molded plastic)?
- What geometrical aides can be used to achieve maximum stiffness (beads, cross beams, gussets)?

The actual design would strongly depend on the number of systems produced.

Air Supply

There are some obvious requirements for the pneumatic system:

- It should be able to quickly inflate and deflate cushions.
- It should control the pressure in each cushion separately.
- It should deflate quickly if the pressure in a cushion gets too high.
- Cushions should hold pressure if the air supply system is detached.
- Pressure lines should be easily attachable to plate (probably automatically).
- Components that are necessary for every system must be inexpensive.

Design issues arising from these attributes are:

- Does air have to be sucked out actively (with vacuum ejectors)?
- What design is detachable and can hold pressure in all cases?
- Does every plate have to have a connector for electric power in addition to the air connectors?

Control System

The system will have to be controlled in some way. Therefore, one needs some sort of input signal to determine the tilt angles. The control system is required to:

- Measure the angles accurately and continuously;
- Be able to get the tilt information from one or more sensors and control the air flows with one continuously running program.

With these demands, it was decided that a dual axis tilt sensor should be used to measure the tilts instead of an absolute height measurement of three or more points of the upper inflatable pallet. With that decision, the following concerns arose:

- Is the plate stiff enough so that the measurement is accurate enough?
- How should the system be controlled (PC, PLC or simple microcontroller)?
- Should small slopes of the plate be taken into account for the box stacking robot? (If you have a slope of $2°$ and a box height of 20 in., the top of the box will be 0.7 in. horizontally off the theoretical position, for possible problems with clearances.)

If the whole stacking process is automated, a data interface between the leveling system and the robot would make sense. Besides the angle information,

FIG. 9.28   Upper pallet with inflatable cushions

data of preferred stacking positions for certain cartons due to the pressure states in the cushions could be sent.

9.7.3. Self-Leveling Pallet Concept

The chosen concept is shown in a CAD model in Fig. 9.28. Here the upper pallet has four inflatable cushions attached on its bottom. In Fig. 9.28 all four cushions are equally inflated, but in actual use this would never occur.

A lower pallet is shown in Fig. 9.29, with the expected uneven top layer. The upper pallet with the pneumatic hoses and level sensor attached would be slowly lowered onto this lower pallet. As the upper pallet starts to tip (while still partially constrained by the forklift), the level sensor would instruct which cushions should be inflated.

Figure 9.30 shows the cushions inflated to adjust to the lower topography. The lower pallet with its mixed load is not shown for clarity purposes.

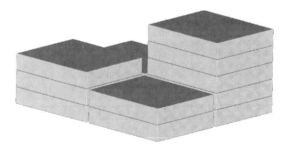

FIG. 9.29   Mixed load pallet

FIG. 9.30   Upper pallet adjusted for lower topography: alternative view

FIG. 9.31 Upper pallet on mixed load pallet

In reality, the cushions would not be inflated as shown while the upper pallet is being held by the forklift above the lower pallet. Figure 9.31 shows the complete implementation. When the level sensor has been satisfied that the upper pallet is properly aligned, the pneumatic hoses and level sensor are removed. This removal most likely would be part of the forklift devices and not performed manually.

9.7.4. Self-Leveling Pallet Prototype

The purpose of the mockup built is to show the principal feasibility of the underlying concept. The leveling system (Fig. 9.32) consists of a pallet, four inflatable and deflatable rubber playground balls usually used as toys for children, a double axis inclinometer, a control board for user inputs, several valves to control the air stream and a PC to control the valves.

The cushions are ordinary rubber balls (Fig. 9.33). The main problems with the balls are their needle valve fittings. Since the opening diameter of the ports is

FIG. 9.32 Leveling system overview

FIG. 9.33 Rubber ball attached to pallet

extremely small, the air stream flowing in or out is very small, even at high pressures. Therefore the pallet's tilt angles cannot be changed too quickly. The lateral stability (stiffness in horizontal direction) of the semi-inflated balls is rather small. Therefore the control program is designed that at least one corner of the system is always resting on the boxes the system is placed on. This increases the stability due to friction.

A device consisting mainly of two aluminum plates is used as a port for the air supply. One plate is attached permanently to the pallet, one is detachable (Figs. 9.34 and 9.35). The inclinometer is resting on three screws that are attached to the removable aluminum plate. With these screws, the tilt sensor can be calibrated in both axes.

On this plate are also two clamping levers to attach it to its counterpart on the pallet side. One half of a multiconnector is screwed to each plate with two aluminum elbows (Fig. 9.36). Air streams through the multiconnector to or from the cushions. The plastic guide of the multiconnector ensures that the detachable part will be centered relative to its counterpart.

Inclinometer

A dual axis inclinometer has been used (Figs. 9.37 and 9.38). Its measuring range is $\pm 20°$ with an accuracy of $\pm 0.06°$ for angles less than $10°$. To measure tilts, the electrical resistance of a body containing an electrolyte is determined.

FIG. 9.34 Plate with tilt sensor attached to pallet: top view

This resistance chances with the angles. Owing to this measuring principal, the inclinometer has a relatively long settling time of 500 ms. Its advantage is the very competitive retail price of about $100. The inclinometer does not only consist of the sensor itself, but also of a hybrid signal conditioner. This microprocessor can communicate with a PC via an RS232 interface.

FIG. 9.35 Plate with tilt sensor attached to pallet: front view

Sensors

FIG. 9.36 Pallet part of interface with multiconnector half: front view

Pneumatic System

The system must be able to inflate and deflate several cushions at the same time. Furthermore, the cushions must hold the pressure when the air supply is detached. Another goal is to have most of the pneumatic components not as a part of the leveling plate. This way, the whole system is cheaper since these components will only be required once. The leveling plate will not be too heavy either.

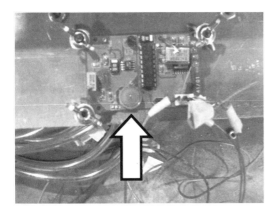

FIG. 9.37 Dual axis inclinometer (lower center on PC board) on leveling system: top view

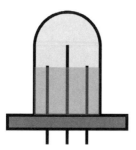

FIG. 9.38 Inclinometer cross-section

The prototype is using two sets of solenoid valves (Fig. 9.39). The first four valves have a common manifold. They are connected to an 80 psi pressure line. Although they are two-position, four-way valves, they are used as on/off valves. Normally they are closed. Opened, air can flow and the cushions can be either inflated or deflated, depending on the state of the second valve.

9.7.5. Experimental Findings

Experiments show that even with the simple setup, an accuracy of $\pm 0.2°$ per axis can easily be achieved without the system becoming unstable. If the tilt sensor is detached from and reattached to the system, the angles stay constant within $0.1°$. Assuming that the tilt sensor is positioned correctly within $0.1°$ and that the accuracy of the sensor is $\pm 0.06°$ for small angles, the absolute accuracy is within $\pm 0.5°$. The medium deviation should be much smaller. This accuracy is achievable for unloaded to heavily loaded states. The system should therefore already be good enough for normal applications (Fig. 9.40).

The needle valves on the rubber balls are deterring the process a little bit. In the prototype it takes about 20 seconds to get from both angles being bigger than $\pm 3°$ to an even surface. Although this is still acceptable, custom-made cushions would need a small standard port to accelerate this operation. A 0.25 in. port should be suitable.

9.7.6. Conclusions

The information gained shows that the concept of a pneumatic pallet leveling system is viable. The comparatively simple idea of a pallet with cushions attached to it that are inflated or deflated depending on the two pitch angles is suitable to create a system that increases load stability without greatly decreasing volume utilization. The key sensor was the inclinometer.

Sensors

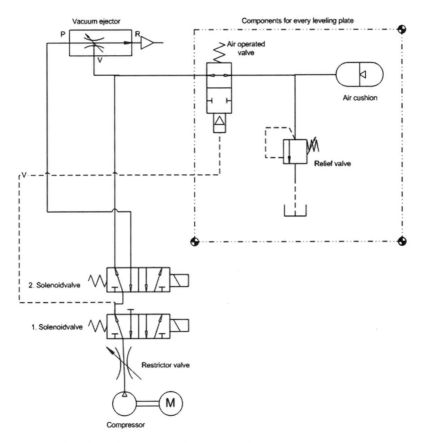

FIG. 9.39 Schematic of pneumatic system used

The system is able to create even surfaces within narrow tolerances, take sufficiently high loads, and it provides an effectual lateral stability. Cartons on which the system is placed are not exposed to stresses they could not tolerate. The cushions are able to hold their pressures when the air supply is detached. The tendency to push out the boxes on which a system is resting is very low for spherical cushions and should virtually be nonexistent for cylindrical cushions. This could only be a problem if extremely small cartons are used.

9.8. CONCLUSIONS

It can take a great amount of engineering to attempt to replicate the sensory input of a human being. Not all of a human's senses need to be duplicated for every automation application, but often more than one would first expect.

FIG. 9.40 Loaded and inflated pallet

The types and numbers of sensors keep growing. There are even entire trade shows dedicated to vision and sensing. And with all of these advances, do not let anyone tell you it cannot be sensed just because the solution was not out there last year.

Before deciding upon a sensor, understand its positives and negatives and assess them with respect to the environment and application.

PROBLEMS

1. You are baking chocolate chip cookies and the recipe states that they will be done in 8 to 10 minutes. What kind of sensors would you use to determine if they are done? What kind of testing and calibration are you most likely to have to perform?
2. You are going sailing on your favorite lake in Maine. The rope that holds up the sail looks a little frayed. What kind of sensor will inform you that the rope is starting to fail, and that the sail is likely to hit you in the head?
3. You are creating an automatic machine to sort and pack apples. What kind of sensors would you use to determine if the apple is ripe or is green? How would you detect bruises?

4. You need to create automation to break an egg for your bakery. What type of gripper and sensing would you use to determine how the egg breaking progress is occurring? How would you detect if any egg shells get into the liquid? (In production, large bakeries purchase liquid eggs and have them delivered similar to other chemicals.)

PROJECT ASSIGNMENT

1. For your project create a list of required sensing for both input signals to the controller, and monitoring output devices for feedback and motion assurance.
2. Investigate the probable impact of using contact sensors (limit switches) vs. noncontact sensors (optical switches). Also investigate using several digital sensors vs. a continuous sensor.
3. Create a matrix to assess the strengths, limitations, and unknowns for the sensors found in parts (1) and (2), and select the best choice, understanding that you will have some unknowns and risks.

REFERENCES

Bringmann, B. (2002). Conception of a Pneumatic Pallet Leveling System. MS Thesis, Rensselaer Polytechnic Institute.

Davis, J., Chen, X. (2001). A laser range scanner designed for minimum calibration complexity. Computer graphics lab, Stanford university. In: Proceedings of the Third International Conference on 3D Digital Imaging and Modeling, 3DIM 2001.

Graham, J., Meagher, J., Derby, S. (1986). A safety and collision avoidance system for industrial robots. *IEEE Trans. Industry Applications*, 1A-22(1): 195–203.

Dr. Ing. Porsche Aktiengesellschaft, Stuttgart. Vacuum mattress, preferably for rescue vehicles. U.S. Patent No. 4,254,518. (1988).

10
Control

This chapter on control is the last chapter of this text concerning the design and build sequence of automation. Chapters 2–10 present the core material machine builders or equipment specification writers need to know on how the process can and should work. The remaining three chapters will add information with larger case studies of new automation products (Chapter 11), specifications (Chapter 12), and the larger group of packaging machines (Chapter 13).

Control is the part of automation from which many mechanical engineers shy away. The often-heard comment is, "let the electrical engineers do that." This comment and attitude are unfortunate, since with today's control options, almost any type of engineer can create a standard level of control implementation.

The concepts within control can be seen from the following situation. Imagine that you have had a bad week, made up of many bad days (perhaps as outlined in the beginning of Chapter 9). You are home on a Sunday afternoon, and you want to forget about your "wonderful" job of stacking boxes. You want to enjoy a mindless three hours of football on your television. After all, you deserve a break.

So on the table on your right you have a bowl of chips and a glass of your favorite cold beverage. On your left are the remote control and your cell phone. The remote is important since your two favorite teams are playing on two different channels at the same time, and you want to follow them both. The cell phone will let you call your friend who roots for one of the teams challenging one of

Control

your favorite teams. The television is in front of you, and you have ordered a pizza for delivery around half time. Any distractions (dog, kids, and spouse) are out shopping or whatever. Life seems under control. What do you have to do?

- Place more beverage in the refrigerator to be cold when you need it.
- Turn on the television.
- Lapse into an "eat a chip then take a sip" mode with your right hand.
- Keep the remote in your left hand to flip whenever the mood strikes or there is a commercial.
- Decide if your team is beating your friend's team enough so as to call and gloat.
- Listen to the doorbell for the pizza delivery person.

This does not seem like work, yet the decision process you are undergoing is probably more demanding than that of your job. This can be seen by a more detailed comparison of these fun tasks to an automatic machine.

Placing more beverages in the refrigerator represents the need to plan ahead. Similarly, you hopefully purchased enough chips the day before the game. If the supply of raw materials into an automatic machine goes to zero, then the machine will waste available production time. Some control programs monitor the production rate and help to inform the inventory control program so as to electronically order more raw materials from the supplier in a Just In Time (JIT) mode.

Your turning the television on emulates the controller signaling a required external device. In this example you only have to do it once, but if you somehow ignored this fact, the afternoon would be wasted watching a blank screen. Automation machine controllers can interface with almost any other device if needed. The number of devices and how they are to be signaled (the number of digital and analog controller output lines) will help determine the size and type of the controller and its options.

The "eat a chip then take a sip" mode is the normal operational mode of your afternoon. In this example, there is no penalty or problem if you happen to eat two chips before sipping, of if you grab more than one chip at a time. For most automation processes, this would not hold true. The order would be more explicit, but this does represent a normal controller process loop where, if the supplies hold true and there are no interruptions, the loop would continue forever. The loop gets broken if and when there is an interruption.

The flipping of stations with the remote is a second independent program loop that is not coupled to eating chips and sipping beverage. Whether or not you have just eaten a chip or taken a sip will not determine if and when the game production staff determines that a block of television commercials needs to be run. And no matter how hard you may scream for your teams, the timing of flipping between games cannot control the games' outcome. Automation controllers can handle more than one program loop, but as an engineer adds more loops to the controller's tasks, the time between successive executions of the same loop becomes longer.

The rest of the afternoon's tasks are interruptions. Some of these are good interruptions, like the door bell signaling that the pizza has arrived. And some could be bad, like when you are out of chips, beverage, a bathroom break, or your team is losing and your friend calls on the cell phone to gloat. The order and the magnitude of these interruptions need to be assessed. Most likely the door bell for the pizza will take precedence over refilling your glass. You do not want to have the pizza delivery person wait too long, or they might leave. Similarly in automation, one needs to plan out all of the parallel scenarios, such that an operator opening a door that trips a safety circuit immediately executes a machine shutdown, rather than worry about signaling the supplier to send more parts to be assembled. Safety has a higher precedence over the machine being out of supplies for 10 minutes more.

The cell phone also represents a device that can be used by you to send a good signal when the conditions so trigger (your team is up 14 points and it is time to gloat to your friend), and a device that receives the news that you do not want to hear. In automation, some interfaced devices work with two directional communications. These devices can be additional controllers or processors that are specifically meant to unburden the main controller.

So our enjoyable afternoon is actually a set of challenging tasks and interrupts handling situations. If it was not so much fun, it would be thought of as work! But to a machine controller, there is no fun or work, just a set of signals to accept, process, and send.

10.1. TIMING DIAGRAMS

When an automation process is supposed to occur can and should be documented in one of several forms. It is only when the order of steps and the related input and output signals have been laid out on paper or the computer monitor that the simplicity or complexity of a controller can be assessed. The example concerning football watching could be represented by a flowchart similar to a computer program, but it will be more instructive to look at a series of machines.

If one looks at the standard home washing machine, the traditional controller is a timer motor-driven set of cams that trigger the different devices. The timer motor does not turn as the washer is being filled with water. A pressure sensor confirms that the water level has been reached, and then the timer motor can proceed. The washer then agitates for a prescribed amount of time, followed by a pumping out of the dirty water, a rinse (with a small amount of water and a short agitation), and a spin. Some washers have additional operations, but they will be ignored for now.

Control

Figure 10.1 shows a timing diagram of these operations and the sensors or motors either triggering events or causing motion. The sensor input is considered "low" or 0 if there is no signal, or "high" or 1 when the signal occurs (as when the water level has been reached). The agitation motor is at rest when noted as 0, and in operation with a 1. (The same motor is used to both agitate and pump out water in real life, by operating the motor at two different speeds. But this will be ignored here.) There are several variations of timing diagrams used, but they all state the same information.

It should be noted that in Fig. 10.1 the X-axis is a nonlinear statement of time. Some events like "Turn On" take a fraction of a second, while other events like "Fill" can take five minutes or more. The "Drain" operation repeats during the "Spin" cycle so as to remove the water that has been wrung out of the clothes.

Historically, there is another answer, when there is relatively no input signal to be processed. A paper tape can be punched with a series of holes to demark when the various output devices should be actuated. The only variation a paper tape is capable of processing is that of stopping the tape until something else has occurred. This was also used in old-time player pianos (seen in cowboy western movies), where popular songs were encoded into many rows of holes in paper rolls often 8 in. in width. The rolls were set in motion, and then the holes would allow for compressed air to escape. When such an escape occurred, the corresponding key on the piano was played. The song would sound the same each time played, and

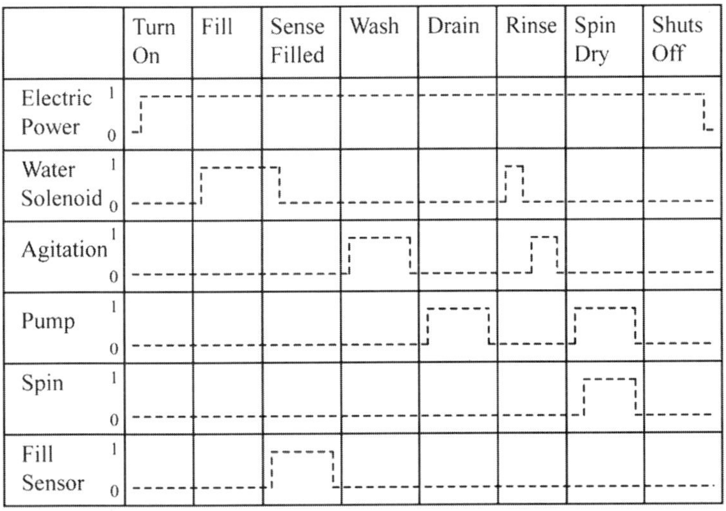

	Turn On	Fill	Sense Filled	Wash	Drain	Rinse	Spin Dry	Shuts Off
Electric Power	⎡‾‾‾	‾‾‾‾	‾‾‾‾‾‾	‾‾‾‾	‾‾‾‾	‾‾‾‾	‾‾‾‾‾	‾‾‾⎤
Water Solenoid		⎡‾‾	‾⎤			⎡‾⎤		
Agitation				⎡‾⎤		⎡‾⎤		
Pump					⎡‾⎤		⎡‾‾⎤	
Spin							⎡‾‾⎤	
Fill Sensor			⎡‾⎤					

FIG. 10.1 Timing diagram for simple washing machine operation

there are no options. The songs are hard coded for life unless one pokes an additional hole with a pencil (or a bullet in the old western movies).

The implementation of a timing diagram for most automation today can be done on a PC, a Programmable Logic Controller (PLC), or other dedicated processors. As mentioned before, most older standard washers do not use any of these options, but essentially wrap the output timing diagram curves each onto its own cam surface. Then each cam has a limit switch that follows the cam surface due to light spring forces. The input timing diagram signals will stop or start the timing motor and cams, as per the water level sensing operation. It is the cam limit switches that are the most likely element to fail in a washing machine, but since they are integrated into the cam and timer motor, are not inexpensive to replace!

Another option less likely found in automation but more likely found in consumer products is that of a programmable controller chip. An automatic breadmaking machine can be purchased for less than $100. It does operate an electric motor and turn on the heating coil. User-operated switches select options. It does not need nor can it afford a low end PLC costing $200. The controller chip used can and does only cost a few dollars at most. There is no requirement to be able to reprogram your breadmaker to do other variations (though it does sound like fun!) so the control code can be burned into the chip once and for all.

10.2. PROGRAMMABLE LOGIC CONTROLLERS

Many control applications require switching (on or off) of various outputs as a function of a number of inputs. This type of control is known as logic control or switching control. The relative simplicity of switching control makes it attractive for use in automatic machines where there is a requirement for the machine to follow a set sequence of operations. The use of switching control to cause a machine to go through a set sequence of operations gives rise to the term sequential control.

A PLC is an industrialized dedicated microcontroller whose operation is based primarily on sequential control of a process. Its Real Time Operating System (RTOS) scans a program in a repetitive cycle and actions are taken (outputs set) depending on logical relationships (specified by the programmer) between the inputs.

10.2.1. Are PLCs Necessary?

If one had an automatic machine with only a few motors, and one wanted to avoid controllers at all costs due to one's silly fear of electronics, one could control the machine with a bank of switches (Fig. 10.2). Now an operator would be needed to

Control

Fig. 10.2 Bank of switches

flip all of these switches when things looked appropriate, rather than the operator performing the task themselves. Now if it was the operation of a heavy capacity overhead crane, this makes all the sense in the world, but if it is a machine to tie bows, this borders on stupidity.

So the banks of switches needs to be replaced with some electric relays, which themselves are activated electronically by a controller. Now no matter what kind of controller is used (even paper tapes) the concept of a relay still prevails. The relay's electromagnet holding a metal arm making an electrical contact is usually replaced by a solid-state version. If it is to conduct a larger power source, a relay can now be a dedicated device several cubic centimeters in size. But if it is a computer-type 5 V TTL signal, a single IC chip could process 10 or 20 relays internally.

The physical use of these older electromechanical relays has diminished, but the concept of the relay lives on. So it is important to understand the older notation, not just for historical reasons, but as a basis of present-day understanding. Figure 10.3 shows the four basic elements, the relay coil, which acts as an electromagnet, the normally open and normally closed switches, which can get activated by the relay coil, and the lamp, which lights up to show that a circuit is closed. These elements are replaced by present-day solid-state devices, so one might not hear the click as one did when an electromagnetic relay closed.

These four elements do not facilitate any human or environmental interaction. Any On/Off or Emergency switches, or any sensor input, needs to be included. Figure 10.4 shows the representation of the six basic switch types. For example, a complex optical sensor should state somewhere (perhaps even on the sensor body itself) what type of switch it behaves like.

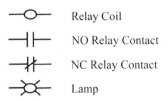

Fig. 10.3 PLC programming elements

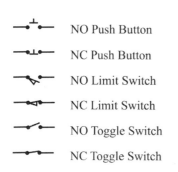

Fig. 10.4 PLC programming elements

Since all of these switches either close a circuit or open a circuit when they are activated, all of the switch types are reduced symbolically to the two representations in Fig. 10.5. They simply state the continuity of the local part of the circuit (allowing current to flow or not), and the engineer is left to understand the physical reality of the situation. This may sound more difficult, but when these symbols are used in larger programs, simplicity is a must.

10.2.2. Ladder Logic

A PLC, whether it is the size of a small brick or a desktop PC, will be programmed in one logic language or another. Most PLCs have several rows of connections for input and output wires. Programming is done with either a teach pendant containing a keyboard full of specific buttons and a simple LCD display, or by a laptop PC. The standard language for years has been called Ladder Logic. Ladder Logic is predicated on the extensive use of relays in the past. A symbol-based solution used to translate electrical schematics into program instructions the PLC can execute. There are standards (IEC 1131 is the modern standard) on what should and should not be in Ladder Logic, but many manufacturers do have some extensions that make their models more powerful than the competition. So beware if you have Brand A code and you hope to download it into Brand B PLC (or even Brand A model 2) without some changes.

One of the key concepts as one writes larger amounts of Ladder Logic in a single executable program is that of Scan Time. Scan Time is the time to make

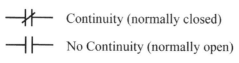

Fig. 10.5 Continuity

Control

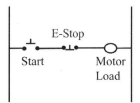

FIG. 10.6 Single rung starter motor circuit

one pass through the program, and one pass can be thought of as one PLC cycle. During each cycle, inputs are sampled, logic is evaluated, and outputs are set. The Scan Time is an indication of how often the logic functions are evaluated.

A written expression of a program using Ladder Logic is called a Ladder Diagram (LD). Here is where the term "ladder" makes sense. Each step in the program is written as a rung on a ladder. One starts at the top rung, and cycles downward, then returns to the top and repeats until ended or interrupted. There are a few general rules on what should or should not appear on a Ladder Diagram:

- Inputs may be repeated in a diagram.
- Only one output should occur in any rung.
- A specific output should appear only once in a diagram.

This will be understood best by stepping through a series of LD examples.

Figure 10.6 shows a single rung that is supposed to start an electric motor. Note the motor symbol is the same as a relay coil, something that causes action when a current is passed. The Start switch is a momentary NO switch that in real life is a 1 in. diameter green button. The NC E-Stop is usually a red mushroom button that is easy to identify on all machines. When the Start button is pressed, the current can flow from the left ladder side to the right ladder side, but the Start button does not stay depressed when one releases it. So the motor starts to run only as long as the button is pressed. This is not very useful in factory automation.

So, to make this a useful LD circuit, a load relay is added in parallel to the Start button (Fig. 10.7). When the Start button is pressed for a fraction of a second, the energy of the motor also closes a relay, which will keep the motor running after the Start button is released. The power to the load relay is disconnected when the E-Stop button is pressed for a fraction of a second. If the E-Stop button was not included in this LD logic, there would be no method of turning the motor off, short of disconnecting the power source.

This logic is how the older controller systems could have functioned, but it would not be convenient to add a load relay to every motor one wanted to control. So the same logic is applied to only internal PLC elements (Fig. 10.8). The Start and E-Stop switches have been updated to the continuity symbols, and the motor

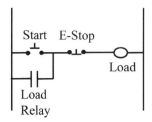

FIG. 10.7 Corrected single rung starter motor circuit

relay has been replaced by an internal relay, both the energized coil and the element in parallel to the Start switch. This is an easier method to implement and change the logic if need be. The only difference is that, without some sensor to confirm that the motor has started and continues to run, it could be possible for the PLC to think that the motor is running because the internal relay says it is, and perhaps due to a loose wire, the motor has stopped running an hour ago.

10.2.3. Model Train Example 1*

A model train can be configured as in Fig. 10.9. This could look like what one would set up on the living room floor, but if one has ever run a model train that has an inner and outer loop, one knows that not only does switch S1 need to be set, but switch S2 has to match it (inner or outer loop) or the train will derail. This may be fun to watch now and again, but it does mean one has to reset the train.

So a PLC can be set up to monitor the situation to guarantee that there will be no derailments. A possible scenario would be:

- The train should start when the NO start switch is pressed;
- The NC stop switch should stop the train;

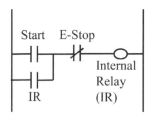

FIG. 10.8 Two-rung software ladder

- Train should only run when BOTH switches (S1 and S2) are in the same position.

Control

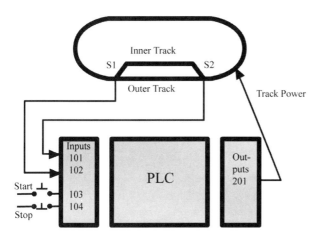

FIG. 10.9 Model train example 1

This means that for the few seconds one has changed one of the switches (S1 or S2) and not the other, the train power will be off. This is not exactly useful, but will be improved in the next example.

One of many possible Ladder Logic Diagrams is given in Fig. 10.10. The 100 numbers are reserved for inputs, the 200 numbers for outputs, and the 300 numbers for relays. The upper two rungs will function as the previous motor starter example, except that the powered up device is only the relay R301. Then in the lower two rungs, the output power to 201 only happens if the relay R301 is energized, and the two switches are in the same direction.

This example does seem a bit highly engineered for an application for a toy, but the logic it demonstrates is a good stepping stone. The author is not suggesting that one would want to implement this in real life.

10.2.4. Model Train Example 2*

To learn more about Ladder Logic, more constraints will be added to the model train layout. Now suppose the following:

- Add the proximity sensors A (105) and B (106) to the setup (Fig. 10.11).
- State that the train should behave as in Example 1.

---

*This example supplied by Dr. Ryan Durante.
*This example supplied by Dr. Ryan Durante.

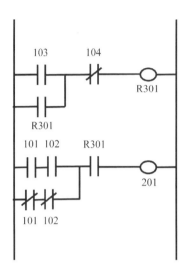

FIG. 10.10 PLC solution for train example 1

- But it should stop if it has crossed sensor A four times without crossing sensor B or crossed sensor B four times without crossing sensor A.

On the surface, one might ask:

- Why are we doing this?
- Does this not make the human operator a slave?

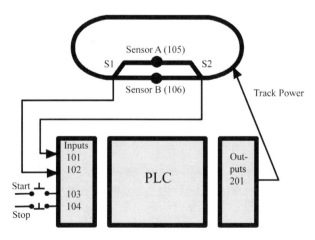

FIG. 10.11 Model train example 2

Control

We are doing this to create the need for counters, and to show the next level of difficulty to writing Ladder Diagrams. And yes, the human has a specific set of tasks to count how many times the train has crossed each loop, or the train will cease to run. In real life one would surely never want to implement this problem! It would be better to have the PLC operate the switches and take the human out of the loop (Fig. 10.12).

So we will need two counters and some elements to compare a counter to a fixed target number. The Ladder Diagram starts to become quite a bit more complex, and this example is only two loops of a model train. The rungs from the top are:

- Top two rungs to energize system relay R301.
- Third rung increments the counter C10 when sensor A is triggered.
- Fourth rung increments the counter C11 when sensor B is triggered.
- Fifth rung resets/enables the counter C10 when system is powered up or the other loop is transversed.
- Sixth rung resets/enables the counter C11 when system is powered up or the other loop is transversed.
- Seventh and eighth rungs compare counters C10 and C11 to power relay R302 when both true.
- Ninth and tenth rungs check for the switches S1 and S2 and allow power if relay R301 is energized and relay R302 is not energized.

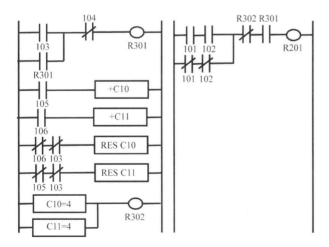

FIG. 10.12 PLC solution for train example 2

10.3. OTHER PROGRAMMING OPTIONS

As one can see from the simple model train examples, the Ladder Logic can become complex very quickly with only a handful of inputs and outputs. Writing the Ladder Diagram on paper can be cumbersome. So there have been other methods developed to set up the logical relationships:

- Function Block Diagrams (FBD): Program elements appear as blocks, which get wired together in a manner resembling a circuit diagram. Blocks contain algorithms and wires indicate data flow. Programs like National Instrument's LabView fall into this category.
- Sequential Function Charts (SFC): Uses flow charts to develop the logical progression and different paths.
- Structured Text (ST): This method uses a high-level language (if-then-else, Case of, etc.). Examples would be C++ and Delphi.

Other directions are based on the existing knowledge of PLC programmers. One uses a PC interface to program the PLC (Fig. 10.13). A second option is Allen Bradley's SoftLogix 5, a Windows NT based soft control

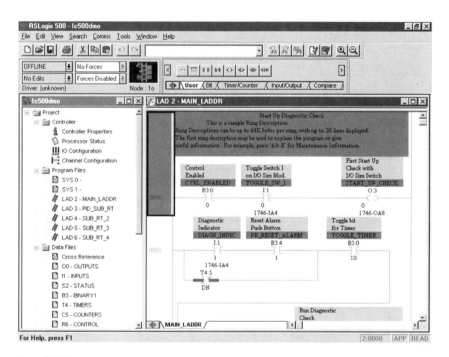

FIG. 10.13 RSLogix 500 programming environment (from Durante, 1999)

engine that offers the openness and power of PC-based control, combined with the functionality of the PLC-5 processor. It also provides hardware and software compatibility with Allen-Bradley I/O, communications and Operator Interface devices. The SoftLogix 5 Controller is ideal for applications with large memory requirements or the need for a high level of information integration.

10.4. CASE STUDY NUMBER 1: AGILE AUTOMATION CONTROL SYSTEMS — THE HANSFORD ASSEMBLY FLEX PROJECT*

Flexible Automation needs more than a robot to be functional. The supporting feeders and conveying system must change quickly, as well as the software that coordinates the control. This is not a simple task if a company plans on creating a commercial product that will service the needs of a range of customers. Hansford Automation, Rochester, NY, made one of the early commercial attempts in the late 1990s.

This case study reviews the Hansford Assembly Flex project conducted at RPI (Durante, 1999). Assembly Flex is a modular automation assembly system in the truest sense. In most cases it can be reconfigured (re-tooled) in less than 20 minutes and go from building water pumps to car stereos. Much of this flexibility stems from its unique software control system developed by Dr. Ryan Durante, the author, and others.

The New York State Center for Automation Technologies (CAT) at Rensselaer Polytechnic Institute (RPI) was contacted by Hansford Automation Systems, for the purposes of entering into a cooperative effort to assist Hansford in the development. The CAT's role in the project was to assist Hansford in developing initial functional specifications and the corresponding software for the Assembly Flex system. Hansford had already determined the mechanical hardware configuration of the system. Figure 10.14 shows the three assembly modules (SCARA, Cartesian, and a Pick and Place) assembled to form a small demonstration system.

The mechanical concept was to have assembly modules (Fig. 10.15) that could accept up to six docking process modules (Fig. 10.16) to form a workcell. The number of process modules was limited to six because of the experience gained from the National Center for Manufacturing Sciences (NCMS) Light Flexible Mechanical Assembly (LFMA) project on which Hansford was the prime integrator. More than six stations per workcell was

---

*This example supplied by Dr. Ryan Durante.

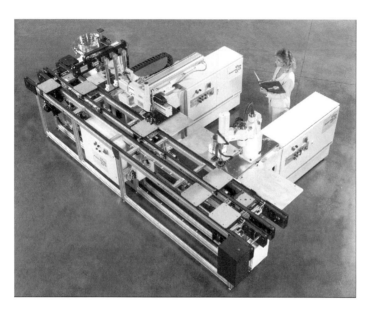

FIG. 10.14 Three cell Assembly Flex system (from Durante, 1999)

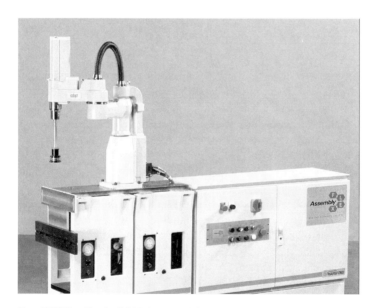

FIG. 10.15 Single SCARA Assembly Flex assembly cell (from Durante, 1999)

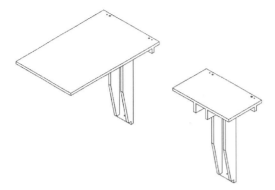

FIG. 10.16 Long and Short Process Modules (from Durante, 1999)

found to be too complicated and caused too much congestion (Molnar and Rezsonya, 1997).

The process modules mount via a three-point kinematic mounting mechanism, which provides highly accurate registration and repeatability. Hansford knew that they wanted the system's mechanical design and controller architecture to be open and modular. They also realized that this would not happen overnight and accepted that the first generation of the Assembly Flex system would be Adept-centric, that is, the robot assembly cells would use Adept MV controllers. They also required that all device-level communication would be via DeviceNet — an up and coming (in the late 1990s) open standard for field (device) level communication.

10.4.1. Agile Control System: Release 1.0

The first official release of the software had to meet the following specifications:

1. All screens must be active.
2. Configuration was to be channel based (described in Durante, 1999).
3. The following modules were supported:
   - SCARA Assembly;
   - Cartesian Assembly;
   - Pick and Place Assembly;
   - Bowl Feeder;
   - Tray Stacker;
   - Intelligent Interface Module (IIM).

The final Assembly Flex Control Suite is composed of three applications: the System Executive, the Event Viewer, and the Configuration Editor.

System Executive

Figure 10.17 shows the main screen of the System Executive. The screen shot shows a system navigator on the left side of the screen. The system executive uses the multiple document interface (MDI) for displaying cell information while maximizing screen real-estate. Four cells are tiled in the client area. The screen is divided into five key sections: client area, cycle control, system navigator, status bar, and event viewer. Depending on the level of security given to any user, they may be able to change the system configuration and/or the robot program, or they may be limited to simply observing the data and permitted to only hit the system stop button for safety reasons.

Figure 10.17 has four module inspectors displayed. The MDI interface allows for displaying multiple module information simultaneously while guaranteeing that other salient features of the human machine interface (HMI) are never obscured from the operators view, particularly the Cycle Control and Event Viewer. It also allows the user to tile and cascade the windows automatically:

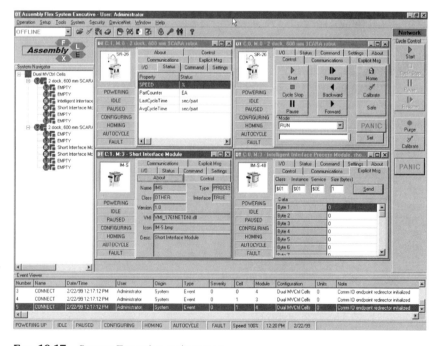

FIG. 10.17   System Executive main screen

Control

A module inspector exists for every module defined in the system. The module inspector is the user's interface into a module. It allows control at all levels, data monitoring, and atomic-level control of all discrete I/O. It displays state information, and allows the user to send explicit messages to the target module. All I/O points are tagged here and displayed. The user can toggle all discrete I/O from this single screen.

The Cycle Control provides system-level supervisory control. It provides Cycle Start, Cycle Stop, Pause, Resume, Purge, Calibrate and Panic Stop functionality to the user. A Cycle Start sends the start command to all workcells to initiate the build cycle. A Cycle Stop will instruct all workcells to stop after their current cycle is complete. This type of stop leaves the system in a known state. This can be contrasted to the Panic Stop, which will force the system to shut down as rapidly as possible. In this case, the system is left in an unknown state and must be calibrated before beginning the next build cycle. The Pause button puts the system into the "pausing on" state. This allows the user to single step through each assembly step for debugging and pilot runs. Resume returns the system to the autocycle state. Purge instructs all cells in the system to completely purge all products from their workspace. This would most likely be done after a Panic Stop.

Event Viewer

The Event Viewer provides the user with a log of all events from the current session including the event number, name, date/time stamp, the user, the origin, the type, severity, the source module, the configuration, the units, and a generic text notation. This provides the user with a quick look at recent events without opening the Event Viewer application.

Configuration Editor

The Configuration Editor is a design time tool that allows drag and drop configuration of the target system. The main screen is shown in Fig. 10.18. The design window is broken up into four major sections: the system information box, the module palette, the module information box, and the workcell design box.

The system information box gives the user system-level information such as name, author, creation date, revision, and description. This information corresponds to what is presented in the system information window of the System Executive.

The module palette utilizes a two-level tabbed worksheet. The top level splits the available modules into assembly (base) and process modules. The assembly modules are then broken down into SCARA, Cartesian, Pick and Place, Conveyor, and Other. Process modules are broken down into feeder, handler, and other (Fig. 10.19).

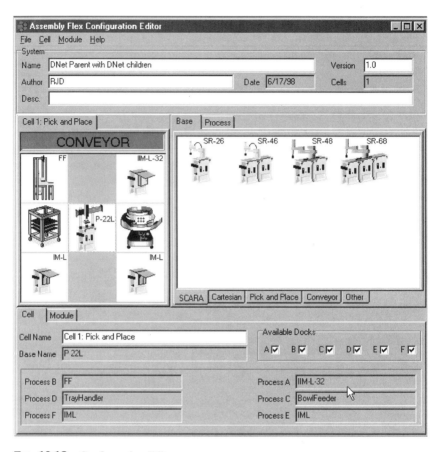

FIG. 10.18 Configuration Editor

The workcell design box is where the target system is constructed. Figure 10.20 shows a two-cell system with a SCARA-based workcell displayed. This workcell shows three active process modules, a tray feeder, a bowl feeder, and an intelligent interface module. The box is tabbed for each cell in the system. Assembly modules can be dropped into the center cell and process modules connected to it. All communication links are created automatically. This corresponds to the physical placing of process modules on either side of the robot, and the proper connection of the standardized cabling. The user no longer has to be concerned about the software logical assignments. The software has performed that task for the user.

Control

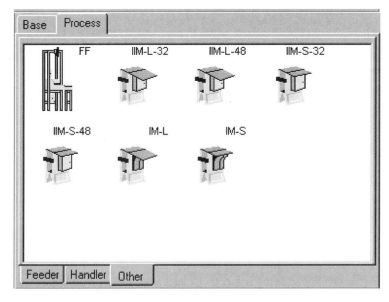

FIG. 10.19 Module Palette

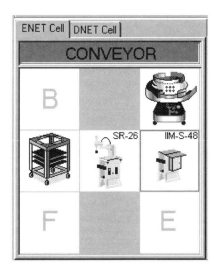

FIG. 10.20 Workcell Design Box

10.4.2. Control System Architecture

The first prototype of this architecture was demonstrated at the 1997 Assembly Technology Exposition (ATE) show in Chicago. It had to meet the following requirements:

1. The Ethernet/DDE link had to be in place and functional (to support the following two tasks).
2. The System Executive had to be able to start/stop the system.
3. Main displays had to include "live feedback" including robot status, PLC information, and conveyor status.
4. All other screens were to be present but not functional.

Figure 10.21 shows the DeviceNet communications scheme in more detail.

This was not the ideal configuration but was driven by the technology available at the time. When the new DeviceNet scanner cards became available, it was possible to have a direct physical connection between DeviceNet and the PC. This did away with the SLC 504 PLC/scanner card and the RS232 serial link. Also, since Allen Bradley has recently released the new Micrologix PLCs, it is now possible to do away with the DNDB CAN interface module. The configuration that will be implemented in the final architecture is shown in Fig. 10.22.

Figure 10.23 shows the finalized robot communications scheme in more detail. The prototype that was developed enjoyed good acceptance at the ATE show. After taking into considerations suggestions and criticisms from the show, work began on the analysis and design of the "real" software architecture.

10.4.3. A Second Assembly Flex Implementation

The power of Assembly Flex as a control software architecture was not fully appreciated until it was implemented on a second automation configuration. A linear track based tray handling robot (Fig. 10.24) had been built by American Dixie Group, but the controller was not implemented. The tray handling robot was outfitted with Assembly Flex by Dr. Ryan Durante in a matter of weeks.

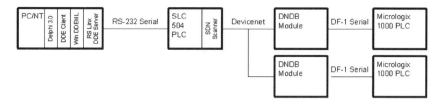

FIG. 10.21 DeviceNet system used for prototype

Control

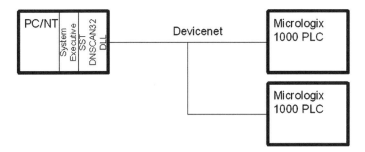

FIG. 10.22 Ideal DeviceNet configuration

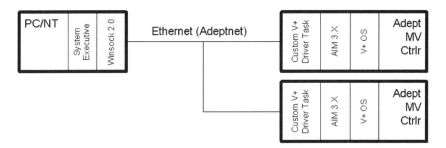

FIG. 10.23 Robot communications implementation

The actual robot (Fig. 10.25) shows a gripper with eight large vacuum-powered suction cups (located directly above the roller conveyor) on a dual arm implementation. There are a tremendous number of vacuum and pneumatic hoses, sensors and electrical wires around the vertical arm. This system was truly a prototype. The mail containers are not shown in this figure for clarity.

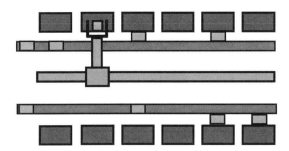

FIG. 10.24 Linear track based single arm tray handling robot concept

FIG. 10.25 Dual arm track based tray handling robot

The controller box was 5 ft high by 6 ft wide (Fig. 10.26), and the electrical power being used was significant.

The implementation of Assembly Flex was straightforward. A customized Human Machine Interface computer display was designed (Fig. 10.27) to represent the robot and the mail container locations. The underlying drivers for the specific electronic motion controllers were updated from the original implementation, similar to the updating of drivers for your PC when you purchase a new model printer.

Control

FIG. 10.26 Controller for dual arm track based robot

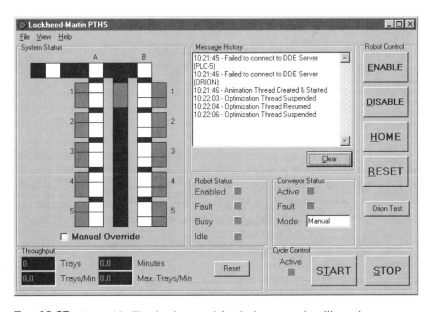

FIG. 10.27 Assembly Flex implemented for dual arm tray handling robot

10.4.4. Project Conclusions

The Hansford Assembly Flex system was shown at the 1998 ATE as well as the IMTS (International Manufacturing Trade Show). The control was accomplished by a fully functional agile automation control system. Final cut over of the software base and documentation was done in December 1998. The straightforward implementation for the tray handling robot application demonstrated the usefulness of modular software.

This system has demonstrated the successful use of Agile Automation Control Systems in flexible automation. Users of all levels of security clearance can use a drag and drop interface with the appropriate levels of system information and control to quickly change over a flexible workcell. The object-oriented approach will allow for quick implementation of new robots and workcell elements.

As the first flexible automation commercial entry into the field, the Assembly Flex system was met with some resistance. Consumers were not accustomed to thinking in terms of flexibility and its longer term financial benefits. As the economy for automation cycled through lean times, Hansford Automation was purchased by DT Industries, and the Assembly Flex project tabled as personnel were dispersed. The next generation will learn from the successes and failures of Assembly Flex in terms of its market and customer base.

10.5. CASE STUDY NUMBER 2: OMAC AUTOMATION CONTROL

A large effort to have improved automation machine control is being conducted by the Open Modular Architecture Controls (OMAC) organization. Started in the 1980s by General Motors, factory floor technical managers were finding that each and every machine from multiple vendors had unique controllers and/or software implementation. Technicians were going to training classes for each machine, and the retained knowledge when a worker moved to a new site or a new company was too limited. So the OMAC effort was initiated. It grew into the packaging automation area in the late 1990s and is flourishing. It is likely that many of the big and medium customers will be insisting on OMAC compatible controllers in the near future, so this can impact on much of the automation industry.

OMAC's packaging working group's Mission Statement is to "enhance the value of packaging machinery and systems by promoting the use of digital motion control and OMAC guidelines for open control architectures." The stated goals (OMAC, 2000) in 2000 were, over a three-year horizon and with no increase in capital or operating cost, potential improvements that should approach 50% for:

- Delivery lead time;
- Startup time;
- Machine/system footprint;
- Product throughput;
- Product changeover time;
- Material loss;
- Machine reconfiguration/overhaul time;
- Machine downtime;
- Mean time between failure.

These goals are very aggressive, and many people in the automation industry (including the author) wonder aloud if they are all possible, but they are worthy goals and should be pursued. Giant customers like Hershey's (Campbell, 2000) see that these goals are needed to stay cost competitive, and that some new packaging machines are already making headway. They see:

- Smaller footprints;
- Faster startups;
- Flat to lower cost per function completed;
- Higher speeds;
- Improved efficiency;
- Faster changeovers;
- Better quality package;
- Reduced waste;
- Improved reliability.

Hershey's also sees that redesigning packaging machines with a clean sheet of paper using Mechatronics can make a big impact:

- Fewer moving parts;
- More precise movements;
- Software is major performance and cost component;
- Machines are more modular;
- Functions added by module/software;
- Fewer machine platforms;
- Performance changes within a single machine platform by changing controls/software.

10.5.1. OMAC Working Groups

There are a variety of subcommittees within the OMAC Packaging Workgroup effort. These subcommittees have changed over time to dynamically mold into the best methods for success, and will likely continue to change in the near future. So the specifics of the various subcommittees will not be detailed here. The reader is encouraged to check their website for the latest results.

However, there are a few nuggets of wisdom that will most likely not change over time. The fact that most automation machine control programs can be written a thousand different ways, having someone other than the original code writer try to understand someone else's code can take forever. There are basics that need to be agreed upon so as to create some sanity in the situation.

So independent of the type of controller, the language used, or any specific brand, one of the goals of the PackML subcommittee of the OMAC Packaging Workgroup is to define two classes of mode of operation:

- User Selectable: Two types of Mode exist within the User Selectable Class: Procedural (in which there are three possible modes of operation: Automatic, Semi-Automatic and Manual) and Equipment Entity (in which there are two possible modes of operation: Automatic and Manual).
- Machine: Machine modes provide the control framework in which the User Selectable modes become available for selection. These can be transparent to the operator. The two proposed modes are ESTOP and IDLE.

The user selectable modes are selected by the machine operator in order to perform relevant procedures and operations (Fig. 10.28). These are similar to most robotic controllers and many automation machines, but some manufacturers have in the past created their own branching and unique definitions that can be confusing to the majority of the workforce, and sometimes even a safety risk, so standards like this are not universally accepted by default.

Machine modes provide the control framework in which the User Selectable modes become available for selection (Fig. 10.29). This concept comes from the need for engineers determining the control sequences to define the

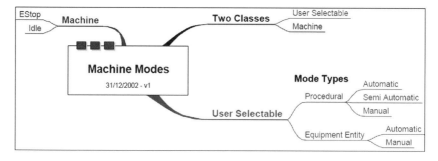

FIG. 10.28 OMAC machine modes (from OMAC PackML Team, 2003)

Control

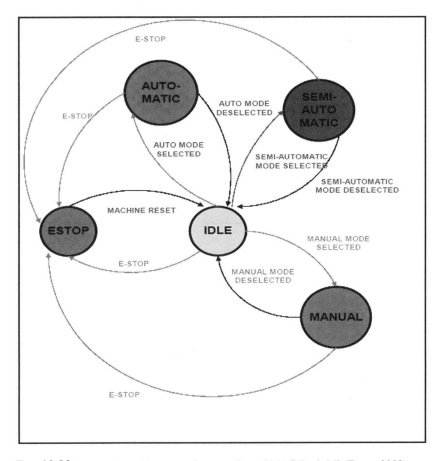

Fig. 10.29 OMAC machine state diagram (from OMAC Pack ML Team, 2003)

possible states the machine can be in at any time. Another way of approaching this issue is to consider at any point in the automation machine's operation, what are the conditions of the following:

- sensors;
- actuators;
- moving members;
- incoming parts;
- processed parts;
- facility supplied pneumatic lines and electrical power.

If any of these fail, due to loose connections, power variations, or the unanticipated (a stray animal decides to nest in your machine for the winter), what happens to your machine control? Does it totally hang up and freeze? Does it send the motors running at full pace and create an internal crash? Do you have to reboot like one does sometimes on a PC?

It is difficult to anticipate all of life's possibilities on your machine going wrong, but many of these are predictable. So if they happen, what state does the machine go into? Or will it just hang while running, which can be a safety nightmare. Figure 10.29 gives a framework to use while sorting out these issues, which is consistent with many of the other new machines being built today.

10.5.2. Conclusion: Do You Use OMAC Principles?

Depending on your customer base, you may not have a choice. Over 100 big customers are on the list of OMAC users that will require compliant automation controllers, or they have promised to buy your competition's machine, even if it costs 10–25% more. The OMAC compliant controller will save them time and money in the long run.

10.6. CONCLUSIONS

Controllers have become almost easy to use. Any engineer, no matter what their background, can learn a specific type of control, whether Ladder Logic, PC based, or microcontroller. The competition within suppliers and the consistency demanded by OMAC will help everyone win.

PROBLEMS

1. Modify Model Train Example 1 to operate the system if one has added two actuators to replace the human intervention for switches S1 and S2, and create the Ladder Diagram.
2. Create a Timing Diagram for making a submarine sandwich (roll, meat, lettuce, cheese, sauce, etc.). What is the impact of changing the order of operations?
3. Analyze eating an ice cream sundae and draw a block diagram of what you do and under what conditions. What options do you have and what are the impacts?
4. Plan a party for you and some friends, but model it with a Timing Diagram. What are the inputs and outputs?

PROJECT ASSIGNMENT

1. Write a list of all inputs and outputs for your project.
2. Draw a Timing Diagram of your machine's normal operational mode.
3. Create a Ladder Diagram.
4. Specify a PLC that would work for controlling your machine. Compare to a PC-controlled system.

REFERENCES

Campbell, K. (2000). OMAC at Hershey Foods Corporation. 2000 Pack Expo Conference, November 3–8, Chicago, IL.

Durante, R. (1999). A Design Methodology for Developing Agile Automation Control Systems Architectures. PhD thesis, Rensselaer Polytechnic Institute, May.

Molnar, M., Rezsonya, T. (1997). Results Summary Report for the NCMS Light Flexible Mechanical Assembly Program. NCMS Internal Memo, May 9.

OMAC (2000). Open Modular Architecture Control for Packaging Machinery — A User's Perspective. Pack Expo Solutions 2000 Conference, November 7.

OMAC Pack ML Team. (2003). *Machine Mode Definition V0.3*. www.omac.org.

11

Bringing New Automation to Market

This chapter will lead the reader through the design and development of three new automation products. These machines were the invention of the author over the past 15 years, and they were all created to fill a specific automation need. One is based in the area of precision assembly and fiber optic alignment, while the other two address needs within the packaging community. There was a range of exciting, good, and mixed events along the way, as well as some major frustrations, but there is a wealth of experiences here to be shared to assist automation designers and entrepreneurs.

11.1. CASE STUDY 1: PRECISION AUTOMATION

11.1.1. The Market Need

The precision of an industrial robot has been a research and development challenge ever since they were invented in the 1950s. There was a desire to make robots fast and therefore as lightweight as possible, yet precise for many applications. One only has to look as far as a standard multi-axis milling machine to see a very rigid precision structure that weighs a great amount, so it does not move extremely fast without requiring significant costs. Robot designers were and are concerned for balancing speed, stiffness, arm and payload weight, and precision.

Bringing New Automation to Market

Robot precision, measured as repeatability and accuracy, has been tested and discussed by industrial robot manufacturers and user groups for many years. Each generation of robots did improve over the previous generation, but many were not sufficient for 0.001 in. or even 1 micron operations. When required, more intelligence was added to the robot and controller, including more sensors for better feedback. Or robot users would add several smaller robot joints and links, almost like a mini robot, to the end of the large robot where the normal robot hand or gripper would be. Attached to the mini robot would be a mini gripper.

The mini robot would use sensor feedback to move or servo to the correct desired position. The sensor could be force feedback, as the large robot and mini robot serial chain were possibly inserting a peg into a hole (the peg and hole analogy is often used as a model of many specific robotic assembly tasks; rarely does one set up a robot system just to place pegs in holes). Or it could be a computer vision system that had some reference marks or cues to align the peg before it was inserted into the hole.

However, this combination of large and mini robot has several limitations:

- Resolution of large robot encoders;
- Deflection of robot arm members;
- Deflection of bearings;
- Gravity effects;
- Servo control system design;
- Friction at joints;
- Dither to overcome friction.

Dither is a common strategy for servo controlled mechanical systems that on the positive side, produces a quicker response. Since the friction at the robot joints is larger at rest (static) than what it is dynamically (kinetic), the frictional effect is minimized by always having the servo motors, and thus the robot arm, moving back and forth on an almost microscopic scale. This can be best observed by feeling the powered up robot vibrate at "rest" with one's hand (touching a robot while its drive system is powered up may be in violation of industrial safety standards). This vibration is the dither effect.

However, the dithering also means that the larger robot holding the mini robot is always shaking. One can see that the mini robot now has a greater problem to solve. The base of the mini robot is moving and twisting in many directions, and the sensor system for inserting the peg has to compensate. If the natural frequency of the larger and mini robots and the control system are not accounted for, the mini gripper might not stand a chance of ever inserting the peg!

So in summary, it is difficult and expensive for a large range of motion (1 or 2 m) and having the robot gripper achieve 1 micron precision. From a machine design view, one is mixing too many orders of magnitude.

11.1.2. The Closed Loop Assembly Micro Positioner (CLAMP) Device

The first embodiment of the author's invention is called the Closed Loop Assembly Micro Positioner or CLAMP device (Derby, 1989a, 1990). How it was invented is similar to what many inventors will tell you. It was a combination of:

- Seeing a market need;
- Understanding the current solutions;
- Making a few or even hundreds of attempts;
- Divine inspiration.

The author, after using robots to try to solve some assembly challenges for industry, was looking for a better solution for several years. Then one day (while riding the train to the big city) the proverbial light bulb went off. Why not learn from a photographer's tripod? It creates a rigid reference base for taking pictures. And it can be relocated to a new location with relative ease. This is also similar to bracing one's wrist if one has a hand tremor, often found in people past their youth.

In many assembly tasks, the precision work (i.e., the peg into the hole) is not over a large range of motion. The mini robot could be place under the "tripod" and perform its tasks. The tripod would steady the mini robot, and somehow isolate it from the large robot's dither and other challenges. So the large robot could have the limited task of moving the "tripod" from one location to another, and could have significantly less precision.

However, to obtain a repeatable precision base, simply letting a tripod touch a work surface or table top would not be sufficient. Forcing the tripod on the work table surface would be better, but not enough. There needs to be some physical references to align the tripod in a repeatable fashion, so thus the second part of the CLAMP invention, the docking process. To use more commonly found industrial terms, the tripod is called the robot's end effector, and the mini robot a micro positioner.

The CLAMP end effector uses a method of forcing its three legs into three hole features (the process is called "docking") found in the work table surface to achieve a very high precision with respect to the assembly workcell. The most common implementation is to use a conical hole, a conical slot, and a flat plate as the three hole features (using three conical holes would be impractical, since it would mean that the three legs of the CLAMP device must be machined and assembled almost to perfection). The conical hole determines three degrees-of-freedom, the conical slot two degrees-of-freedom, and the flat plat the remaining degree of freedom, thus determining the total six degrees-of-freedom between the end effector and work table surface. This aligning process of two rigid bodies has been referred to as a kinematic mount for many decades.

Bringing New Automation to Market

The kinematic mount must be precise so as to allow the micro positioner to be as accurate as possible. The ends of the legs can either have a conical surface or preferably a spherical ball. Mating conical elements defines a set of line contact surfaces, and requires very tight machining tolerances. However the set of point contacts of a ball into a conical hole or slot is less demanding, and hardened metal ball components are commercially available (Fig. 11.1). This use of spherical ball ends on the three legs is also useful in that the CLAMP device can be docked into the same set of docking features in three different orientations, 120° from each other. This docking process for a high-precision robotic end effector has been recently duplicated (Würsch et al., 2001).

When the three legs are docked, there is need for some freedom of motion or compliance between the larger low precision robot and the three docking features, otherwise there will be significant wear on the contact points. The CLAMP docking end effector incorporates the use of mechanical springs and several alignment features to assist with the docking and undocking process (Figs 11.2 to 11.5). Conical disks are held into conical holes on the top mounting plate by a series of compression springs. When the large robot pushes down to properly seat or dock the CLAMP device, the springs are compressed and thus allow the lower CLAMP end effector to float freely into the docking features. When the CLAMP device is lifted from the docking features, the springs compress the CLAMP device back into registration to the top mounting plate. This assures that the CLAMP device will be able to successfully dock into the next set of features, and eliminates free vibrations if the coupling was to be simply a set of springs.

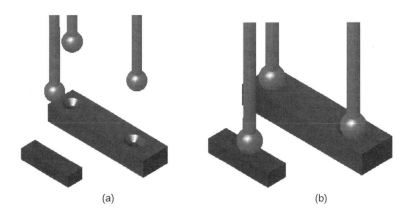

FIG. 11.1 (a) Three legs above docking features; (b) three legs in conical hole, slot, and flat plate

Fig. 11.2 Conceptual CLAMP end effector with two finger micro gripper

CLAMP End Effector Performance

The repeatability of the docking process was measured with a laser interferometer. Measurements were taken while docking with a range of misalignment of the large robot, a range of docking force (and therefore compression of the springs in the top mounting plate), and a range of angular misalignment. The composite results of the CLAMP device are given in Fig. 11.6. These results of less than 1 micron (0.000,04 in.) for repeatability give an excellent foundation for micro positioners and grippers to manipulate.

One limitation of the CLAMP system comes from needing to place a set of docking features around the place where one needs to perform the task. One option is to place the docking features on an XY table's moving surface created

Fig. 11.3 Skeletal CLAMP device docked into docking features

Bringing New Automation to Market

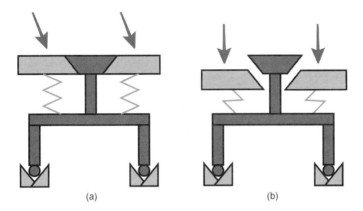

FIG. 11.4 Conceptual top mounting plate compliance device (two-dimensional model)

above the desired working surface (Derby, 1989b). With a hole in the moving surface of the XY table for the micro positioned and gripper to access the working surface, flexibility can be achieved. Another precision implementation would be the coupling of two robot arms. With a CLAMP device on one arm and the docking features on the second arm, a relatively stable precision task can be carried out in free space (Ducoste and Derby, 1990).

11.1.3. CLAMP Production and Market Acceptance

The author was awarded a U.S. Patent (No. 4,919,586, April 1990), and had formed a startup company (GRASP, Inc.). The author remained a faculty member, but hired full-time employees. He hired a talented former graduate student of his, Mr. John McCarthy, as the lead engineer and machinist. John, being the rare

FIG. 11.5 Actual top mounting plate compliance device

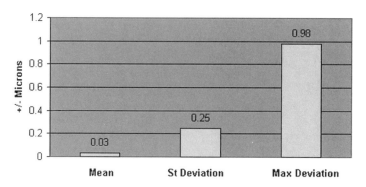

FIG. 11.6 Docking repeatability test summary

combination of Electrical Engineer, Mechanical Engineer, and machinist by education and practical training, was able to design and produce the first CLAMP units.

Since the docking process needed to only hit the conical targets, the larger robot precision was required to only be within 0.25 inch. This was actually a very loose performance specification for robots in the late 1980s, so GRASP also designed and built a moderate speed, lower cost SCARA robot arm. Its major design feature was that the arm members were entirely an exoskeleton frame, allowing for a simplistic manufacturing and assembly procedure. Electric servo motors with harmonic gear boxes were used to drive the arm. It was referred to as the CLAMP robot (Derby, 1992).

The CLAMP end effector was built from aluminum and used off-the-shelf hardened steel balls for the docking legs. It had stepper motor drives for each direction of motion, either translational or rotational, or both, depending on the application. It could be intermittently removed from the larger robot by use of a robotic quick change adapter and a standard gripper temporarily installed for transporting large items.

A PC, several PC interface boards, and some power supplies and amplifiers controlled the CLAMP robot and docking end effector. The PC code was written in the C language, and a user interface written. Mr. Albert Santos, the second full-time employee, handled the control software.

Several CLAMP robots and end effectors were manufactured. They were exhibited at several major assembly robotic trade shows. Interest was good but mixed, since the CLAMP device was a totally new approach to solving precision automation. Much time was spent with each visitor on how the process worked, because no-one had seen this before. The third full-time employee,

Mr. John Cerveny, was in charge of marketing, and had his hands full trying to educate the potential customer base that as encountered.

However, since sales were slow, and all of the GRASP employees were fresh out of school, a seasoned salesman was hired, Mr. Randy Fields. He knew how the sales cycle worked at our potential customers' factories, and got the list of potential sales to grow to several dozen serious buyers, but timing is everything!

Now it was early 1991, and the Gulf War was under way. The recession of 1991 was kicking into high gear. So despite our sales leader's effort, the company lost one to two customers per week. Not to the competition, but to the fears of recession, and their own potential plant closings. The automation market hit a very soft season, not just for GRASP, but for most of the automation equipment suppliers. Existing major robot companies that had been profitable for years also closed their doors in 1991. There was little that could have been done differently at GRASP that would have meant success. A larger amount of investment dollars would have prolonged the company, but the downturn lasted several years. So the CLAMP device was mothballed for many years, and the author encountered the experiences of company bankruptcy.

11.1.4. The Dockbot

The setback of GRASP did slow down the entrepreneurial effort into automation for the author. But another day of inspiration occurred in the late 1990s, and he was tasked to develop several new inventions. The first of these is the Dockbot (Derby and McFadden, 2002).

The CLAMP end effector is limited in that a larger robot must hold it in place while the smaller micro positioner and gripper does its assembly task. Thus one large robot cannot service additional smaller robots (CLAMPs) simultaneously. It also becomes difficult to coordinate the larger robot to an incrementally moving work surface, such as a pallet conveyor. The novelty of the new docking and locking solution (the Dockbot concept) is the addition of:

- Method of locking the legs of the existing CLAMP device;
- Requirement for the wrist exchange;
- The ability for the larger lower precision robot or automation device to be able to re-aquire the Dockbot, since the wrist interface may have shifted during the compliance motion while docking;
- Device no longer obtains electrical control signals through the larger lower precision robot or automation device via the wrist;
- Now uses connections from worktable for control signal/power; or
- Uses wireless control means (i.e., IR, RF signal);
- Uses stored power;
- Uses onboard microprocessor for control;

The docking process is very similar to the CLAMP device. The locking process can use:

- Ball and detent methods;
- Spring-loaded rollers similar to many kitchen cabinet latches;
- Permanent magnets and steel plates;
- Electromagnets and steel plates;
- Threaded socket latches — rotary.

Additionally, the preferred method of the locking process does assist in improving the docking process. The spring-loaded rollers option also guides the spherical ball ends of the legs into the docking features, softening the docking process (Figs. 11.7–11.9).

Various Dockbots Solution

Since the larger lower precision robot is not required to be connected to the Dockbot continuously, the lower precision robot may service multiple similar or different devices. This system may be configured as a single SCARA (Fig. 11.10) or six degree-of-freedom (DOF) robot moving multiple Dockbots within its reachable workspace. The Dockbots are not limited to having the docking features parallel

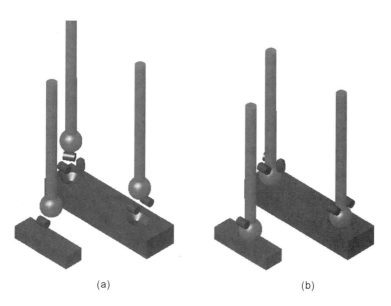

(a)   (b)

FIG. 11.7 Spring-loaded rollers above docking features: (a) legs above; (b) locked in

Bringing New Automation to Market

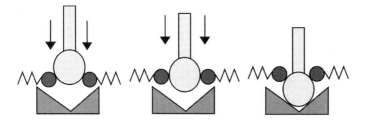

FIG. 11.8 Two-dimensional version of spring-loaded rollers and docking/locking process

to the floor. They could be at various angles, on a vertical surface, or even upside down.

The multiple Dockbots could each have different functions internally, with various micro grippers or operational processes. Each could get its own internal microprocessor program via RF or IR communication, for true distributed robotics. The Dockbots can be effectively parallel processing, where the SCARA or 6 DOF robot is the distribution and allocation device.

The Dockbot's docking features can also be located on multiple pallet surfaces, where the SCARA or 6 DOF robot essentially leapfrogs one Dockbot over a series of other Dockbots to the next available pallet.

11.1.5. Micro Positioners

Micro positioners have jumped from experiments in the laboratory to commercial products. Mechatronic solutions have been very useful in making them a reality.

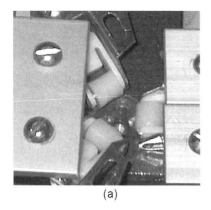
(a)

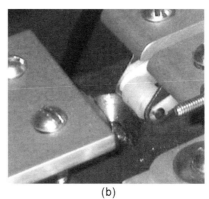
(b)

FIG. 11.9 Spring-loaded rollers: (a) above conical hole; (b) above conical slot

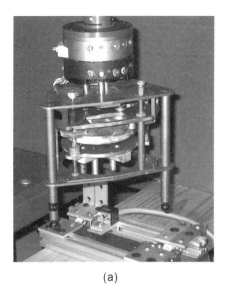 

(a) (b)

FIG. 11.10 Dockbot being docked by SCARA robot: (a) above (b) in docking features

Many of these devices were developed for aligning fiber optics, which are used for Internet and phone communications. Most have either one or two degrees-of-freedom or direction of motion. The enabling technologies include:

- Micro Stepper motors;
- Voice coil technology (CDROM uses this);
- Piezoelectric actuators.

There are piezoelectric actuators that have one or two axis stages available, with 2.5 cm travel in X and Y directions and 0.1 micron resolution (Del-tron, www.deltron.com/newproduct.htm). There are micro stepper motors available with an accuracy of 0.01 mm per 25 mm of travel and a repeatability of 0.0025 mm (NanoMotion Ltd., www.nanomotion.com). There are also three-axis stage systems (National Aperture, Inc., www.nationalaperture.com) available (Fig. 11.11). A micro gripper would need to be added to all of these for useful work to be performed.

11.1.6. Dockbot Production and Market Acceptance

The effort to bring the Dockbot to market was different from the CLAMP device and the two other machines to be discussed later in this chapter. The goal was to find a technology partner to license the technology, as opposed to full development and marketing within the author's company. There was only limited

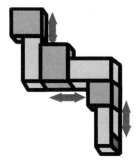

FIG. 11.11 Nanopositioner placed underneath Dockbot: without micro gripper

prototyping performed, and various presentations and conference talks given. The markets explored included:

- Laboratory automation;
- MEMS fabrication;
- Fiber optic alignment.

Those conference attendees handling new technology received the concept of the Dockbot with excitement. Many exhorted the author for the novel approach, and that the industry had suffered from a lack of creativity, the constant cloning of everyone else's machines. But poor timing saw the Dot Com world crumble (fiber optics) and the drug discovery field (laboratory automation) lose money. The MEMS area was slower to take off than was predicted. So the author is still looking for the right partners.

In the future, it might be possible to extend this approach to have the micro positioner of today's Dockbot move an even smaller Dockbot to its docking location to manipulate a nano positioner. This is an area of future research.

11.2. CASE STUDY 2: PALLETIZING

Palletizing is the process of stacking items on a pallet for shipping. The items can be bags, cans, or loose-formed product, but the great majority is cardboard boxes or cases. These cases can have finished product ready to place on the market shelf, such as a dozen cans of beans or 10 boxes of corn flakes. Or they can be filled with assorted products such as letters for the post office.

11.2.1. The USPS Mail Tray Project

It was the palletizing of cardboard trays of letters that first drew the author's attention to this market. In the mid 1990s, the United States Postal Service (USPS) was looking to automate the material handling of these trays. Every evening, hundreds of postal processing facilities around the country sort out the first class mail and place letters in reusable cardboard trays bound for various locations (Figure 11.12). The trays are approximately 12 in. × 24 in. × 6 in. in dimension. Filled trays have a cardboard sleeve slid over them, and a plastic band is strapped around the length of the tray to keep the sleeve and tray an integral unit.

The USPS ships most of these trays on several different models of carts (Fig. 11.13). These carts are approximately 6 ft high, and can accommodate three trays stacked side by side per layer. The trays are reused many times, and can become quite ugly looking after only a few uses. The stacks of trays are somewhat unstable, often looking like they could fall over, and the carts' loading was, and in most cases still is, done manually.

The first reaction by the author and many other engineers was to replace the cardboard trays by some more durable and rigid material. France's mail system was automated during the 1990s by changing to a reusable plastic tray with an interlocking lid. This changeover was the logical choice for the USPS also, but being a joint effort of the Government and private industry, the USPS often finds change very difficult. So the automation was to be done with a mix of good and relatively poor trays (in many industrial applications, this would be considered foolish).

The automation is to be configured in an area of the USPS facility that is potentially different in every site. Depending on the location of the site, the normal outside transportation routes for the mail carts, and other factors, the number of mail carts to have trays loaded into them would vary from 8 to 20. The available floor space was also unique in many sites due to the desire to not modify the overall building framework and support columns.

This meant the automation system should preferably be modular (to adapt to every USPS facility) and reconfigurable (to adapt to a new mail route) so that the machine did not have to be scrapped. There was also the internal love at the top administration of the USPS for existing commercially available robots,

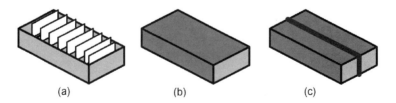

FIG. 11.12 (a) Tray with mail; (b) tray with sleeve; (c) strapped

Bringing New Automation to Market

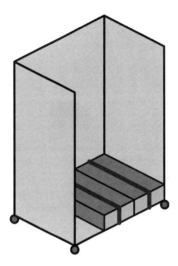

FIG. 11.13 USPS mail cart

similar to what one sees on commercials for automotive companies welding car bodies.

However, this preconceived notion that the solution must take a certain form severely limited the project's success! As will be seen, those commercially designed robots only work best in a certain configuration.

11.2.2. Commercial Robotic Prototype Implementations

The pedestal robot has its maximum reach in a circle (Derby, 1981) (see Chapter 5). To palletize on the maximum number of carts, they too must be in a circle. When the prototype automation system was located in the developmental lab space of a commercial automation house, the circular layout did not seem to be deficient, but when a second unit was actually installed in a USPS site, problems arose. The existing building had support columns that the workcell must accommodate, and the overall traffic flow of people and carts on the floor was a rectangular grid. The need to load and replace carts radially from the circular layout required even more space claims than was originally planned. Simply put, it was trying to fit a round peg into a square hole. The corners of the square were unusable and therefore wasted space. Many of the other future sites had

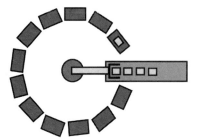

FIG. 11.14 Pedestal robot workcell: top view

support columns in less accommodating locations, so that many other site managers were reluctant to implement the workcell.

Another problem was the number of carts that could be reached by the pedestal robot. A limit of 11 or 12 was all that was possible. However, many potential sites required that more than 12 carts needed to be filled at any one time, so either a second robot system was needed, or manual loading of carts beyond the limit would still be required.

A second prototype robot workcell was configured by the same commercial automation house around an overhead gantry unit. Looking somewhat like an independently supported overhead crane similar to those found in many factory high bays, two column supported rails allowed for travel by a moving horizontal beam. The beam supported a vertical telescoping sliding member that had the gripper attached to its lowermost point.

On first inspection, this robot workcell configuration seemed superior. The supported rails could be as long as needed, so as to load any number of carts. Only more column supports would be needed. However, several problems arose:

- The robot workcell cost became much higher due to the massive robot structure.
- The cycle throughput became an issue, since the gantry supported beam had to travel from the pickup point at one end to the farthest cart in the same cycle time.
- Many of the existing industrial sites did not have the available ceiling space to install the robot.

Both of these prototypes had their champions within the USPS, and during a stretch of many months, a contract of 100 of each automation system was given to two different firms. Neither of these systems was successful due to costs and/or physical constraints. The remaining approximately 7000 possible system locations remain manual to date.

11.2.3. First Implementation: Linear Track System Design

There were two different answers to this problem invented by the author. The first one was developed when the system throughput requirements were an average of six trays per minute, 10–12 carts, with a 20 lb maximum payload (The Evolution of Geometry and Its Impact on Industry, S. Derby, C. Cooper, Fifth National Applied Mechanisms and Robotics Conference, University of Cincinnati, Cincinnati, OH, paper AMR97-056, October 1997). The fact that the system specifications crept seemingly every few months or so (30 lb payload, followed by 7–10 cycles per minute, followed by 14–16 carts, etc.) all helped to give automation system designers ongoing headaches.

The general automation design strategy used to determine the linear track robot configuration was to:

1. Understand the performance specifications.
2. Look at the various cart/robot configurations.
3. Do analyses based on cycle time and floor space.
4. Create new solutions to the remaining problems.

The most challenging specification was the cycle time.

The various possible configurations of the carts and "robot" were explored. Since the boxes needed to be stacked at all corners of the carts, it was decided that either a linear slider configuration or a SCARA arm design would be needed. The slider was chosen as giving the greatest access to all regions of the cart, and as being the most flexible for redesign. This flexibility proved later to be quite valuable. A second consideration was the option of rotating the linear arm in the horizontal plane, or having a dual arm system. Although higher in costs, the dual arm method was again selected to meet a potentially higher throughput, which was eventually required.

The options of either a linear track or a pedestal base were simultaneously explored. All of the data from the industry, as well as the authors' experience, showed that the linear track system would be the more efficient due to the rectangular world similar to most machining centers. The dual arms were configured to be side by side, in opposite directions, and moved by the same vertical post on the same horizontal track body.

Any potentially higher costs of the robot could be easily balanced by the lesser cost for site preparation and the cost of floor space. The cycle time was somewhat equal, assuming a limit of 12 carts (as per the earlier pedestal robot configuration).

The author's first design was a linear track robot, but with ultimately an improved strategy on the tray pickup locations. The actuators used are similar to those found in industrial robots. The PLC allowed for reprogrammablility, one of the criteria in defining any automation as a robot. This reprogrammablility allows for the workcell to be installed in many of the customer's sites, without the

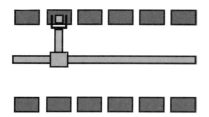

FIG. 11.15 Linear track robot: top view

need for massive plant floor reorganization. Figure 11.15 shows a simplified layout, with the robot loading one of the carts.

The later problem in the design specification process arose from the shortened cycle time. In order to facilitate this requirement, a pair of conveyors was added between the robot and each row of carts, as well as a staging location in front of each cart (Fig. 11.16). The trays were picked up closer to their final cart destination. This made the process the following:

1. The tray barcode would be scanned.
2. The tray would be sent down the proper conveyor, in front of the destination cart.
3. The tray would be pushed by a lever into the staging area in front of the destination cart.
4. The robot would determine an optimum sequence to lift the tray from its staging area and load into the cart.

See Chapter 10 Case Study Number 1 for more details.

This change in plan meant that the vertical reach of the robots had to be doubled. It was easier at this point to increase the length of the slider member than it would have been to add length to a pair of members to a SCARA design. Longer members in a SCARA design would have had interference problems with the cart walls and previously stacked trays.

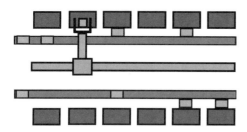

FIG. 11.16 Linear track robot with staging conveyors

The use of the pair of conveyors and the staging areas increased the throughput so that the desired processing of 14 trays per minute was achievable. Without these additions, it was physically impossible to meet the demand due to excessive speed and practical limitations of material handling.

Prototype Built and Tested

The author worked with American Dixie Group, formerly located in Albany, NY, to design and build the linear track system. Most components were readily available off the shelf. Some combinations, like the 80 ft of hardened steel bearing rails for the main linear track motion, meant that alignment and assembly were not simple. Electric servo motors were used for the three different linear motions. Rotary motion of the servo motors was transformed to linear by use of fiber reinforced rubber timing belts and matching pulleys. Smaller rotary motions that were only 0, 90, or 180° were carried out using pneumatic rotary actuators that have internal programmable stops.

Since the tray and sleeve unit was to be picked by vacuum cups, and this tray/sleeve unit was often somewhat abused from many uses, a powerful vacuum supply was needed. Instead of supplying compressed air to the robot arms and then using vacuum generators, it was determined to have a pair of dedicated vacuum pumps placed on the moving arm base. This was rather unique, since most if not all commercial palletizing robots use vacuum generators due to their lower weight.

Weight was a concern after the linear robot system was built and debugged. The resulting design was a tremendous moving mass running from end to end, and the throughput rate of 14 trays per minute was now demanded to be for more than 12 carts! So the linear acceleration was becoming extremely fast, so much so as to make the technicians worried. So at each end of the larger linear travel direction, a huge spring dashpot/dampener was added to stop any potential high-speed runaway motions.

The unit worked as predicted, but its performance was stretched by the increased specifications. This design became less and less desirable with every new demand. So, the author was not satisfied that the right design was now being used, but what should the new design be?

11.2.4. Second Implementation: The Stackbot Design

Some days an inventor can try as he might, and never come close to anything unique or novel worth patenting. And if one has a new idea, such as what happened when inventing the CLAMP device, one can say it was a very good, exciting day. However, on the inspirational day in the late 1990s when the author invented the Dockbot, he also came up with a novel idea for palletizing mail letter trays, called the Stackbot (Derby and Kirchner, 2000).

Assume that most palletizing will be done with similar sized boxes for many hours/shifts without changeover, such as the letter trays into a cart. With this assumption, there are only a limited number of positions in X and Y (parallel to the floor) to which the box must be moved. Why have an expensive set of automation components (servomotors, linear slides, etc.) that allow for unlimited X and Y position access when only a limited number is needed?

The novel concept here is to create a device that uses a selectable multiple cam track plate. The cam track plate's function is solely to determine the X and Y position. It does not carry any of the load of the payload (box or tray), or any vertical load of the device itself. It has a mechanical switch that selects which cam track the payload should travel. The tracks and switch function similar to a railroad train yard, with the various spurs to allow for train car loading or unloading. A dedicated unpowered arm would support the payload weight and give overall rigidity. This arm's sole function is for support, even though it may look similar to a powered robotic arm.

Initial Testing Model

When an inventor conceives of a possibly new automation design, one most likely does not sit down at a CAD system and make production-ready drawings. One must build an initial testing model, or set of models, to prove to one self and any potential investors that the design has merit. Without this proof of principle demonstration, very few people will open their wallets.

Extruded aluminum structures and the various connectors available from websites and catalogs are a great asset. Although the costs are not cheap, one can fabricate many things in one's basement or garage using them with only a few tools. One brand is called 80/20, and it includes rotary and sliding joints that make such concept models straightforward. 80/20 markets itself as the "Professional Erector Set," and the author would have to agree.

An air cylinder was used to power the linear motion. In order to keep things compact, a rodless cylinder was used for the initial testing model. The initial track plate was made from fiberboard, and then a Plexiglas one was cut on a CNC machine when the fiberboard geometry proved to be correct.

There was no capacity in this model to move the track plate or gripper up and down as would be required in practice, but since the concept was to slide either the track plate or gripper using conventional components, it was not seen to be necessary to make this active in the model.

This model, which was almost the correct scale, enough to load three trays side by side in the mail cart, had a trackplate that was quite huge (36 in. × 36 in.). Perhaps it is not practical if one wanted to access an entire 48 in. × 48 in. pallet.

This model did its job, both technically to prove to the author and others in the material handling arena, and to excite a company to invest. Morrison Berk-

Bringing New Automation to Market

shire Inc. (MBI) in North Adams, MA, was looking to expand its product line into new markets. They licensed the design from the author, and with the author's help, designed a commercial unit.

The Commercial Product

Mr. Ned Kirchner, VP of New Product Development of MBI, and Mr. Bill Lyons worked to transform the initial concept to a viable product. Many concepts and variations were sketched and analyzed. Since catering only to the USPS seemed like a hit or miss proposition, the marketing efforts of MBI determined that the Stackbot should also be able to palletize on 48 in. × 48 in. pallets, and if it was to make a good entry into the material handling industry, it should be able to access two pallets at once. (Very few existing commercial palletizing machines work with more than one pallet at a time.) The required trackplate geometry proportions seem unwieldy.

This is where Bill Lyons was clever. When looking at the range of existing mechanisms that have been used over the years (some of them for more than hundreds of years!!), one will find the mechanical linkage called a pantograph (Fig. 11.17). Although mentioned in many Mechanisms textbooks to copy small pictures to be larger or vice versa, it never seems to be used in practice, but this was a powerful combination with the trackplate concept.

The pantograph proportions determine the ratio of input to output. Chosen for a compromise of size and force loading was a 5 to 1 ratio. So the motion of the pin in the trackplate is magnified five times by the vacuum cup gripper, while the trackplate cam tracks see five times the loading of the arm and payload dynamic motion. Figure 11.18 shows a second table top model, with its pantograph.

As can also be see in Fig. 11.17, the forearm link closest to the gripper that transfers the box changes its angle along the cam track path. This means that if the Z-axis connected to the gripper is fixed rigidly to the forearm, the box will be rotated by an amount approximately proportional to the arm's travel. This

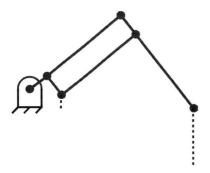

FIG. 11.17 Generic pantograph linkage: cam track path on left, gripper path on right

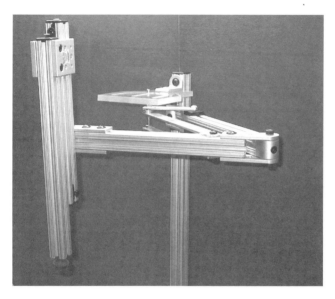

FIG. 11.18 Second Stackbot model

makes things very difficult in placing boxes in a tight, bricklike pattern on the pallet. So the Z-axis needs to be able to rotate with respect to the forearm, and it should keep an absolute net rotational relationship of $0°$ at all times.

The simplest solution, and the one chosen, was to learn from the old-fashioned drafting table attachment that would allow the user to move the "L" shaped drafting guide anywhere while staying parallel to the drafting board edge. This was done in this older implementation usually by cables, pulleys, and sometimes springs to keep the cables taut. A similar setup was done for the Stackbot by using steel belts and pulleys, but the steel belts showed signs of corrosion in a very short time, and actually snapped. Fiber reinforced timing belts and matched sets of pulleys were used and have stood up well (Fig. 11.19).

Then the *XY* position along the cam track can be traversed by applying a linear motion that drives the pin in the cam track of the plate. An electric stepper motor/linear drive unit from Superior Electric was chosen so as to be able to stop at more than one location along the cam track. This is useful when placing boxes along a row on the pallet. A pneumatic rotary actuator was initially chosen as an index device that operated the cam track-switching device. A pneumatic shot pin is energized and locks the track-switching device into the desired directional slot depending on how long the rotary actuator took to move. This was found to be good enough for our factory testing, but in a Beta site testing facility (a bakery) the oils generated from baking were vaporized in the air and coated the Stackbot

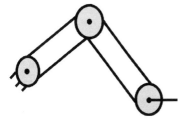

FIG. 11.19 Z-axis anti rotation timing belt and pulleys: top view

enough to increase friction and change the performance of the rotary actuator. So, a Geneva indexing mechanism was implemented instead. It was a higher cost solution, but its output was more determined and reliable.

The required vertical Z-axis motion for taking a box from its pickup point and stacking it at its place on a pallet can be achieved several ways. The first is to raise and lower the entire cam track plate and unpowered arm assembly from the pickup point and lifting it to the required stacking height. This motion requires a means to have a constrained vertical motion and a method to power it. An electrical motor would be one example, but this has a large moving mass and was rejected.

The second option is to keep the cam track plate and unpowered arm assembly at the maximum height required to stack the top layer. The gripper (suction cups were used) holding the box would have a vertical motion to lift it from the pickup point. An electric motor could be used, but was found to be too expensive. Initially a pair of telescoping pneumatic cylinders with 80 in. of travel was used to lower and lift the gripper, but this proved to be very costly. And due to the fact that the cross-sectional area of the cylinder's internal faces that react to the air pressure are different depending on the direction of travel, the results were not favorable either.

Finally a design for the Z-axis that used a standard pneumatic cylinder, and a pair of sliding members and a set of timing belts, created a device that amplified the air cylinder's motion by a factor of two. This gave the results used in production (Fig. 11.20).

The Stackbot can be used to service something beyond the dedicated USPS mail carts if the cam track plate is removable, so as to allow for different sets of palletizing locations. Exchanging the cam track plates is done manually. Track plates are milled using CNC machines. These plates would be held in place by quick release pins, so different boxes or trays could be palletized with a small changeover effort.

Control of the overall device was carried out with a Seimens PLC and a Superior Electric motion controller for the linear stepper motor unit. Incoming

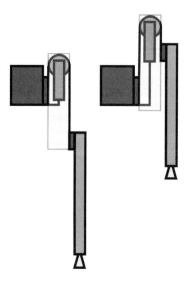

Fig. 11.20 Improved Stackbot Z-axis design

box or tray sensor signals would be processed to determine the proper X, Y, and Z height location on the pallet for placement.

In summary, there is less flexibility in the overall Stackbot concept than a standard robotic palletizer, but with the limited number of components and each set of components doing a single task, creates a device that costs less to produce.

What was almost surprising was that a U.S. patent was awarded (Derby, S., and Lyons, B. U.S. Patent no. 6,394,740, May 28, 2002). Not that the author and others had seen the design before, but the invention was a combination of traditional existing mechanisms. There was no enabling technology reason for where the Stackbot concept (Figs 11.21 and 11.22) could not have been invented 75 years ago.

11.2.5. Multiple Stackbots Solution

Another version of this concept would be to create a system of the Stackbots as modules, each servicing its own pair of pallets or USPS mail carts. This creates a parallel material handling system. Many existing palletizing equipment or robotics workcells attempt to cost justify themselves by servicing many pallets or containers at once, but the throughput speeds required to cost justify become unreasonable and unsafe. Stackbots are designed to move more slowly, and be inherently safer.

Bringing New Automation to Market

FIG. 11.21 Stackbot

A single or set of multiple conveyors would transport the box or containers to the appropriate Stackbot and pallet location. Each of the individual Stackbots would function in parallel to the other Stackbots (Figs. 11.23, 11.24, and 11.25). The parallel throughput can be much higher than a single piece of equipment or robot, and would work at a slower and safer speed. And being modular in design, additional Stackbots could be added or removed from a system, and configured to handle another task. Controller functions on each Stackbot module would be linked to a central processor, perhaps a PC. This concept can be called Distributed Palletizing (Derby, 2002).

Distributed Palletizing

Palletizing is often carried out throughout the factory by manual means. Many companies palletize at a central location, since their palletizing systems are

FIG. 11.22 Stackbot cam plate tracks: pin in motion

costly. Some use robotic workcells, and some used dedicated palletizers, which also can be costly.

Chapter 13 will cover many of the Packaging machines including robotic and traditional palletizers, but it is useful to see how the Stackbot, which was conceived and designed to stack letter trays, can be used to revolutionize an existing factory.

A generic sketch of a three-line factory is given in Fig. 11.26. This sketch represents a great number of actual factories visited by the author. A star marks the end of each production line. One could argue that this layout is by definition illogical, but as the author has seen during these site visits, not uncommon. It is

Bringing New Automation to Market

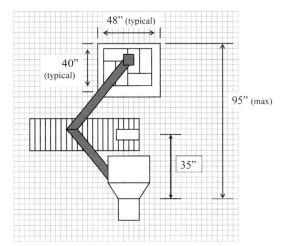

FIG. 11.23 Single Stackbot pallet layout

too costly to reorganize the factory floor and move everything. A production manager has several options for palletizing:

Option 1. One could have a person manually palletize at the end of each line. Depending on the rate of case output and the length of accumulating

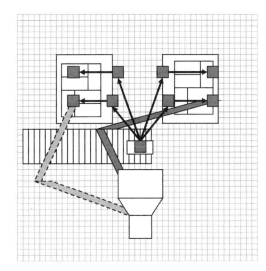

FIG. 11.24 Dual Stackbot pallet layout: with path of box shown

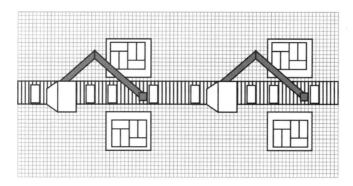

FIG. 11.25 Modular Stackbot layout: Stackbot bases not shown

conveyors, the manpower needed can range from one person per line to one person walking back and forth to cover all three lines. The latter approach is usually unlikely. When a pallet (located at each of the three stars) is filled, a forklift is needed to move it to the shipping area. A path wide enough for pallets is needed (Fig. 11.23).

A suboption here is locating several pallets in the end of the line area, so that the operator does not have to wait for a forklift to change pallets (Fig. 11.27). One is able to keep palletizing on a neighboring pallet. To wait would require an accumulator, and the operator would have to be able to work fast enough to catch up to the production rate.

Option 2. A second option is to install a conveyor system that takes the output of the three lines and either elevates them to a second floor material handling suite, or moves them to a neighboring room. The author has seen both

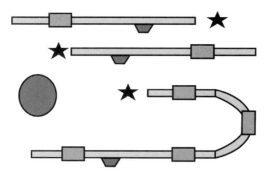

FIG. 11.26 Generic factory floor: stars mark ends of lines

Bringing New Automation to Market

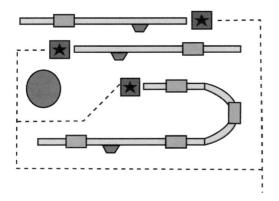

FIG. 11.27 Pallet locations and forklift paths (dashed lines)

approaches used widely. Figure 11.28 represents the conveyor system with the paths marked by dashed lines.

The material handling suite would contain some buffers for accumulation of product, a tracking system if the product becomes mixed as implied in Fig. 11.28, or a set of buffers if each line output is to be kept segregated. A high-speed palletizer would be used to create a pallet load of each product upon receipt of a pallet's worth. A standard robot system would not necessarily have the required throughput.

Option 3. A third option would be represented again by Fig. 11.27, where a palletizer, possible a Stackbot, is placed at the end of each line, around the marked star. This approach becomes practical when the price of the

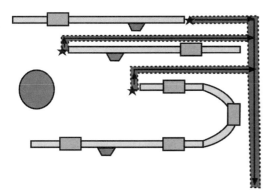

FIG. 11.28 Conveyor system (dashed lines) moves products to a material handling suite located on another floor or in another room

appropriate capacity palletizer is cost effective. This factor depends a great deal on the rate of product production, and the size of the product.

This product to be palletized could be a small baked good that is first placed in a carton, and then multiple cartons placed into a box. Or it could be a personal health or beauty item, which again is relatively small. In these situations, the resulting palletizing is performed at a reasonably slow rate. These rates may not justify a standard robotic system.

The same suboption as in manual palletizing is then available; that is, using a large accumulator (if needed) for a buffer, so as to switch out filled pallets for empty ones. Alternatively, does one use a pallet conveyor? Or a palletizer that itself can actively handle two pallets?

Use of the third option does define what "Distributed Palletizing" means. Several factories that had implemented a complete material handling suite (Option 2, Fig. 11.28) have removed the complex system due to operational difficulties. Although one plant was not willing to publicly give all of the reasons, it was stated that it "was not worth the operational costs and the headaches." This plant had reverted to Option 1, manual palletizing at the end of each line.

It is reasonable to assume that a centralized system such as Option 2 should be able to operate successfully and cost effectively, so one has to wonder what the other contributing factors were. However, Option 2 does create a complex system that requires a great deal of product tracking and accumulating.

A distributed palletizing system with units handling two pallets at a time allows for the suboption of not having a significant accumulator. This is important when there is little floor space available. Forklifts need to be scheduled only when a pallet is full. The servicing window is much larger and thus production can continue smoothly.

Table 11.1 summarizes this comparison, pointing out the benefits of a distributed palletizing approach in some factories. It should be noted that, in other factories, each line's output could justify a larger palletizing system.

11.2.6. Stackbot Production and Market Acceptance

The development efforts at MBI were on again and off again, depending on what other products they were trying to develop and bring to market. There were some performance clauses in the agreement with the author, but not strong enough to drive the Stackbot development as fast as the author really wanted. Obtaining stronger performance clauses was not simple during the initial agreement negotiations, and neither side envisioned things becoming as slow as they did.

An initial Beta site was set up in a bakery company in Eastern Massachusetts, and perhaps the relationship was started too early in the development

TABLE 11.1 Palletizing options requirements comparison

	Option 1: Manual	Option 2: Centralized	Option 3: Distributed
Palletizing labor quantity	High	Minimal	Minimal
Forklift labor quantity	High	Moderate	High
Forklift activity location	Distributed	Centralized	Distributed
Accumulator requirements	None if handles multiple pallets	Very large, depends on number of production lines	None if handles multiple pallets
Conveyor Requirements	Limited	Very large	Limited
Floor space requirements	Nominal	Need space for conveyors and accumulators	Nominal
Product tracking requirements	None	Needed to keep correct product on correct pallet	None
Integrated system control	Only signal to get filled pallet removed	Very demanding	Only signal to get filled pallet removed

process. The first Stackbot to go out the door was not tested for any significant stretch of time at MBI facilities. Marketing was pressuring to get a Beta site so as to have a working unit out there to tell potential customers. The Beta site was over two hours away, which limited engineers just driving over to check things out as they should, and the bakery company personnel were not reporting performance issues with the Stackbot on a timely basis.

The Stackbot prototype was also shown at two trade shows to garner customer feedback. Adjustments in minor functions and programming transpired. A sales representative firm was contracted with in conjunction with MBI's dedicated sales force. It was hoped that this rep firm would get into more prospective factories. Things were still progressing slowly. The economy was already slowing down, and very few potential sales were lost to the competition.

Then MBI made some internal corporate decisions to eliminate many of the new products it was bringing to market for another seeming brighter market to hopefully dominate. The Stackbot design was brought back into the author's company. Then a production agreement was established with Cambridge Valley Machining Inc. (CVMI) in Cambridge, NY. (A relationship with CVMI was ongoing from the Trackbot machine, the third case study, to be discussed later.)

Several automation machine features were determined to be a possible deterrent to future product sales, and a team of CVMI engineers addressed them. These included (problem and resolution):

- Overall arm shaking and perceived lack of stiffness — stiffeners welding onto tubular arm structural members.
- Z-axis motion being too violent — improved pneumatic valves and control to better time the air flow amount and direction.
- Troublesome track-switching device — replacing the pneumatic rotary actuator with a Geneva mechanism, as discussed earlier.

The time to perform additional debugging, testing the system to find the weakest links, and in general hardening the automation system to bring it to commercial readiness took much longer than the author and his team had envisioned. These types of commitments are easy to write in a business plan, but harder to actually do, and thus the costs go higher than planned.

But perhaps the biggest challenge to automation builders is the market. The Stackbot was taken to the Pack Expo Las Vegas trade show, held September 10–12, 2001. During the first day, September 10, 100 sales leads (which are a significant number!) were generated. Potential customers were quite excited, and some wanted to buy the unit off of the floor.

Then Tuesday was September 11, 2001, a day that no-one will forget. The 100 leads from the day before might as well have been burned. What was a marginally downward market for automation turned into more than 24 months of being stagnant. Coupled with the economic woes of the USPS, the market for the Stackbot palletizer became much less rosy than what had been envisioned a few years before!

11.3. CASE STUDY 3: POUCH SINGULATION

Singulation is the process of obtaining a single item from a random pile of similar product. It is often found as a bowl feeder presenting one screw after another for automated assembly. Or it can be one of the most challenging automation problems if it is a nonrigid item such as a pouch (Fig. 11.29). As a human assembly worker, it is not too much trouble to grab into a bin and select a pouch and place it in a box (unless the box is moving, as discussed later!), but to replicate this action with standard automation machines is far more difficult.

A machine vision system viewing into the bin or conveyor may see a jumbled mess. Making decisions of where one pouch starts and where the next ends is challenging due to image quality, lighting, reflections, angles of product alignment, and so on (Fig. 11.30). Researchers have been working on this problem for decades, but not cost effectively as far as industry is concerned.

Bringing New Automation to Market

FIG. 11.29 Pouches of chocolate cocoa powder and snack mix

However, to better understand this problem, it is best to back up a step or two and review parts feeding in general.

11.3.1. Standard Parts Feeding

The most commonly found parts feeder for automated assembly is the vibratory bowl feeder. Depending on the part's geometry, and the likelihood of parts being entangled with each other, some bowl feeders can work fantasticly while others are a thorn in system operator's sides, but a vibratory bowl is not at all flexible for

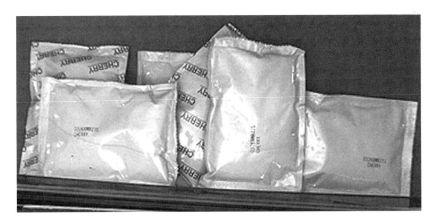

FIG. 11.30 Pouches as seen with vision camera

quick part changeover, and is often not satisfactory for the feeder of flexible parts like a pouch.

One could make a limited argument that a standard conveyor is a flexible parts feeder, although it does not determine a part's position and orientation without additional mechanical hardware and/or vision system assistance.

In order to compare the abilities of these standard feeders to more complex feeders, a table is given (Table 11.2). This table distinguishes between required degrees-of-freedom that use servo or stepper motor and control, from degrees-of-freedom that are simple air cylinders with an on/off binary control. Implied is a significant cost difference.

Adept Flex Feeder

One of the commercial flexible parts feeders available uses a combination of conveyors and actuators with a fully functioning SCARA robot (Carlisle, 1997). This design (Fig. 11.31) recirculates parts under a vision camera, so as to use known part orientations when picking up by the robot.

The robot may simply move the picked up part to a registered position on a conveyor or parts pallet, or it may perform the actual assembly process. If it does the former, it becomes a great amount of automation to perform the moderate task. Table 11.2 documents the system for comparison.

Robot and Bowl Feeder

This system uses a vibratory bowl feeder to move parts to a set of pickup points (Fig. 11.32). This mix of older technology and robotics can be used for less total complex automation costs, but is not as flexible. Bowl features would need to be changed to handle other parts (Kohno et al., 1985).

Programmable Reconfigurable Parts Feeder

A different yet functionally similar variation of the Adept Flex Feeder is the Programmable Reconfigurable Parts Feeder as shown in Fig. 11.33 (Gordon, 1994). It uses controlled conveyor belts, a vibratory table and a vision system to singulate parts from each other, and distributes them in random orientation. The robot arm then picks up parts with known orientation. Again, the robot can perform a highly demanding assembly task, or it may simply place the part into a holding template. It is not commercially available as far as the author can determine.

11.3.2. Feeding Pouches

Today many products are packaged in pouches. These pouches are often just one component of a finished product. The cheese sauce for macaroni and cheese, screws in a "some assembly required" kit, and seasonings for instant soup are

TABLE 11.2 Feeder Types and Degrees of Freedom

Feeder type	Servo DOF	On/Off DOF	Constrained DOF	Degree flexible	Vision	Useful
Bowl Feeder	0	1 Bowl	3 Position 3 Orientation	Very low	None	Yes
Conveyor	0	1 Conveyor	2 Position? 2 Orientation?	High	None	No
Adept Flex Feeder (Carlisle, 1997)	4–6 Robot 2 Conveyors	1 Bucket	3 Position 3 Orientation	Very high	Moderate	Yes
Robot/Bowl Feeder (Kohno et al., 1985)	4–6 Robot	1 Bowl	3 Position 3 Orientation	Moderate	Moderate	Yes
Programmable Reconfigurable Parts Feeder (Gordon, 1994)	4–6 Robot 2 Conveyors	1 Conveyor 1 Vib Table	3 Position 3 Orientation	Very high	Moderate	Yes
Flexible Pouch Feeder (Paster, 1998)	2 Rotational	6 Pistons 1 Conveyor 1 Rotate Bin	3 Position 3 Orientation	Moderate	Limited	Yes

DOF, degree-of-freedom.

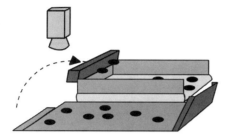

FIG. 11.31 Adept flexible feeder

all examples of pouches packaged together with other components. Often, this pouch is not produced at the location of final packaging. This means that the pouches are shipped, usually in bulk containers, to the packaging site.

Most manufacturing plants today rely on manual labor to place these pouches into their parent container. This is because there is no system for automatically separating and handling them. Classic methods for handling bulk parts, such as vibratory bowl feeders, gravity feeders, and escapement mechanisms, cannot be easily applied to pouches. They are not rigid and deform easily. Methods that work for "hard" parts cannot accept pouches with bent corners or edges, and often cause these defects themselves. Attempts to use bowl feeder technology by industry to singulate pouches have had poor results for these reasons.

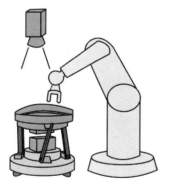

FIG. 11.32 Robot and bowl feeder

Bringing New Automation to Market

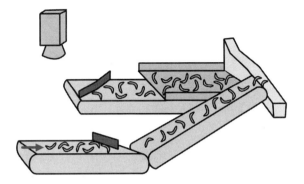

FIG. 11.33 Programmable reconfigurable parts feeder

Feeder Design Strategy: Robot or No Robot

The general strategy of existing commercial feeders, of which almost all would handle flexible pouches satisfactorily, is to manipulate the product, look at it with a machine vision system, and pick it up with a robot. The robot could be either a SCARA or a 6 DOF design. Given, the use of the robot here does allow for more complex operations than a simple loading found in packaging operation, but the requirement to always have a robot may be costly.

These current designs are based on the need to almost always select a product with every robotic motion cycle. If this requirement is removed, then another strategy can be developed. If one examines the operations of the commonly found vibratory bowl feeder, it is based on the "shotgun" effect of letting a larger number of products be cycled by some sorting process, assuming that the output of the bowl feeder will never be starving the next manufacturing operation. This needs to be done while not deforming the pouches.

11.3.3. Vacuum Pickup of Pouches

The use of a vacuum to pick up pouches from a large bin was tested. A vacuum proved an effective means of raising a pouch from the storage bin. The success of this method prompted extensive testing (Norman, 1996). Two suction cups were mounted on a robot to run a series of repeatable tests. The suction cups were lowered into a large bin filled with pouches. Mounting the suction cup with a spring and controlling the distance the cup was lowered controlled the downward force.

The spring was necessary to dampen the impact on the pouches and account for the varying depth of the pouches. As the bin was emptied, the level of pouches would drop in the bin. This downward force was important because too low a force would not ensure a good vacuum seal. Too high a force could dent or puncture the pouch.

A series of tests were conducted using two types of pouches, two levels of vacuum pressure, and three types of suction cups. The effect of vibrating the bin to help settle the pouches and prevent dead spots from forming was also examined.

It was difficult to distinguish an optimum suction cup. Larger cups had a higher rate for single and multiple picks, and were more effective at picking heavier pouches due to the increased area. Flexible suction cups were more suited for some pouches than other pouch types. A higher vacuum pressure raised pick rates, but also improved the chances of a multiple pick. A lower pressure was shown to reduce multiple picks, especially for heavy pouches, where the reduced pressure was not capable of supporting multiple pouches. The choice of suction cup was demonstrated to be relatively pouch type dependent. An overall success rate of 50–85% was achieved (Norman, 1996).

The only serious problem in singulating the packets using a suction cup was multiple pickup. When two packets were overlapping and the suction cup hit the edge of the upper packet it was possible to form a good seal across the packets and pick them both up. In some tests, 3–6% of pickup attempts yielded multiple packets (Norman, 1996). This occurrence had to be addressed by the flexible pouch feeder.

11.3.4. Pouch Feeder from a Modified Bowl

The author and his Master's student, Aren Paster, designed an initial prototype pouch feeder that looked like a modified vibratory bowl feeder (Paster, 1998), but it has many additional automation components and functions. It was the first generation automation from the testing performed at Rensselaer Polytechnic Institute.

Flexible Pouch Feeder Design Considerations

The goal of the flexible pouch feeder (Fig 11.34 and 11.35) is to dump a box of packets in, and get a stream of orientated, singulated packets out. Suction cups proved an effective way to manipulate them. The remaining hurdles are to deliver the pouches to suction cup, or vice versa, to orient the pouches properly, and to get them out of the machine. Breaking these requirements down into discrete, individual steps yields the following short list of actions needed to singulate and orient pouches from bulk:

- Spread packets out in large bin;
- Rotate suction cup assembly;
- Lower suction cup assembly;
- Detect 0 vs. 1 vs. 2 packets;
- Detect orientation of packets;

Bringing New Automation to Market

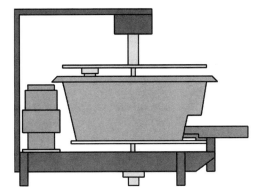

FIG. 11.34   Flexible pouch feeder overview

- Orient packets correctly;
- Feed packets to exit conveyor;
- Operate completely automatically.

Spread of Packets in the Bin

The packets will only be picked up at five discrete repeatable locations around the bin. If nothing is done, all the packets at these five locations will be removed and although many packets remain in the bin, "dead spots" of no packets will develop below the pick stations.

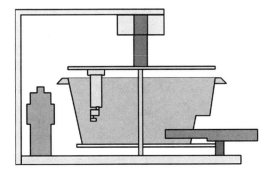

FIG. 11.35   Flexible pouch feeder cutaway

The floor on the bottom of the bin could be rotated. This would not redistribute the packets greatly, but would move the dead spots away from the pick stations and ensure a more even removal of packets. Implementation of this concept is simple, as the motion need not be tightly controlled. By rotating the floor in the opposite direction of the suction cup assembly, the dead spots will not "follow" under the suction cups. Packets that are added to the system, whether it is from a hopper or those discarded by the vision system, will not accumulate at a constant location in the bin.

Rotate Suction Cup Assembly

The suction cups will be suspended above the bin floor on a rotating disc. This disc will index the cups to six stations equally spaced about the bin. The first five stations will give the suction cup a chance to pick up a packet. At the last station the machine will determine if there is only one packet, if it is oriented correctly, and then send "good" packets out of the system.

A servo motor used to produce the rotary motion would provide the highest level of control for the machine. This system requires a computer, but provides excellent control of the motion. Acceleration and deceleration of the assembly could be precisely controlled, and feedback used to guarantee positional accuracy.

Raise and Lower Suction Cups

The suction cups must be lowered down into the bin to the level of the packets. This will allow the cups to form a seal with the packets, and "grab" them with vacuum pressure. Before the assembly indexes, the suction cup will be raised to reduce the chances of knocking the packet off.

An air cylinder could be used to raise and lower the suction cup. This would require a source of compressed air, either shop air, or a compressor on the machine. Varying the air pressure to the cylinder can control the extension and retraction speed. The force with which the suction cup hits the packets could also be controlled. This force is directly related to the air pressure and thus the speed. Owing to this relationship, to hit the packets lightly, the cylinder must move slowly. This force is important since a high downward force could damage a weak packet. The cylinder could be spring actuated in one direction or air operated in both directions.

If the suction cup successfully picks up a packet at the first pick station, there is no need to try again at the remaining four stations. Lowering the suction cup into the bin may knock the packet off. If the suction cup does not grab a packet at the first station, it should descend at the next station and keep trying until it does pick up one. This logical action needs to be controlled, so the machine must sense if the suction cup is holding a packet.

When the suction cup is not covered and vacuum pressure is applied to it, the pressure in the line is slightly below atmospheric pressure. As soon as a packet covers the cup, a seal is formed between the cup and the packet. This lets the vacuum pump suck all the air out of the line, lowering the line pressure significantly. A vacuum sensor could detect this pressure drop, which indicates the presence of a packet. Unfortunately, if valves are used with the suction cups, this may prevent the concept from working. With six suction cups and only one vacuum pump, if one pump was covered, the vacuum force would be transmitted to the five uncovered cups instead of being split evenly. This would provide insufficient force to pick up a packet if too few of the cups were covered. To prevent this from happening, special valves are used just upstream of the suction cups. When the cup is uncovered, a small amount of air is sucked into the cup. As soon as the cup is covered, the valve opens. This exposes the suction cup to full vacuum pressure and it will hold onto whatever began to cover it. This helps to maintain full vacuum pressure in the line and improves the chances of picking up a packet.

Detect Number of Packets

If two packets overlap on the bottom of the bin, and the suction cup hits the packets right on the edge of the upper packet, it is possible to pick up two packets at once. Heavy packets present too much weight for the suction cup, but a lighter packet poses the problem of multiple pickups. This is clearly not desirable, as the goal of the machine is to singulate pouches. This situation must be detected so that multiple packets can be rejected.

When multiple packets are picked up, they rarely overlap more than 30% (Paster, 1998). This means that the area of multiple packets is considerably bigger than a single packet. A vision system could determine the area of the packet, compare this value to a preset size, and easily determine if the suction cup had picked up a single packet or multiple packets. This idea is even more attractive because a vision system would be able to determine the orientation of the packet in addition to the size.

A computer vision system could be used to look at the suction cup and determine the size and shape of the packet(s). This is an expensive option for what needs to be done. A simpler vision system could be built of inexpensive components. An infrared (IR) light source was suspended above a grid of IR detectors. When a packet passes between the grid and the detectors, the detectors are shielded from the IR light. The detectors, which no longer detect light, are in the shadow of the packet. This shadow represents the size and orientation of the suction cups payload.

Detect and Correct Orientation

Once a single packet has been removed from the bin, it is desirable to have it in a known orientation before exiting the machine. A rectangular packet destined to be placed inside a box may not fit if it is rotated 90°. To provide the packet in a known orientation, the flexible pouch feeder must determine that the current orientation is correctly using sensors, or force the packet into the correctly orientation using "blind" physical manipulation.

Both the computer vision and the simpler vision systems mentioned earlier can sense the orientation of the packet. The straight edges and flat nature of the packets make this a simple task. This information can then be used to rotate the packet to the desired orientation. A second servo motor would be engaged to the suction cup arm at this point, enabling the controlled rotation to be achieved.

Feed Packet to Exit Conveyor

After a single packet in the correct orientation is produced, it must leave the system. It will be placed on a conveyor and carried out of the flexible pouch feeder. How this will be done will depend largely on how the packet is held or located when orientation is complete.

Machine Control

The machine must have some program or code to control the many different actions, inputs, and outputs required to function automatically. Solenoids or air cylinders must be actuated, motors controlled, and vacuum pressure turned on and off. Logical decisions need to be made based on input from sensors and time constraints. A PC was chosen.

The computer control program "sees" what the packet looks like by examining which detectors are in the shadow of the packet. A simple count of how many detectors are covered allows the program to determine the size of the packet. If too many detectors are covered, the suction cup picked up two packets. If only one packet is present, the program checks its orientation. The program compares the length and width of the shadow to the preset packet dimensions. If the packet is crooked or bent, these dimensions will not be correct. The packet is then slowly rotated while the computer rescans the detectors over and over again. When the dimensions of the shadow match those defined in the control code, the packet has been orientated and it stops rotating.

The rotary assembly now indexes the suction cup to a position above the exit conveyor. If the suction cup was holding only one packet and it was able to orient it correctly, it is dropped onto the conveyor. If this did not happen, the suction cup will continue to hold on and drop the packet(s) into the bin immediately before descending at the first station. This allows the machine to give up on a packet if it cannot guarantee that it was aligned and unbent. The packet is returned to the bin so that it may be successfully picked up and orientated later. After some time, a collection of bent packets might need to be removed by an operator.

Pickup Efficiency

Two different pouches were used during the experimentation. The first measures 2.5 in. × 4 in. and weighs approximately 0.4 oz. The surface is smooth and consistent and is not prone to wrinkling or puncturing. The pouch is small, lightweight, and solid. The center of mass of the pouch is nearly constant. However, this pouch is very thin, and its small size and flat shape could pose problems during singulation.

The second pouch is produced for a soup product. It measures 4 in. × 5.4 in. and weighs approximately 1.6 oz. It is substantially heavier and larger than the first pouch. The contents are also rather stable, but the center of the pouch is considerable higher than the edge. This bowing around the perimeter provides a difficult spot for the suction cup to establish a good seal. The second pouches have a flexible, wrinkled surface compared to the smoother first pouches.

The suction cup was lowered into the bin with packets beneath it. Tests were run to determine the ability of the suction cup to pick up single pouches (good), multiple pouches (bad), and deformed pouches (good). A number of pouches were deformed to determine the limits of "acceptable" pouch geometry.

The results of these tests were very promising. The concept of handling flexible pouches using vacuum-powered suction cups as demonstrated in previous research (Norman, 1996) has been further proven. These results indicate that the valves used on the suction cups have a significant impact on the success rate.

Usefulness of the Machine

Although the picking of pouches proved promising with the suction cups, the physical layout of the machine became too difficult to pack all of the needed components into the bowl area. The system architecture was just not right. The effort at RPI was placed on the shelf until a better idea would come along.

11.3.5. The Trackbot Invention

Some days are those you remember forever. The same day in the late 1990s when the author invented the Dockbot and the Stackbot, he also invented the Trackbot! There were several other inventions he invented that day that he has yet to develop. It was a very productive day.

The Trackbot design used some of the benefits of the previously described invention, but was more flexible. The earlier concept was always trying to pick pouches at predefined intervals (60° in this case), but the pouches were randomly located. After a pouch was picked, it might be placed into a package or box, something that may require the suction cup device to rotate the pouch to get the desired orientation. The desire to pick up at positions defined by sensors or vision systems required a system architecture that was flexible.

Standard Robot Motion

The world of industrial robot arms has been living with the standard robot motion cycle. Defined by some as the "Adept Cycle," a robot and gripper will move a part as shown in Fig. 11.36. The up, over, and place motions, referred to as the "Work Motion," produce a useful operation. The other up, over, and down motion, referred to as the "Return Motion," is wasted time and energy. It is required as the robot and gripper must return to complete the motion cycle. And except in limited cases where two SCARA arms can coordinate their motion, the "Return Motion" also occupies three-dimensional space such that no other robot can move the next part during that time.

This can lead to a very inefficient use of time in many production situations. Or it can lead to the designing of very fast robots (using accelerations and decelerations of up to 10 g) that often move or place the carried part so roughly that the part flies out from the gripper, misses its target, or gets damages by the high dynamic (multiple g) loading.

The design and performance of a novel robot system called the Trackbot will now be discussed. It was invented by the author, and is patented (Derby, S., and Smith, J. U.S. Patent 6688451, Feb 10, 2004). It has many unique design characteristics, and these characteristics yield a set of motion strategies unlike most single robots today (Tsai, 1999; Groover et al., 1986; Craig, 1986).

The general application area is to be able to pick up a product part from a collection of parts along a conveyor. There are many other variations to this description, and some of them will be addressed later.

Trackbot Design

When one hears the word "track" associated with a robot, it is usually a conventional robot placed on a linear track to extend the effective reach. Some tracks are

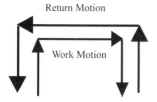

FIG. 11.36 Standard robot trajectories for most current robot configurations

based on the floor, some overhead (Kato, 1988, 1992; Jerue and Walsh, 2000; Vogt et al., 2001). The system described here is nothing like these standard configurations.

The Trackbot robot system (Derby et al., 2002) has a rigid track to carry the multiple individual robot heads, or Bots. A constantly moving drive chain is used as the motion source so as to not have a drive motor on each Bot, thus saving weight. Figure 11.37 shows the overall concept.

The product parts are represented as simple discs (Fig. 11.38), but in practice can be any size and shape part, potentially a mixed group. A mixed group might be handled best by having dedicated Bots for each type of part, that is, one Bot has a suction cup while another has a gripper to pick up long thin parts. The overhead vision camera represents the possibly use of multiple cameras in order to have a field of vision large enough to recognize the location of the parts on the conveyor.

The moving drive chain runs at a constant speed, while each Bot has the mechanisms to "grab" or "release" from the chain, conceptually similar to a ski lift.

Although perhaps obvious, this design does not permit one Bot to pass the one ahead of it, and direct contact of Bots would constitute a collision. The overall system computer needs to calculate where each Bot will be over time, so as to avoid these collisions. So the actual embodiment of this concept has many concerns.

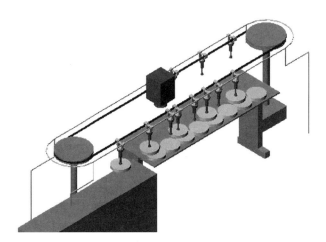

FIG. 11.37 Overall Trackbot system layout. Eleven Bots are on the oval track. Product parts being moved are shown as discs. Many discs are located on the input conveyor. The output conveyor is presently empty. A video camera is located at the midpoint between the two chain drive wheels.

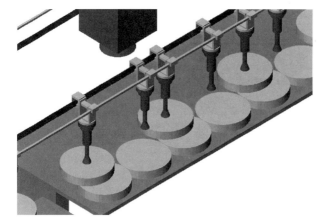

FIG. 11.38 Conceptualization of the Bots being held on the lowermost track and being propelled by the neighboring (thicker) drive chain. The disc products would most likely be more jumbled in a true application.

11.3.6. Initial Test Model

Similar to the early development of the Stackbot, an initial test model of the Trackbot was developed (Fig. 11.39). It was a benchtop unit that had a single straight section of track, and a timing belt on pulleys, much smaller in diameter than the proportions shown in Fig. 11.37. The simple Bot has an onboard Basic Stamp microprocessor, and it was programmed to wait, grab the chain for a few seconds, then release. It literally pinched the timing belt using rubber contacts to obtain motion. The Bot's motion was limited to 2 ft of travel, and had to be manually reset to the starting area after each run. What it demonstrated was that pinching a chain or belt would be unsatisfactory for any length of service. The pinch points showed excessive wear with limited use.

Prototype Production Trackbot System

Shown in Fig. 11.40 is the prototype Trackbot robotic material handing system. Shown are four Bots on one of the straight sections of an oval track. The overall system is 13 ft long, 5 ft wide, and 6 ft high.

Each Bot has a telescoping Z-axis of motion with a vacuum cup at the bottom. Each Bot has its own internal Pentium-based single board computer. An IR communication device is located near the top of each Bot, allowing wireless communication. Upon the passing of the IR station, the programmed instructions for each Bot are updated. After receiving such an update, each Bot operates autonomously.

The rail and chain system is shown in Fig. 11.41. It consists of two metallic conductive rails and two chains. The conductive rails allow for 12 V power to be

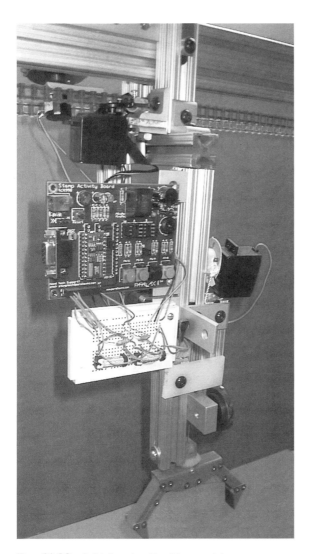

FIG. 11.39 Initial testing Trackbot model

supplied to each Bot through their conductive wheels. The upper chain remains stationary, while the lower chain moves at a constant speed. Each Bot has two cogs that are always engaged into the chains, and are actuated by a clutch pack. The lower cog is clutched to restrict its rotation to enable movement along with the chain. The upper cog is rotating while engaging the upper chain, giving positional information. Releasing the lower cog's clutch and

FIG. 11.40 Actual Trackbot material handling system. This application is using suction cups to singulate filled pouches. Multiple pouches are randomly located on the input conveyor, while two singulated pouches have been placed on the output conveyor at the left.

FIG. 11.41 Original Trackbot rail and chain drive system. The upper and lower rails constrain the Bots. The lower chain moves and is used to propel the Bots. The upper stationary chain is used for braking and linear path position feedback.

Bringing New Automation to Market

clutching the upper cog achieves stopping. This applies a braking action in a repeatable way. Engaging the lower cog produces the acceleration and eventual matching of the drive chain. The overall robotic repeatability is designed to be ± 0.25 in. since the prescribed tasks do not require any significant precision. Any backlash due to the use of a cog and chain is insignificant.

11.3.7. Design Considerations

The reasoning behind this design was the following:

1. It was assumed that vacuum cups on the moving Bots would still pick up pouches. This meant that one either had to:
   (a) supply compressed air, for a vacuum generator;
   (b) supply vacuum directly; or
   (c) generate a vacuum on the Bot by a locally placed electric powered motor.

   Choices (a) and (b) require hoses that could connect the Bots to a central hub, while choice (c) added more weight and space claims. No off-the-shelf small lightweight vacuum generator could be found, so that eliminated choice (c). Having extensive tubing handle a vacuum (choice (b)) is not usually recommended, since the hosing can get sucked flat if not properly designed and installed, and it takes longer to achieve a vacuum, since the entire volume of the hose needs to get to the 8–10 psi pressure drop. Choice (a) was the best of the group. It still required a custom rotary hub to be designed.

2. The Bots needed to have electrical power supplied to them. The most effective amount would be 12 V DC. The three choices were:
   (a) have an electrical cord tethered to each Bot;
   (b) use conductive rails similar to many commuter trains;
   (c) generate electrical power from the motion of the Bots.

   Choice (a) requires an electrical slip ring, which has a history of contact wear, electrically noisy connections, and maintenance woes. So this choice was left as a backup. Choice (c) would be only intermittently generating power while the chain drive was propelling the Bots. It would be difficult to implement, but it could work. Again weight and space claims ruled it out. So choice (b) required that the rails be electrically isolated from each other, and supplied with a 1000 W 12 V power supply for the 12 Bots. The 1000 W amount was a worst case loading determined during the Bot design. There were still many questions at that time, and it was later found to be excessive.

3. The Bots then needed to pick up electricity from the rails. The original design (Fig. 11.41) included round cross sectional rails. This would

allow for the wheels (Fig. 11.42) to stay in contact whether the Bot was traveling on a straight piece of track or a curved portion. The Bots had a suspension system to hopefully assist with the dynamics. Contact on both sections of track is not a trivial matter, since the wheel carriages must float while keeping the cogs of the clutch pack engaged with the drive and breaking chain. The initial idea chosen to implement were wheels made of electrically conductive silicon rubber. Although they worked surprising well electrically, the Bots basically "blew out the tires" and shredded rubber lay under the track after only a few laps. Then the wheels were made from copper, which did not exhibit the best wear characteristics. Copper filings filled the track too often.

4. The Trackbot prototype was designed to be both an evaluation and testing machine, and a machine to bring to a trade show. A series of internal conveyors was integrated into the machine, solely to recirculate the pouches after picking. No one wanted to have to baby sit the machine for five days at the trade show by moving pouches from one end to the other every three minutes.

5. The vacuum cups would work well for lifting chocolate cocoa powder pouches, but after many reuses in testing and at trade shows, the pouches would break down. Chocolate powder was found on all of the recirculating conveyors, and did get sucked into the vacuum generators. It was desired to grab heavier pouches with a slippery metallic cover, so a switch was made to a more powerful and better filtered vacuum generator from Piab (Hingham, MA, USA).

6. The chain drive system required a pair of chains, one for propulsion, and one for braking. But both chains required the same pair of 3 ft diameter cogs that form the loop. Concern was raised that the propul-

FIG. 11.42 Original wheel profile

sion chain would receive more wear, and would be likely to stretch. So since the entire design of the clutch pack was reversible in its operations, the Trackbot was designed to be able to swap the roles of the chains. After so many days of operation, a set of toggles would be switched to change which chain and pair of cogs would be the driver and which for the braking. The only real additional component costs were that the Bots would need to have rotary encoders on both halves of the clutch pack, and that the Bot's software would have to intelligently determine which chain was the motion source.

7. The initial concept demonstrated at the Pack Expo trade show in Chicago, November 2000, had a set of line sensors to determine if pouches were present. As seen in Fig. 11.43, this is similar to a multiple implementation of how most grocery market conveyor belts operate. This was found to be insufficient, since for pouches of any real size or weight, the center of the pouch needed to be obtained, and the sensors only indicated a pouch's presence.

8. The Bot's Z-axis rotation was initially seen to be not important to implement. After the successful picking of a pouch, it could be placed on an output conveyor and rotated thereafter, but too many potential applications suggested that this operation was not the desired one. So the Z-axis rotation was added to the production design.

Redesign Directions

The round rails, wheels, and Bot suspension proved to be unsatisfactory. The wheels did not stay in contact 100% of the time. The transition from straight

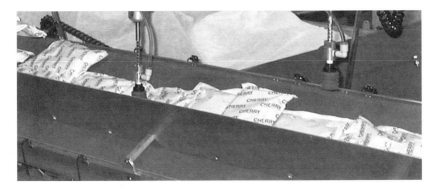

FIG. 11.43 Multiple sensor operation on the conveyor: note holes in metal plate

motion to curved motion at 4 ft per second causes many an electrical supply gap. Even by added large capacitors to the Bots, the power would drop enough that the onboard CPU would drop and reboot, a definite problem!

So the rails were made to be of rectangular cross-section, and the wheels replaced by commercial cam followers (Fig. 11.44). These followers also conducted electricity.

So unlike a servo motor drive unit found in most robotic actuators, this dual clutching drive system (Figs. 11.45 and 11.46) gives a repeatable but non-variable response. It does also provide a much more cost-effective solution. However, because of this dual chain drive method, some unique opportunities arise for motion analysis.

11.3.8. Trackbot Motion Strategies

To better understand some of the desired motion strategies, a brief look at some of the general uses is needed. In Figs. 11.47–11.49, the square blocks represent the Bot locations. The solid rectangle represents the infeed conveyor and Bot pickup locations. The striped rectangle represents the output conveyor. Figure 11.47 shows the use of one side of the Trackbot, while Fig. 11.48 shows the using of both sides. Figure 11.49 shows six infeed conveyors, each containing a different product to be combined at the striped output conveyor. This process is often referred to as "kitting."

FIG. 11.44 Revised Bot wheel profile

Bringing New Automation to Market

FIG. 11.45 Revised track rail system

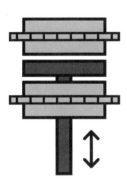

FIG. 11.46 Chain, cog, and clutch pack design

Motion Assumptions

The motion of a single Bot will now be discussed. For most applications of the Trackbot the product pickup point requires that the Bot come to rest. A generic velocity vs. time plot is shown in Fig. 11.50. Although the configuration of the two

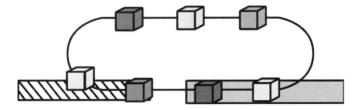

FIG. 11.47 Conceptual representation of seven Bots, rail, and chain drive, input and output conveyors. Bots are moving clockwise

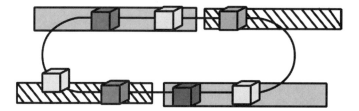

FIG. 11.48 The addition of a second input and output conveyor

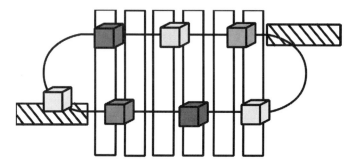

FIG. 11.49 The addition of a second output conveyor and the use of six input conveyors used to create product kits

chains is one constantly moving and the other fixed, the engaging of the cogs' clutch packs (Fig. 11.46) produces a noninstantaneous effect. This is fortunate since the desired Bot motion is smooth rather than screeching to a halt or to full speed.

It should be noted that for most of the applications in material handling concerned in this paper, the product can be placed while the Bot remains in motion. So for one productive looping of the track, the Bot would come to a stop only once. When a second Bot has been added to the same track, the timing of this one stopping location becomes the challenge.

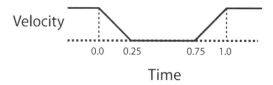

FIG. 11.50 Velocity profile along track vs. time. Zero velocity segment corresponds to Z-axis travel time to grab product, 0.5 sec. Maximum velocity is 4 ft/sec.

Bringing New Automation to Market

FIG. 11.51 Path of two Bots on the same track. Second Bot does not enter deceleration, activity, and acceleration zone until first Bot is back up to speed

It is this strategy of two Bots where the improvement over standard robots is seen. Where the time and space of a standard robotic motion as shown in Fig. 11.36 has the wasted return motion, during that time a second Bot can be performing its task and increasing the overall system throughput.

Part of the overall strategy is how conservative a Trackbot system programmer wants to be. If a Bot is going to stop at the location represented by the star in Fig. 11.51, and the entire zone initiated by the signaling of the lead Bot to begin braking, and concluded by the Bot again achieving the overall chain speed, is to be kept as a "No-Bot" zone, then the rule base is graphically given by Fig. 11.51. If, however, the system programmer is less conservative, the next Bot can get closer, more representing bumper-to-bumper traffic on a congested highway as seen in Fig. 11.52.

After some review of these constraints, a preferred strategy to handle a multiple number of Bots is to operate them in a wave. The four Bots in Fig. 11.53 are programmed as a wave. The absolute distance between each consecutive pair of Bots may be adjusted within some higher level of programming. Bot-to-Bot

FIG. 11.52 Path of two Bots on the same track. Second Bot does enter deceleration, activity, and acceleration zone before first Bot is back up to speed. No factor of safety

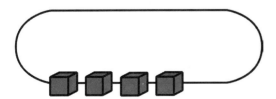

FIG. 11.53 Wave of four Bots on the same track. These Bots are coordinated to keep together to maximize throughput

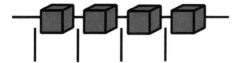

FIG. 11.54 Zones of one wave of four Bots. Each Bot is programmed to go a targeted position within its zone

collisions must be avoided by the overall system computer, but the average over time is a constant.

It will be shown that the maximum number of Bots and waves will be dependent on track length, chain speed, and stop time.

The actual Trackbot Bots are 6 in. wide. A nominal Bot-to-Bot centerline distance is kept at 12 in. Small variations of ± 2 in. within each Bot's zone (Fig. 11.54) can be programmed to pick up randomly located parts on the input conveyor.

11.3.9. Motion Analysis

But how does this compare to a traditional robot and when is the Trackbot a better option? To address these questions a comparison of a robot to a track full of Bots will now be presented.

Single Bot

For comparison purposes, the discussion will continue to base its assumptions around the existing prototype Trackbot. One of the current mechanical design limitations is a minimal radius of curvature of 10 in. (the actual prototype has an 18 in. radius).

Another limitation for comparison in this paper is the assumption that the Bots do stop to pick up a part, but deposit it at the current maximum speed of 4 ft/sec. Given that:

- 4 ft/sec speed chain;
- No-Bot zone of 1.5 ft;
- No-Bot segment takes 1 sec.

A circular track of 5.5 ft circumference makes sense. The 5.5 ft includes the 1.5 ft of the No-Bot zone. During the first second the Bot travels 4 ft, the following second the pickup sequence. The part is to be placed on the opposite side of the circle from the pickup point. The results are shown in Table 11.3.

Bringing New Automation to Market

TABLE 11.3 Throughput Comparison

Bots vs. robot	Track length/Robot path (ft)	Throughput rate (/min)
Trad Robot	3.2	40
1 Bot	5.5	30
2 Bots	5.5	60
Trad Robot	3.8	35
2 Bots	10.5	37
3 Bots	10.5	55
2 waves of 2 Bots	10.5	74
Trad Robot	3.8	35
2 waves of 3 Bots	10.5	110
Trad Robot	3.8	35
2 waves of 4 Bots	11.5	137

Standard Robot

A standard robot must be programmed to make a somewhat similar motion in order to make a fair comparison (Fig. 11.55). This motion is the same whether a XYZ Cartesian, SCARA, or pedestal style robot. The robot travels down to the infeed conveyor, grabs a part, and moves to a location represented by the average dropoff location of the Trackbot. For comparing to the single bot above, the robot will have covered 18 in. replicating the No-Bot zone, leaving 22 in. to be covered at 4 ft/sec. The robot also must decelerate and reverse direction, approximately 0.5 sec. The 22in. will increase to 30 in. when comparing for waves of multiple Bots. Given that:

- 4 ft/sec max velocity;
- up/down motion 1 sec as per No-Bot zone;
- 22 in. remaining horizontal motion (0.5 sec).

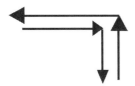

FIG. 11.55 Traditional robot's similar motion

Two Bots

If one places two Bots on the same circular track, one Bot is in the No-Bot zone while the second Bot covers the four linear feet. The total distance has been chosen such that entering the No-Bot zone is not relevant. Given:

- 4 ft/sec speed chain;
- No-Bot zone of 1.5 ft;
- No-Bot segment takes 1 sec.

Two Bots on Oval Track

The next logical embodiment of track would be an oval. Two circular cogs on each end tension the chain. A standard 2 ft diameter wheel size is used. And since the two wheels cannot interfere with each other, the straight section of chain must be 2.2 ft long. This produces a total circumference of 10.5 ft.

Given:

- 4 ft/sec speed chain;
- No-Bot zone of 1.5 ft
- No-Bot segment takes 1 sec.

This produces a less than optimal arrangement. The extra distance takes more time and thus overall efficiency is reduced (Table 11.3). With only two Bots, the "next" one is nominally never close enough to the first one to gain any cycle time.

Three Bots on Oval Track

By adding a third Bot to the exact same track and conditions, almost the same level of performance of two Bots on a circular track can be duplicated. At first inspection, the additional size and Bot seems a waste, but the distance covered by the part has been increased. Table 11.3 compares a traditional robot's minimum distance. But the dropoff point can be much further away from the pickup point with the Trackbot with no additional cycle time penalty. A standard robot trajectory cannot be extended without cycle time penalty.

Two Waves of Two Bots

If the same oval track is used, a pair of two Bot waves can be placed effectively. The No-Bot zone becomes 1 ft longer because of the Bot pair.

Given:

- 4 ft/sec speed chain;
- No-Bot zone of 2.5 ft;
- No-Bot segment takes 1.25 sec.

Bringing New Automation to Market

Two Waves of Three Bots

For two waves of three Bots, the same track length can be used. The No-Bot zone has increased to 3.5 ft, and the duration is now 1.5 sec, but the remaining 7 ft of track will be covered in 1.75 sec, so the system is compatible.

Two Waves of Four Bots

Two pairs of four Bot waves can now be investigated. Since the No-Bot zone is now 4.5 ft taking 1.75 sec, the remaining track length is only 6 ft, and at 4 ft/sec and 1.5 sec, the 10.5 ft track length is not compatible. The track would need to be increased to 11.5 ft overall.

The use of waves as the most efficient method becomes clear when one looks at the practical size of the No-Bot zone and the density possible if the wave of Bots loosely act as a unit. This is similar to congested highway driving, where if drivers are alert enough to stay on the tail of the car ahead of them during stop and go driving, the maximum throughput can be achieved.

The closed-loop track-based multihead robot discussed in this paper can have a significant impact on overall cycle time. For short pickup/dropoff distances, the system can be more efficient than a traditional robot. For longer distances such as 11.5 ft work and return total path length, the system can reach a factor of four times the throughput.

11.3.10. Distributed Control

First Control Architecture

Control and processing for the Trackbot system is distributed among multiple computers and processing units (Derby and Brown, 2002). This distribution of control allows for the quick response times necessary for real-time control and allows for simpler programming of the individual units. It also reduces the processing performance needs of any one individual unit. The Trackbot uses the following control and processing units:

- Host computer;
- Remote computer;
- Bots;
- Video acquisition and processing unit.

Overall control of the system is performed by the host computer, a standard PC running Windows NT (Fig. 11.56). On initialization the host establishes TCP/IP connections to both the remote computer and the video processing unit using standard Ethernet connections. After establishing the connections the host passes initialization and parameter information to each unit. By doing this, the host retains complete control and oversight of the entire system. In addition,

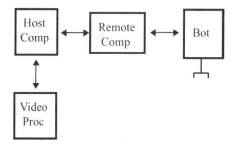

Fig. 11.56 Overall control block diagram

the host establishes a second TCP/IP connection with the remote unit to allow the remote unit to communicate urgent information to the host at any time. The host also has control of the various mechanical parts of the system, including the chain drive, air supply, and conveyors, through an I/O board. If an emergency condition is detected, the host can immediately stop the Bots.

When a user starts the processing operation the host starts a new control and processing thread. This thread determines where a wave of Bots should be sent to pick up product from the input conveyor and passes that information to the remote computer. Position information is sent for a complete wave of Bots each time, so that minimum Bot spacings are maintained to avoid collisions. The host is continuously updating the position information to the remote so that when a wave of Bots is ready to receive new commands, the most recent information is available. Attempting to have individual Bots plan their own buffering or obstacle avoidance is not possible with this architecture.

The remote computer is responsible for control of and communication with the individual Bots. The remote computer receives Bot parameter and product position information from the host over the TCP/IP connection established by the host. It then passes that information to individual Bots as they pass by the IR communication link. The remote computer is also responsible for maintaining spacing between individual Bots and waves of Bots. This is done by calculating and passing to the Bots additional delay times for the Bots when they stop to pick up product. The remote computer has approximately 125 msec to communicate with an individual Bot. This assumes a 4 ft/sec chain speed and approximately a 1 ft window in which the IR link between the remote computer and an individual Bot is allowed. During this time the remote computer must:

- Determine that a Bot has entered the IR window;
- Establish communication with the Bot;
- Confirm expected Bot identification;
- Calculate Bot spacing and delay, if any;

- Request status information from the Bot;
- Pass any command, control, and parameter information to the Bot.

The remote computer detects that a Bot has entered the IR window when a flag on the Bot breaks a through beam sensor. A simple handshake and communication protocol is employed for the IR link, which includes a checksum for each message passed (Fig. 11.57).

Each Bot has a single board Pentium computer, flash memory, and an IR communication device located on top. Although a dedicated Pentium computer might seem overkill, the relatively cheap cost of the single board system, plus the ability to add a local vision system and have it processed efficiently, and the ability to add other functions, makes it more open for further expansions than a simple Basic Stamp processor or the like.

Each Bot operates independently from every other Bot, receiving its commands via serial IR from the remote computer. The Bot executes as a state machine as shown in Table 11.4. After initialization the Bot enters a state where it is waiting to establish an IR link with the remote computer. When this link is established, the Bot can receive requests for status and can receive updated parameters and command information. A command to pick up a product consists of a position and a delay time. After receiving a command the Bot looks for a flag that indicates the home, or zero, position. A position encoder on the

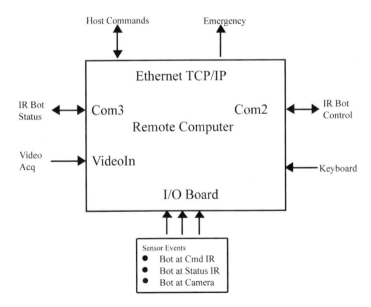

FIG. 11.57 Remote unit functionality

TABLE 11.4 Bot Processor Functions

Thread	Description	Waits on	Signals
Main	State machine to perform Bot actions, dynamically update Bot parameters, communicate with remote via IR	IR input Sensor events Decoder event	
ISR	Sensor input interrupt routine	Sensor inputs Decoder counter	Sensor events Decoder event

moving cog is used to determine track position and in turn when to apply braking. The actual position at which the Bot finally stops is compared to the desired position in order to perform adaptive braking. This allows for variations in clutch slippage among the individual Bots to be accounted for and for accurate stopping positions to be possible. In a similar way, acceleration parameters are calculated when the Bot starts moving again. This allows for accurate times between when a Bot first applies its brake and when it finally achieves full speed again after stopping. This is important for maintaining Bot spacing and to avoid collisions between Bots. After the Bot has picked up a product it then drops it at the location it had previously received from the remote unit. The Bot processing then returns to the original state of waiting for an IR link. In a future system the Bot may also have a video input. This would allow the Bot to perform more precise positioning as well as rotating a product into a desired orientation before dropping the product off.

A vision system is used to determine the location of product on the input conveyor. An intelligent video acquisition and processing unit is used. This unit allows custom processing algorithms to be loaded and executed. The video unit continuously acquires new images and performs the processing to determine product location. The most recent locations are passed to the host computer on request, where they are subject to additional restriction criteria based on Bot spacing requirements. By determining product location in the video processor, the host computer only needs to analyze a small number of points (usually less than 10) as it requires them, as opposed to needing to continuously process 640×480 image points. The exact algorithm performed by the video unit can be modified and loaded onto the unit depending on the product being handled by the Trackbot.

11.3.11. Lessons Learned

The initial control scheme had some major difficulties. These needed attention before product commercialization. Most of the control problems were around

Bringing New Automation to Market

TABLE 11.5 Host Processor Functions

Thread	Description	Waits on	Signals
Main	User interface	User	
Processing	Determines pickup position, transfers pickup request to remote, starts conveyor when required	Sleeps between remote queries	Conveyor timer
Emergency	Receives status from remote in event of emergency situation detected	Blocking read of emergency port socket	User (Stop chain?)
Conveyor timer callback	Check conveyor, stop it if full, reset timer if not	Timer	Conveyor relay

TABLE 11.6 Remote Unit Functions

Thread	Description	Waits on	Signals
Main (video)	Video processing, keyboard processing	BotAtCamera event video frame done keyboard input	Terminate event
Remote command	Receives commands from host	Blocking read of remote port socket Terminate event	IRCommand event Adds commands to Bot Command Block
IR command processing	Process commands and requests requiring communicating with Bots via IR. This includes commands in Bot Command Block, enumeration and updates to Bot parameters.	BotAtIR event IRCommand event Terminate event IR response Terminate event	
IR comm	Performs low-level IR communication	IR port input	IR response
Emergency	Sends emergency status to Host when signaled	Emergency event Terminate event	
ISR	Sensor input interrupt routine	Sensor inputs	BotAtIR event BotAtCamera event

the restricted communication between the remote computer and the Bots and the issues of Real Time Control.

The IR communication "window" between the remote computer and a Bot was short compared to one lap around the track. The use of the separate remote computer to communicate with the Bots was chosen in order to minimize the real-time response time when a Bot passed the IR window and maximize the possible communication time. Even with this, the amount of information that could be passed between each Bot and the remote computer during that "window" was limited. As development continued it was found that additional status information was required from each Bot as it passed by. This information needed to be processed by the remote CPU and commands sent back to the Bot based on it. This included the need to anticipate what each Bot was going to do and to adapt based on how the Bot actually responded. There became a significant amount of information to transfer and confirm in a very small amount of time.

The brake control needed to handle the slipping of the drive clutch. As the hardware was broken in, the braking distance and acceleration characteristics changed. This, in turn, changed the positioning restrictions that needed to be accounted for by the remote CPU. The success or failure of a Bot in picking up product could also indicate problems with a particular Bot that would need to be handled as well.

The variable speed drive on the track chain also created some electromagnetic interference issues with the remote CPU and the IR serial communication device. This would sometimes require that information be re-sent, shortening an already short communication window. The IR device and driver circuitry was not meant to see industrial environments.

Second Control Architecture

Figure 11.58 shows the revised Trackbot Control Architecture. This includes what is really a simplification of all of the tasks and threads in the first architecture. By the improvement of RF wireless Ethernet technology to commercial readiness, the host CPU can talk to any Bot at any time. This eliminates the need for a real-time response when a Bot reaches a particular position (i.e., the IR communication window) and removes the limitation of a fixed amount of time to transfer information to and from a Bot. By doing this the remote CPU can be eliminated. The host computer now has complete control of the system, including the actual communication with the Bots. The Bots can also send status information back to the host (e.g., successfully picking up product), to which the host can respond (e.g., telling the Bot where to drop the product). The ability to communicate with any Bot at any time also allows the possibility of a more interactive command and control architecture between the host and the Bots in the future.

Bringing New Automation to Market

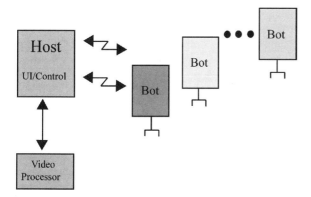

FIG. 11.58 Second control architecture

The Bots were improved with onboard backup batteries to eliminate some issues due to noisy pickups on the rails for the 12 V power supply. Each Bot's single board processor has an interface for an onboard vision camera, which can be added when needed. This allows for the picked up product orientation to be determined. The Z-axis has a rotary motion added to rotate the product for correct placement. The brake control was modified to remember the braking parameters from earlier runs, rather than determining them from initial conditions each time the Trackbot is turned on.

11.3.12. Trackbot Production and Market Acceptance

In early 2000, the author only had the initial testing model built. His company reached a development agreement with Cambridge Valley Machining Inc. (CVMI) of Cambridge, New York, in May of that year. After a very positive marketing visit to a major food producer, a major effort was undertaken to develop the prototype Trackbot in six months so as to make the November 2000 Pack Expo trade show. This is the premiere packaging trade show in the world, and it is held every two years.

So, the author, Dr. John McFadden (CEO of the author's company, Distributed Robotics LLC-DRLLC), and Mr. David Brown (controls and programming consultant) met weekly in design reviews with the CVMI team. This team included Mr. Donald Schneider, Sr. CVMI president, Mr. Merritt Bell, chief designer, and Mr. Jesse Ruppel, engineering intern. The entire group was starting with a clean sheet of paper, since nothing like the Trackbot had ever been built.

The fact that the prototype Trackbot, running at 1 ft/min rather than the target 4 ft/min, did get to Pack Expo was an engineering miracle. At the trade show, the marketing and sales company, American Productivity Group LLC (APGLLC), with Mr. Dwight Carey as principle, handled the contacts. APGLLC had many automation representatives attending the show, and 18–20 of them learned about the Trackbot.

The reactions at the show varied greatly. Some attendees loved the new technology, while some could never seem to understand it. Others wanted it to go much faster, and some were insistent on the use of vision (only the multiple line of sensors was implemented at this time). Still others needed a Z-axis rotation for their application. Some wanted to singulate empty Stand Up Pouches (SUP) to send them to pouch feeding machines, but no-one wanted to place down any amount of dollars to buy the first one. Since it was not a proven technology, no-one wanted to be the first. They all wanted to talk after the company had one in production in someone else's factory. This was a little surprising, since some customers want the first of any new product, but automation and packaging people are generally more conservative!

So the vision system described earlier was implemented, and the author was a speaker at the Pouches 2002 conference in the middle of that year. He visited several pouch equipment producers to gain information for a meaningful presentation on how the Trackbot fitted into the entire pouch material handing market. This conference focused on the SUP market, and was attended by over 300 of the industry users and suppliers. However, mid-2002 was already a slow market; a major machine producer had only sold two units in six months!

So the Trackbot production plans are complete, but the first production unit has not been built. Current DRLLC investors are looking towards the commitment of a Beta site customer, and the improvement in the economy for automation and capital equipment in general before proceeding.

11.4. OVERALL EXPERIENCES

The author's overall experiences of creating a startup company for the development and production of automation have seen ups and downs. Many times it has been exciting, when sharing a new method to solve an old problem; other times when the market goes flat, disappointing!

Here are some observations, many of which will echo generally held opinions, but when one encounters them for oneself, it does drive the message home:

1. Things take longer than expected. The development of the Stackbot at MBI and at CVMI always took longer than one had hoped. Some of this was due to components having 12–14 week lead times, some to the fact that each company was producing other products that had paying customers eagerly awaiting.

2. It takes more dollars than expected. Both Stackbot and Trackbot took more money than initially planned. Increased feature desires from customers as well as early engineering attempts that were not satisfactory were responsible.
3. Business plans are easy to write, but hard to stick to. When in a business plan one anticipates a certain burn rate of dollars, and milestones that seem to slip away too easily, rewriting the business plan to reflect current reality too many times can discourage potential investors.
4. The market is fickle. Case studies of past development projects are good to learn from, but the market conditions can change in a very short time. The United States entry into Afghanistan in October 2001 helped to discourage one potential Stackbot customer, who the previous week wanted 14 units!
5. Which came first, the chicken or the egg? This conundrum lives on for potential investors wanting a startup company to get a few machines in production to feel good about investing. Yet a startup company needs funds to develop and produce the machines to be out there in the first place. DRLLC needs additional capital to make the next steps happen, but the 2002–2003 market has squashed the investment process.
6. Patents last for 20 years. This is a good thing when the market for automation (as well as many other markets) is so cyclical. The author's inventions are protected for many years until things are better, but the high costs of obtaining all of the international patents that might be lucrative have prevented DRLLC from the worldwide protection originally desired. Major automation companies like Adept Technologies, who had $100 million in sales of their robots in 2000, were forced to reduce their head count in early 2003 to around 25% of their peak when sales waned. Patents are a great reserve to draw on when rebuilding.
7. Inventing happens when it happens. One can brainstorm and work hard every day on a problem, and with more design and build experiences, coupled with learning from others' work, one can see the reapplication of some traditional concepts into new areas. However, some divine inspiration does help now and again. That proverbial light bulb going off over one's head, or the little voice from within giving one a vision on how to proceed is priceless.
8. Licensing technology is difficult. A common strategy for inventors is to license the invention to some large corporation and collect royalties. The author was initially successful with the Stackbot at MBI, but that did not last. Licensing the Dockbot has proven to be frustrating.

9. Never say never. After GRASP closed its doors, the author told everyone who would listen that he would never start another automation company, but when the new inventions (Dockbot, Stackbot, and Trackbot) became clear to him, he felt a strong sense to go ahead and do it again. This included raising capital, a fate he and most engineers dread!
10. Do not ever give up.

Conclusions. Outside, uncontrollable factors can have a profound effect on the automation designer.

QUESTIONS

1. Based on the technology of the Dockbot, Stackbot, and Trackbot, what range of devices could be made from the combination of two or three of these? What would their markets be?
2. Based on the mix of technology of the Dockbot, Stackbot, and Trackbot in 1998, what could an alternative approach to machine development and bringing them to market be, if one knows that there will be a giant economic slowdown in 2002–2003?
3. What operations could the Trackbot achieve in the material handling sortation business, as in USPS, UPS, and FEDEX?
4. What are the limitations of the Stackbot's track plate such that it is difficult to handle more than two pallets simultaneously?
5. What is the lower limit of number of Bots one could use on a Trackbot and why? Why would this be practical compared to a standard robot?
6. How would a Dockbot handle a smaller Dockbot for nano-assembly?
7. Sketch how a Dockbot would operate in conjunction with a *XY* table.
8. Sketch what a two-arm robot version of a Dockbot might look like.
9. Design an alternative method of driving a Trackbot Bot clutch pack.
10. What could be done businesswise in 2003 for Distributed Robotics LLC to improve the situation?

REFERENCES

Carlisle, B. (1997). Flexible Part Feeder. U.S. Patent no. 5,687,831, November 18.
Craig, J. (1986). *Introduction to Robotics*. Boston: Addison Wesley.
Derby, S. J. (1981). The maximum reach of a revolute jointed manipulator. *Mechanisms and Machine Theory Journal, International Federation for the Theory of Machines and Mechanism* 16:255–262.
Derby, S. (1989a). The closed loop assembly micro-positioner (CLAMP) end effector for high precision robotic assembly. *Robots 13 Conference Proceedings* 1:9/39–9/50.

Derby, S. (1989b). High-accuracy placement using a hybrid robot/X-Y table system. *Printed Circuit Assembly* 3(8):26–31.
Derby, S. (1990). Mechanical closed loop robotic arm end effector positioning system. U.S. Patent no. 4,919,586, April 1990.
Derby, S. (1992). Selective use of mechatronics when designing assembly robots. *Mechatronics Systems Engineering* 1(4):261–268.
Derby, S., Cooper, C. (1997). "The Evolution of Robot Geometry and its Impact on Industry", Fifth National Applied Mechanisms and Robotics Conference, University of Cincinnati, Cincinnati, OH, paper AMR97-056.
Derby, S. (2002). Case study comparisons: gaining efficient distribution flow. In: Proceedings Warehouse of the Future Conference, Orlando, Florida, June 24.
Ducoste, J., Derby, S. (1990). A docking end effector for a two arm robot. *ASME Mechanism Conference Proceedings* DE-Vol 26:271–278.
Derby, S., Kirchner, N. (2000). Design of the Stackbot Cam Track Palletizer/Material Handling Device. In: Proceedings ASME Mechanisms Conference, Baltimore, MD, Sept 11–13, pp. MECH 14131/1–6.
Derby, S., Brown, D. (2002). Distributed Control of a Track Based Multi-Head Robot. 2002 ASME DETC Conference, Montreal, Canada, Sept. 2002, pp. CIE 34502/1–6.
Derby, S., Lyons, B. Material Handling Device. Patent no. 6,394,740. May 28, 2002.
Derby, S., McFadden, J. (2002). A high precision robotics docking end effector. *Industrial Robot Journal* 29(4):354–358.
Derby, S., McFadden, J., Brookes, K., Brown, D. (2002). Design and Analysis of a Closed Loop Track Based Multi-Head Robot. 2002 Japan-USA Symposium on Flexible Automation, Hiroshima, Japan, July 14–19, pp. 1421–1426.
Gordon, S. (1994). Programmable reconfigurable parts feeder. U.S. Patent no. 5,314,055, May 24.
Groover, M., Weiss, M., Nagel, R., Odrey, N. (1986). *Industrial Robots: Technology, Programming, and Applications*. New York: McGraw Hill.
Jerue, R., Walsh, S. (2000). Material handling robot and rail assembly. U.S. Patent no. 6,059,092, May 9.
Kato, H. (1988). Conveying apparatus for ceiling-suspended industrial robot. U.S. Patent no. 4,726,732, Feb. 23.
Kato, H. (1992). Industrial robot apparatus. U.S. Patent no. 5,161,936, Nov. 10.
Kohno, M., Sugimoto, K., Nakagawa, Y., Part feeding and assembling system. U.S. Patent no. 4,527,326, July 9, 1985.
Norman, J. (1996). Automatic Singulation of Pouches from Bulk. Masters Thesis, Rensselaer Polytechnic Institute, Nov.
Paster, A. (1998). An Automated Device for the Singulation of Flexible Pouches. Master's thesis, Rensselaer Polytechnic Institute, December.
Tsai, L. (1999). *Robot Analysis*. Hoboken: Wiley.
The Evolution of Robot Geometry and its Impact on Industry.
Vogt, F., Lang, W., Ego, T., Jack, K., Erb, H., Zimmer, E., Rohling, B., Wiegand, M. (2001). Device and method for feeding, clamping and processing, particularly for geometric welding of automobile welding components in a work station. U.S. Patent no. 6,170,732, Jan. 9.
Würsch, A., Scussat, M., Clavel, R., Salathé, R. P. (2001). An innovative micro optical element assembly robot. *Robotics Today* 14(3):1–2.

12
System Specifications

The title of this chapter, Systems Specifications, might sound boring to the average engineer. Why not let the administrators and bean counters worry about that. Granted, it is usually not the type of thing that an engineer wants to spend their entire work week on, assuming that they have the option to be designing a machine as an alternative; but let us see how it does impact the engineer, perhaps more than one could ever imagine.

Hypothetically, let us say at a New Year's Eve party your defenses are down and you loan your old friend $40,000. He promises to pay you back plus an additional $20,000 interest ($60,000 total) on the 4th of July, but with the silly condition that you two must meet at the top of the Eiffel Tower at 9 pm. If you do not meet at that time and place, you will get nothing back. Obviously, this example is farfetched, and one's defenses would have to be pretty low to accept these terms. Most lawyers would advise you not to get into such an agreement.

But let us say you do take up this loan to your friend, since it looks far more lucrative than the stock market. So on the 4th of July you are in Paris, France, on top of the Eiffel Tower. But your old "friend" is not there. You take out your satellite compatible cell phone, and give him a call. Guess what, your friend is at the top of the half scale Eiffel Tower at the Paris hotel in Las Vegas, Nevada. You lose!

Now if one is in the business of designing and building automation, a very similar series of events can occur. Let us assume that you are building a $100,000

System Specifications

machine, and at the placing of the order the customer must pay 30% or $30,000. With normal expenditures, you have spent this amount, plus $40,000 more building the machine. You probably used the purchase order to get a line of credit from the bank for the $40,000. The machine is now ready for customer approval, after which you will get 60% more ($60,000). You will get the final 10% ($10,000) after installation. The customer sees the machine run, and surprise, they hate it. It does not move fast enough, it does it the wrong way, and it is even the wrong color. They are very upset and are balking at paying. This is a custom machine, and no-one else would want to purchase it, even if the legalities were clear.

So, just like the farfetched loan on New Year's Eve, you missed the expectations, and are out $40,000 that you had indirectly loaned the customer. The Eiffel Tower scenario is not that farfetched after all. In fact, it demonstrated a very likely scenario if there is a lack of communication and a lack of written system specifications.

12.1. EXPECTATIONS

In so many phases of life, if the reality that happens does not match expectations, then someone is disappointed. Whether one is a diehard Red Sox or Cubs fan, whose lot in life it seems is to suffer, or just having a bad meal at a restaurant, incorrect expectations can ruin short-term successes. And the longer the timeframe, the more the mind forgets the actual conversation (or verbal agreement) and the dreamer inside everyone takes over.

Since automation takes from three months to sometimes several years when developing entirely new processes and equipment, this additional time has several pitfalls:

- Competition — The original market plans can go haywire if the competition comes out with something similar before your machine can make its products.
- Costs pressures — The initial development of your machine might produce results that show more devices need to be added, at a greater cost. So to keep the machine cost justifiable, now it must work faster than the original plans.
- Feature creep — As the customer thinks more and more about what your machine is supposed to do, they start to find other tasks to automate, and do not see why your machine cannot do them also, for the same costs.

The author's 25 years of experience show that almost 80% of the time the problem can be classified as Feature Creep. As time goes by as the machine is designed and built, it is natural for the customer to wait and wonder, almost like a child waiting for a present under the tree on Christmas. The longer the wait, the harder it is to keep the expectations the same as they were when the machine was ordered, and just like on Christmas morning when one does get the toy for which one was so eagerly awaiting, sometimes the toy is only OK, not terrific as one had hoped.

Cost pressures do appear many numbers of times. The experiences of the USPS mail tray handling in Sec. 11.2.3 state that the original throughput rate of 7–10 trays per minute was increased to 14 trays per minute solely due to no vendor being able to create an automation solution for the original price estimate. This price estimate was dreamed up by some managers without any sound engineering reasoning.

Expectations can also creep not just from the customer wishes but from one's own sales staff or independent product representatives. Early excitement over the Stackbot in Chapter 11 by product reps who desired payload capacity of 20 lb was replaced by a need for 30 lb, and then they had to have a 40 lb capacity or it was the end of the world. This came from initial customer enthusiasm when the economy was very strong, and the justification criteria were easy to achieve. However, but as the economy cycled to slower times, the same Stackbot needed to handle two 20 lb boxes at once, not just a single box, in order to be justified. One of the Stackbot engineering team's biggest impediments to bringing a product to market was the changing demands from their own reps.

Potential customers were initially excited over a simple Stackbot implementation, but as sales people discussed the realities of how and where installations could occur, more accessories and safety options were needed than initially thought. The author found a real parallel between education in the classroom and education at a tradeshow or customer visit. Many industrial customers are bright people, but they have not had coursework or training in automation, and they need to be taught the options (this is different than selling an automobile, which 95% of the United States can understand, at least to a reasonable limit). This takes time and patience, and will often whittle down the number of sales leads to less than one had at the start of the day.

So what kind of things can be expected? What features need to be established? The list started in Chapter 4 is a good starting point:

- Throughput rates;
- The processed product's size and weight;
- How to determine out-of-tolerance incoming product and what to do with them;
- Who is responsible to guarantee that the incoming product is within tolerance;
- Available electrical and compressed air facilities;
- Floor capacity — in weight per square foot;
- Floor space available and ceiling height;

System Specifications

- Pillars and other obstacles to be negotiated;
- Available entry passage to desired machine site;
- Expected machine life — Mean Time to Repair;
- Safety procedures at the desired site including relevant ANSI standards.

These features from Chapter 4 on The Automation Design Process immediately state the need for the engineer to set or verify the specifications. If any of these features change, the engineering work done to date is potentially wasted. Data on all of these features need to be in a written document.

12.2. OTHER PROBLEMS BEYOND SPECIFICATIONS

It should be stated that even if one has every specification written down in very clear terms, the customer may still be disappointed. They will admit that you built a machine to do exactly what the feature list states, but somehow something in the product has changed, usually due to marketing. Now if it can be accommodated with an Engineering Change Order (ECO), and you are willing to modify the machine at a price that still makes automation economical, both sides can still feel like winners, but sometimes it can be worse that that.

An automation customer developed a new food product. It was a snack size cup of applesauce like one sees in the market in all sorts of flavors, but it had a cap on top of the applesauce lid with a short but very thick straw. The idea was for kids in grade school to suck up the applesauce as a more enjoyable way to eat a basically healthy food. It was test marketed to several groups of boys (who later were rumored to be the kids of the marketing team), and they loved it. So, an automation firm was contracted to take the applesauce cups and place the bent straw and cap on top of it. The down payment was made, and some custom parts were ordered and delivered (timing screw to space the applesauce cups), and the machine was starting to take shape.

The automation firm had invested significant moneys beyond the down payment when the customer all of a sudden faxed a notice to stop work. It seems that the customer had done additional market tests, and found out two important things. First, they found that girls in general thought that sucking up applesauce was gross. And secondly, lunchroom monitors were not too enamored with the thick straws filled with applesauce becoming food projectile launchers. So the marketing efforts were in panic mode, and the project was halted for good.

The automation company was left with a partially built machine and some large bills that they had to eat (there was also most of a pallet of applesauce cups designated for machine testing that needed to be disposed of). The moral of the story was that the contract agreement needed to have a cancellation clause that was more aggressive in protecting the machine builder. One needs to look at all of the possible angles from both the customer and the builder side to fairly protect both parties. One would like to limit the number of times one is in court over these contracts. One is rarely pleased with the results in court, even when supposedly one has "won."

12.3. EXAMPLE 1: BULK MAIL CARRIER (BMC) UNLOADER SPECIFICATIONS

As part of the Case Study Number 3 in Chapter 4, the RPI team developed a list of specifications that the actual BMC unloader should satisfy. This list was incorporated into the New York State Library (NYSL) bid document, which will be discussed in Sec. 12.4.

This list was reviewed many times over several weeks, since the team was describing a device that had never been built at full scale before. The system it was to be integrated with was not completely defined yet either, but there were key issues that jumped out to the team fairly quickly. The team wanted to let the successful BMC unloader bidder have some room to engineer many smaller details that would not be critical to the operation, but might be more specific to the firm's manufacturing capabilities. However, the team had reasonable thoughts on many tradeoff issues, and wanted to be sure that the unloader would be useful:

- The arrangement of the BMC, receptacle, and elevating conveyor is, to some degree, reconfigurable, to attempt to accommodate various constraints of space in other libraries.
- The actuator for a system was a question worthy of consideration at this stage. Pneumatics would be too fast and unreliable for this application. Hydraulic actuators are too messy. Actuating the unloader at one of the ground pivots was a consideration, but requires enormous torque. The benefit of rotating the load about the center of gravity, and moving it horizontally and not vertically, was to minimize the actuator size.
- The means of actuation that seemed best equipped to deal with this was the motorized lead screw. There was concern about the length of lead screw required for this application (5 ft), although this later evaporated.
- Failure conditions were also considered. It would obviously not be safe for the unloader to crash to the ground in the event of a power failure. Nor would it be safe to have the unloader left in a non starting position during a loss of power. Although the tendency of lead screws to NOT be back-driveable had advantages to the first item, there was some concern on the second point. For the NYSL/TBBL, the potential for loss of power is so minimal that this item evaporated. Since deployment of this device could occur at a facility with not so redundant power backups, this concern did not disappear completely, and resurfaced

System Specifications

during the BMC Unloader builder solicitation. However, ultimately, as long as the BMC did not crash to the ground, maintaining an "up" position until power is restored is acceptable.

12.3.1. Design Specifications

The engineering team placed the above concerns in the discussion part of the unloader bid document, and listed the following items specifically.

BMC Unloader

- The unloader is two identical four-bar linkages mounted on either side of a pallet that secures the BMC for the unloading process. The details of the links of these mechanisms are as follows:
  - The ground link, with a center distance, between pivots, of 22.5 in.
  - The "driver/driven" links, which are identical, with a center distance of 61.75 in.
- The pallet, with a center distance of 30 in. The bottom pivot is located $(-35.5, 20)$ in. from the lower right when the BMC is loaded from the left, the top pivot is located $(-35.5, 50)$ in. from the lower right when the BMC is loaded from the left.
- Precision in assembling the linkage is vital in ensuring successful operation, that is, as exact tolerances to the above specifications as practical.
- The geometry of the pallet is formulated to accommodate several notable features of the BMC, as follows:
  - The BMC has four wheels, two fixed, 34 in. apart, two pivoting 24 in. apart. Consequently, the pallet has slots in its base to accommodate these wheels. The BMC is loaded fixed wheels first, to provide maximum control over manouvering the BMC into the unloader to the individual performing this task.
  - To facilitate loading the BMC into the pallet, the pallet sits 4 in. off the ground. This is accomplished with either the addition of legs to the pallet or a stationary block on which the pallet rests when not active. At no time may such a feature interfere with the motion of the linkage once the BMC is secured.
  - The front of the BMC has a 2 in. lip at the base of the body. To prevent tipping and lateral movement while secured within the loader, a similar lip is added at the top of the facing wall of the pallet to balance the BMC.
- Three devices secure the BMC during unloading, one ledge and two bars. The ledge, in conjunction with the first bar, prevents the BMC from sliding out of the top of the pallet. The second bar holds the BMC into

the pallet and prevents possible motion out of the rear of the pallet. The second bar is secured when the operator has loaded the BMC into the pallet. The first bar is either secured in this manner, or is a permanent part of the structure of the pallet. Creating additional positions for bars is an option, which allows the pallet to accommodate other types of bins.
- A mechanical stop is located (19, 6) in. relative to the right ground pivot, extending 17 in. up and to the left at an angle of 30°. This stop confines the "dumping angle" to 30° or less. Additionally, this stop prevents the pallet from inverting.
- A second mechanical stop is located (53, 64) in. relative to the right ground pivot. This stop aids in confining the "dumping angle" to 30° or less and also contributes to preventing the pallet from inverting. It may consist of a bar mounted in the support structure of the module.
- The equipment footprint is approximately 8 + ft wide × 12 + ft long.
- The maximum height of the machine during any stage of operation is 9 ft.
- Proximity sensors in the "Unloading Area" are necessary as an emergency feature to detect the presence of a person in this area. If interrupted, this must trigger an emergency stop and prohibit any motion of the machine.
- Isolation of moving parts is necessary wherever possible.
- Minimum Actuator Force: 2 × 750 lb. linear force. It is intended that a lead screw attached to the side of the pallet transmit this force.
- Actuation is accomplished at the geometric center of the side of the pallet, desirable for several reasons. Actuation is achieved with a lead screw. The motor for the lead screw is attached to a pivot, mechanically grounded; the other end of the lead screw is free. The nut through which the lead screw passes is mounted at the center of the pallet. This is duplicated on both sides, that is, one actuator for each linkage. The necessary lead screw length is 5.5 ft.

Related Equipment to Supply

The receptacle and elevator conveyors receive and conduct away material from an unloading BMC.

Receptacle Conveyor

- Operating velocity: 20 fpm
- Conveyor-top height: 1/2 ft
- Width: 2.5 ft
- Length: 6 ft
- Belt material: Standard
- Angle: 0°
- Guides

System Specifications

- Inner guide (facing BMC) — several feet in length to prevent undershooting spillage
- Outer Guide (opposite BMC) — several feet in length to prevent overflow

Elevating Conveyor

- Operating velocity: intermittent, 20 fpm
- Conveyor-top height: N/A
- Width: 2.5 ft
- Length: 4–5 ft
- Belt material: Flighted
- Startup characteristics: Standard
- Angle: 45° or more
- Guides
  — Guides are 1 ft in height to confine material to travel on the conveyor without spilling over the side.

12.3.2. Comments

This list of specifications at first glance may seem impressive. The team designing and modeling had a great deal of knowledge, more than what most potential automation customers would ever have. This amount of knowledge even produced a U.S. patent (Simon, 2000). The earlier example about the applesauce cups and straws probably would have listed a throughput rate and overall size constraint, and only a few more items, but note the list given here for the BMC unloader does not mention speeds, cycle times, and a few other features that one would need to know after sorting the given demands thoroughly. A partial reason is that the throughput of the BMC is very slow compared to the rest of the system, so that the speed is almost irrelevant.

Yet there are still more specifications that could have been established. This could include things such as the width of the doorway for moving in the assembled machine, and the floor loading restrictions.

Since the amount of technical material was significant, and the crossed four-bar design not a standard concept, a mandatory bidder's conference was held. Here, invited company personnel attended a morning long presentation and were encouraged to ask questions. State contract rules also dictated that if any company had a question to ask after the conference, the answer would be sent to all attending conference companies, so as to keep a level playing field.

12.4. REQUEST FOR QUOTE

A Request For Quote (FRQ) or a Request For Proposal (RFP) are developed and circulated for potential automation firms to place a bid. The customer can review all submitted bids, and select the best one. What is best is a function of the customer. If the customer is a private firm, they do not always work on the lowest bid. If it is a Government agency, it may have to select the lowest bid, or it may have very specific judging criteria that it must use.

The author has been a part of this process many times, and the bid price is usually only a part of the decision criteria. Other criteria will be:

- Bidding firm's experience;
- Bidding firm's references;
- Perceived quality of work;
- Perceived ability to create safe machines;
- Perceived ability to accommodate RPI team's conceptual design;
- Timeframe of bid work completion date.

The bid price may only constitute 20–25% of the decision criteria.

Depending on the customer (private vs. Government), the accepted bid is usually confirmed with some type of written documentation. Some automation firms dealing with private firms may have a legal form that is only 3–5 pages in length, spelling out terms of payment, specifications, cancellation fees, and the like. Often there is a page of legal boiler plating, or terms that are used for every customer, that do not get a lawyer involved for review every contract, but some customers do not agree to everything on the boiler plating, so one ends up using their lawyer more than one's wallet would like.

When a Government customer writes a detailed RFP for any item, it will often attach its boiler plating to the machine specifications as a way to warn the bidding firms what to expect. This is basically a good thing, since if the rules and regulations are just too many hoops to jump through, a firm that would potentially bid (and maybe even win) will decide that it is just too hard to play according to these rules.

12.4.1. Example 2: BMC Unloader Bid Award Package*

The NYSL/TBBL group was bound to place the BMC unloader out for bid. Being a Government agency, many groups internal to them were involved, and the process was not very quick, but it was quite thorough. Simpler agreements might have left out statements such as who owns the intellectual property (the design of the automation machine) and who has rights to make and sell additional machines to other customers.

---

*This example was supplied by the New York State Library Talking Braille and Book Library Group.

System Specifications

The Bid Award Package is 36 pages long, and only the highlights will be covered here. The original RFP was included and constituted 34 pages. The other two pages were the cover letter, declaring who the winning bid was, stating the overall process of how the review was conducted, and noting if it was the lowest bid and/or highest score on the criteria evaluation.

The RFP had the following sections:

1. General introduction: background statement of who the players were and the overall goal.
2. Confidentiality: all information in the RFP was to be kept confidential. There was at the time the need to keep things secret until the patent application was filed.
3. Quantity and licensing: that NYSL/TBBL would remain the owner of the design. They wanted only one BMC unloader, but stated the possibility that additional unloaders could be marketed, and that the winning bid would have an opportunity to make additional units, but this was contingent on a good machine and fair price.
4. Delivery and acceptance: NYSL people would inspect the unloader at the machine builder's site and would conduct an acceptance test. Then the builder would ship the unit to Albany, New York.
5. Proposal content:
   - Description of the bidding company, capabilities, quality management, and service policies;
   - Describe relevant automation machines company has completed;
   - Five references;
   - Description of technical approach;
   - Utilities required at NYSL site for operation;
   - Cost, payment terms, and conditions;
   - Warranty period and policy;
   - Service rates and policy.
6. Description of operation: explaining the RPI team concept so as to inform the bidding companies how it should work. This included:
   - Operating principles (similar to the Case Study discussion in Chapter 4 and the material in Sec. 12.3.1);
   - Components: definitions of the relevant terms used in the operating principles section;
   - Sample sequence of operation: the anticipated sequence from the RPI team's efforts. It was referred to as "sample" since a bidding company might have a better idea. Or they might have a less than optimal idea but it would reduce costs significantly that it should be considered.

7. Functional specifications: dimensions and relevant facts on all of the products (green cases and other materials) found in a BMC. Also dealt with how a BMC should be loaded into the machine, and space restrictions at the NYSL site. It also included:
   - The mechanical and electrical interfacing requirements;
   - Safety;
   - Paint color and corrosion avoidance;
   - BMC tipping angle.
8. Performance specifications: company required to successfully execute the acceptance test.
9. Documentation: written in WORD and CAD drawings could be in any format. Documentation needed to be supplied in printed form and on CDROM. The documentation consisted of:
   - Operations manual;
   - Maintenance manual;
   - Technical documentation: CAD drawings and Bill of Materials.
10. BMC unloader acceptance test: NYSL would supply a sample of materials usually found in a BMC. The specific number of each type of cases and other products was specified. The machine was to be run in a step-by-step mode, and in an automatic mode, for as long as the people conducting the test desired. Testing included a partially filled BMC, a fully loaded BMC, and an overloaded BMC. Some tests were only conducted a few times, while one endurance test was conducted for 24 hours. There were forms created for each test, with places for the NYSL people to fill in how it performed, and to list any questions or concerns.
11. Proposal evaluation criteria: the different categories and the maximum points allowed for each category are listed. Also included were the evaluation forms. Additional forms for the site review were included.

12.4.2. Example 2 Conclusions

The goal through this bidding process was to select the top company. This included the need to make a site visit to any company that had a proposal that was complete. This process did take RPI people significant time and expense for travel and evaluation, but it did lead to a good selection. For a government sponsored process, it did proceed as well as one could expect. It did take several months longer than the author would have guessed.

Many commercial companies selecting an automation firm might only visit the top two or three sites, to save time and money. Commercial companies may conduct this effort more quickly. In fact, impulsive single owner companies might only visit one company and be willing to write the check for the initial payment, but then the commercial firms might be more likely to change their mind as the market changes.

12.5. CONCLUSIONS

Several lessons can be learned listening to engineering managers who have seen specifications going both well and towards failure. They are:

- Communication is the key, and the more often one communicates the better.
- Use common sense, and keep notes since memories can let you down.
- Trust and openness go a long way in developing success between customer and supplier.
- Stop the design process at some point and get on with building it. Some engineers will refine the design for 10 years if not directed to do otherwise.
- Striving for Win–Win is not well understood today, and holds the key to great success.

Developing system specifications is a required must. They need not be as detailed as the BMC unloader example for the NYSL/TBBL group. That example is at one end of the spectrum, but at the other end of the spectrum, one at least needs to state the basics (throughput, machine interfaces, what is acceptable), and all relevant assumptions one wants from the machine designer in order to avoid finding oneself in the wrong Eiffel Tower!

REFERENCES

Derby, S., Simon, M., Allen, M. Bulk mail container unloading system, apparatus and method. U.S. Patent no. 6,394,736. May 28, 2002.

Simon, M. (2000). A Modular Automated Handling System for the New York State Library: Investigation, Design, and Implementation. MS thesis, Rensselaer Polytechnic Institute.

13
Packaging Machines

As stated earlier in this text, the area of packaging machines is quite large. The consortium of machine producers is called the Packaging Machinery Manufacturers Institute (PMMI). The PMMI has many tasks, including organizing the largest tradeshows each year (Pack Expo) in the United States dedicated to packaging. It also has an education division, which was run for many years by the late Glenn Davis. Glenn wrote many instructional books for PMMI member companies, some for training technicians, and others summarizing the types and range of the various packaging machines (Davis, 1997). His vision lives on at PMMI, and parts of this text are inspired from his writing.

The PMMI tradeshow website (www.packexpo.com) offers a great listing of the types of packaging machines. Table 13.1 shows a high-level view of these types of machines, while the Appendix lists the complete list of over 350 types of machines. This chapter will deal with some of the most common packaging machines as an introduction. The reader is encouraged to read Glenn's work, and to view the thousands of machines in the Pack Expo website underneath the 350 product category listing.

13.1. LIQUID FILLING MACHINES

Most of us use bottles filled with liquids every day of our lives. The two-liter bottle shown in Fig. 13.1 needs to be filled with automation to keep profits high.

TABLE 13.1 Pack Expo Packaging Machine Categories

Packaging Machinery
Adhesive/Glue Applicators
Cartoning Machines
Case/Tray Erecting Machines
Case/Tray Handling Machines
Case/Tray Loading/Unloading Machines
Case/Tray Sealing Machines
Case/Tray Wrapping Machines
Cleaning Machines
Closing Machines
Container & Component Handling Machines Conveyors
Cooling, Warming & Drying Machines
Decorating Machines
Fill & Seal Machines
Filling Machines
Form, Fill & Seal Machines
Group Packaging Form, Fill & Seal Machines
Inspecting Machines
Labeling Machines
Marking Machines
Packaging Support/Specialty

(*continued*)

Table 13.1 *Continued*

Packaging Machinery

Equipment
  Pallet Forming, Dismantling & Securing
Machines
  Pasteurizing
Machines
  Rebuilt/Used
Equipment
  Shrinking
Equipment
  Sterilizing
Machines
  Strapping
Machines
  Wrapping
Machines

There are several ways to fill a bottle, the two biggest being volumetric and net weight. Figures 13.2 and 13.3 show a piston that is used to transfer a fixed volume into the cylinder above the empty bottle. Valves are used to allow the proper flow at the upstroke and downstroke. If the liquid density is to vary at all, this fact is ignored with this method.

For net weight filling, the empty bottle is placed on a scale (Fig. 13.4). The tank has a piston that opens a valve just above the bottle mouth. The scale is used as sensory feedback to guarantee the correct weight, even if the filled bottle looks a little under filled due to product density changes on the heavy side.

FIG. 13.1 Liquid-filled bottle

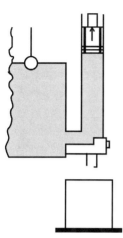

FIG. 13.2 Volumetric filling: piston up

All bottles need to be capped, some by threaded caps that get placed via a cap chute (Fig. 13.5), and are then led to rollers that will spin the cap onto the bottle threads until tight. Other caps are themselves sealed through a heat shrinking process (Fig. 13.6).

13.2. CARTONING AND BOXES

Many products (nonliquid) get placed in a cardboard box or carton. Either name can be used in practice, but the packaging industry usually states that the individ-

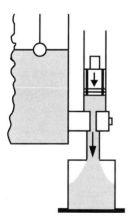

FIG. 13.3 Volumetric filling: piston down

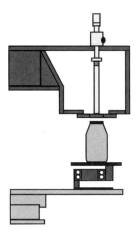

FIG. 13.4 Net weight filling: cutaway view

FIG. 13.5 Cap chute

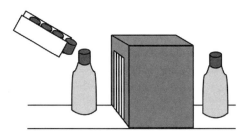

FIG. 13.6 Heat shrink capping

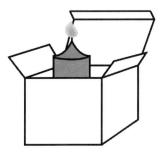

FIG. 13.7 Candle in a box: do not light before packaging!

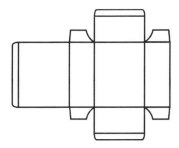

FIG. 13.8 Box in folded form for shipping

ual item is placed into a carton, and a collection of cartons placed in a box or case. When an individual item is to be cartoned (Fig. 13.7), the box first needs to be constructed or erected from a stack of folded carton blanks. These blanks come from the printer folded (Fig. 13.8) and stacked to take up less shipping space.

Carton opening (Fig. 13.9) is usually one of the first steps required in a cartoning machine (Fig. 13.10). These machines usually run with a continuous conveyor belt, and are highly synchronous in design.

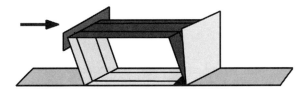

FIG. 13.9 Carton opener

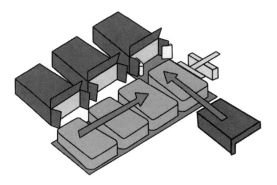

FIG. 13.10 One-shot filler cartoner

Whether it is to be used as an inside liner for a powdered product that ultimately gets cartoned (instant pudding for example), or it is to produce pouches for sale without a cartoner, the Form Fill and Seal (FFS) machine is fun to watch. It takes a roll of wrap and folds it into a continuously moving funnel. The two edges are sealed to make a tube, and then the correct amount of powdered product is dropped in. Then with careful timing, sealing fingers create a seal and cut one pouch from the next. The machine in Fig. 13.11 is shown creating an inner pouch and dropping it into a carton, all at remarkable speeds.

FIG. 13.11 Form fill seal machine

FIG. 13.12 Labeling a product

13.3. LABELING

Another broad area of packaging need is the area of labeling a product. Labels can be glued on, self-sticking (Fig. 13.12), or heat shrink wrapped (Fig. 13.13). Placing labels on flexible products like bags of potato chips is not easy, so much effort has been made to develop great printing processes onto the bag material itself.

13.4. CASES

Cases are made from corrugated cardboard, and like cartons, shipped in their knocked down condition (Fig. 13.14). Cases need to be erected (Fig. 13.15) by pushing the two opposite corners to form a rectangle, and the bottom flaps folded and either glued or taped. A filled case is then closed by folding the top flaps and sealing (Fig. 13.16). The case erecting machine can use tape or glue, depending on the machine's configuration, and the top flaps sealed by either method, depending on what type of box sealer one purchases. Depending on the product being shipped in a case, and perhaps even how far the case is to be shipped, some

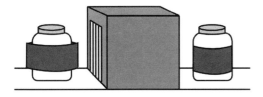

FIG. 13.13 Shrink fit labeling

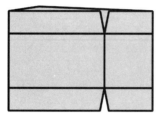

FIG. 13.14 Knocked down case

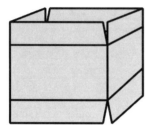

FIG. 13.15 Erected case

users have a personal preference for box sealing that is both logical and emotional. Cases are most often loaded from the top, but some applications use side loading of a single tier or layer in order to achieve the best packaging results (Fig. 13.17).

13.5. PALLETIZING

A pallet of cases can look something like Fig. 13.18. Here the cases are stacked in an alternating pattern on adjacent layers, and have a top layer or sheet to assist with lateral stability. Also in this figure, bands have been applied to lock the product into place. This method tends toward the extreme, and was more likely

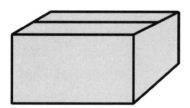

FIG. 13.16 Closed case

Packaging Machines

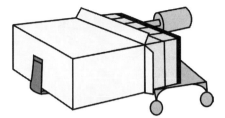

FIG. 13.17 Single tier loading

found in the past. Presently, plastic shrink wrap or stretch wrap is more often used. Wrapping can be done by either a machine or a human, and can be removed without the potential danger of cutting metal bands and watching them fly in all directions with their razor sharp edge that results from the cutting process.

Pallets traditionally are made of wood, but the longer life of molded plastic is becoming more accepted (Fig. 13.19). Wooden pallets are used more than once, but often do not last more than three transfers due to loading and forklift truck operator abuse. Cases on pallets can be placed in packing patterns based on the size of the pallet and the size of the cases. There are dozens of possible patterns, with some of the more basic ones shown in Fig. 13.20.

Pallets can be filled by the row (Fig. 13.21) or by the box (Fig. 13.22). There are all levels of complexity and speeds available depending on product

FIG. 13.18 Pallet of cases

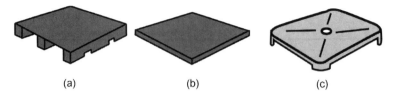

FIG. 13.19 Pallet forms: (a) wood; (b) solid slab; (c) molded plastic

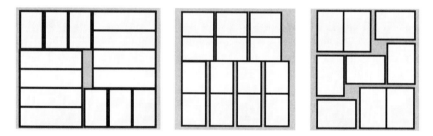

FIG. 13.20 Pallet packing patterns

FIG. 13.21 Pusher bar moving a row of cases onto a pallet layer

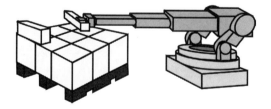

FIG. 13.22 Robotic palletizer

Packaging Machines

FIG. 13.23 Pouch

throughput. A cereal manufacturer making corn flakes will crank out many cases each hour, and can justify a high-speed palletizer that moves much faster than a robotic application.

13.6. FORMING POUCHES

There are many types of pouches in use today. Some of the better designs of yesteryear (Fig. 13.23) are now limited to being placed in a carton (Fig. 13.24). This older style pouch does not stand up by itself on a shelf, and does not grab the buyer's attention. Even fin seals (Fig. 13.25) and lap seals (Fig. 13.26) are old news.

Traditional pouch fillers (Fig. 13.27) have been around for decades, but the new Stand Up Pouch (SUP) has grabbed a lion's share of the market. Each year billions of SUPs are sold in the United States. Machines to fill SUPs have been developed, but the handling of both the empty SUP pouches and the filled SUPs does cause trouble.

13.7. BLISTER PACKS

Blister packs are not new to the consumer, but are still very much in demand and are likely to stay that way. A range of products can be packaged in blister packs

FIG. 13.24 Pouch in a carton

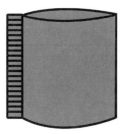

FIG. 13.25 Fin seal

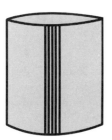

FIG. 13.26 Lap seal

FIG. 13.27 Web feeding for pouch formation

Packaging Machines

FIG. 13.28 Blister packs

(Fig. 13.28), some for internal use inside a carton (medicines and pills) while other products hang on racks using the blister pack as a way to show off the product without exposure to the elements.

The blister pack can be sealed to a card (Fig. 13.29) or it can be sealed with a foil, usually producing a more weather resistant package.

13.8. BAGS

Bags of one form or another have been around for centuries (Fig. 13.30). The paper bag is relatively new compared to the use of animal skins by cavemen. The filling of paper bags produces a product that is not very sexy compared to SUPs and modern packaging, but products like lawn fertilizer and bird food will continue to come in paper bags, at least in large quantities.

Most bags are filled by net weight, and use low-pressure high-volume air to blow the product into the bag's opening (Fig. 13.31). Some bags like lawn fertilizer blow into patented designed bags that actually self seal when they are filled.

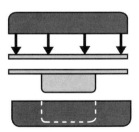

FIG. 13.29 Card feeding

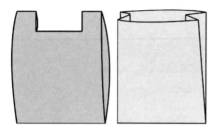

Fig. 13.30 Bags

Fig. 13.31 Net weight filling

13.9. CONCLUSIONS

To many engineers, the package a product comes in is not very critical to one's purchasing decision. An engineer can usually think of what the product is, and because or despite the packaging, decide to purchase it, probably as a repeat customer, but since engineers are only 3–4% of the United States population, and are not a good representative of the purchasing public, packaging plays a huge role in how a product lives or dies.

Packaging continues to become more flashy, high tech, and sometimes even less recyclable, but unless the product beats out the competition and gains market share, unfortunately recycling is not always high on the list of concerns. So, the evolution of packaging has mixed results towards recycling, a travesty in the author's opinion. On the positive side, new packaging drives the automation market for new machines to be designed and built.

REFERENCE

Davis, G. (1997). Introduction to Packaging Machines. Arlington: PMMI.

Appendix A

The following table is the complete packaging machinery equipment tree from www.packexpo.com, the tradeshow website of the Packaging Machinery Manufacturers Institute (PMMI).

- Packaging Machinery
  - Adhesive/Glue Applicators
  - Cartoning Machines
    - Carton Blank Erecting
    - Carton Closing Machines
    - End Flap Carton Closing
    - Horizontal End Load Cartoner
    - Mandrel Bag-In-Box
    - Mandrel Carton Form, Fill & Seal
    - Skillet Erecting
    - Three Flap Carton Closing
    - Top Load Carton Form, Fill & Seal
    - Tray Erect Load & Seal
    - Vertical Cartoner
    - Wrap-Around Cartoner
    - Wrap-Around Sleeving
  - Case/Tray Erecting Machines
    - Case Erecting
    - Division Inserting

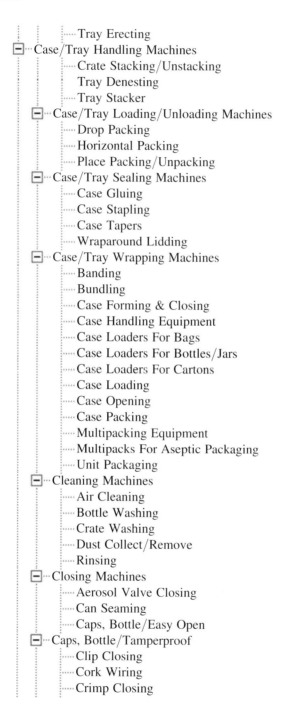

- Tray Erecting
- Case/Tray Handling Machines
  - Crate Stacking/Unstacking
  - Tray Denesting
  - Tray Stacker
- Case/Tray Loading/Unloading Machines
  - Drop Packing
  - Horizontal Packing
  - Place Packing/Unpacking
- Case/Tray Sealing Machines
  - Case Gluing
  - Case Stapling
  - Case Tapers
  - Wraparound Lidding
- Case/Tray Wrapping Machines
  - Banding
  - Bundling
  - Case Forming & Closing
  - Case Handling Equipment
  - Case Loaders For Bags
  - Case Loaders For Bottles/Jars
  - Case Loaders For Cartons
  - Case Loading
  - Case Opening
  - Case Packing
  - Multipacking Equipment
  - Multipacks For Aseptic Packaging
  - Unit Packaging
- Cleaning Machines
  - Air Cleaning
  - Bottle Washing
  - Crate Washing
  - Dust Collect/Remove
  - Rinsing
- Closing Machines
  - Aerosol Valve Closing
  - Can Seaming
  - Caps, Bottle/Easy Open
- Caps, Bottle/Tamperproof
  - Clip Closing
  - Cork Wiring
  - Crimp Closing

Appendix A

- Crown Capping
- Fitment Applicator
- Foil Sealing
- Fold Closing
- Fusion Sealing
- Glue Sealing
- Gummed Tape Sealing
- Heat Sealing
- Induction Sealing
- Nail Closing
- Overcapping
- Plugging; Corking
- Press-On Lidding
- Pump Applicator
- Rivet Closing
- Roll-On Capping
- Screw Capping
- Sewing
- Solder Sealing
- Sonic/Ultrasonic and RF Sealing
- Spin Welding
- Staple Closing
- Strapping
- Tamper Evident Seals & Bands
- Tape Sealing
- Tuck Closing
- Twist-Tie Closing
- Weld Sealing

Container & Component Handling Machines
- Accumulating
- Accumulating & Collating
- Accumulation Table
- Arranging Machines/Feeding, Orienting
- Bag Handling
- Bag Presenting
- Barrel & Drum Handling
- Bottle Handling
- Box Handling Systems
- Can Handling
- Carton Handling
- Case Handling
- Component Unscrambler

- Conveying Machines
- Dispensing Machines
- Handle Applicator
- Inserting
- Key Applicator
- Layer Pad Dispenser
- Leaflet Feeder
- Pallet Dispenser
- Pouch Handling
- Puck Handling, Flow Assistance
- Pucking/Depucking
- Rigid Container Denester
- Rigid Container Orienter
- Rigid Container Single Liner
- Rigid Container Unscrambler
- Robotics
- Sack Seal & Present
- Spoon Applicator
- Straw Applicator
- Tear Tape Applicator
- Top Sheet Dispensing

Conveyors
- Air Cushion Conveyor
- Air Driven Conveyor
- Apron/Slat Conveyor
- Bag Conveyor
- Belt Conveyor
- Bucket Conveyor
- Bulk Conveyor
- Cable Conveyor
- Chain Conveyor
- Converging/Dividing Conveyor
- Drag Chain Conveyor
- Drum Conveyor
- Elevating Conveyor
- Magnetic Conveyor
- Mobile Conveyor
- Monorail Conveyor
- Pallet-Load Conveyor
- Pneumatic Conveyor
- Roller Conveyor
- Screw Conveyor

Appendix A

- Spiral Conveyor
- Steel Belt Conveyor
- Unit & Package Handling
- Vibratory Conveyor
- Wheel Conveyor
- Wire Mesh Conveyor

- Cooling, Warming & Drying Machines
  - Cooling
  - Drying
  - Warming
- Decorating Machines
  - Capsuling
  - Foiling
  - Shrink/Stretch Sleeving
  - Tag Labeling
- Fill & Seal Machines
  - Ampoule/Vial Fill & Close
  - Bag Fill & Seal
  - Blister Fill & Seal
  - Bottle Fill & Cap
  - Can Fill & Seam
  - Cask Or Keg Fill & Seal
  - Cup/Tub Fill & Seal
  - Fill and Seal Machines Unspecified
  - Sack Fill & Close
  - Tube Fill & Seal
- Filling Machines
  - Aerosol Filler
  - Auger Filler
  - Bag Filler
  - Bag-In-Box Filler
  - Bottle/Jar Filler
  - Can Filler
  - Canning Filler
  - Capsule & Tablet Packaging Filler
  - Carton Filler
  - Case Filler
  - Count Filler
  - Cup Filler
  - Displacement Filler
  - Drum Filler
  - Dry/Solid Product Filler

- Feeder
- Filling & Closing
- Flow Meter Filler
- Gas & Vacuum Combo Filler
- Gas Flush Filler
- Gravity Filler
- Gross Weighing
- Liquid Filler
- Net Weighing
- Portion Packaging Filler
- Pressure Filler
- Rotating Chamber Filler
- Sterile Filler
- Timed Flow Filler
- Vacuum Filler
- Vacuum Level Filler
- Viscous/Semi-Viscous Filler
- Volumetric Cup Filler
- Volumetric Piston Filler

- Form, Fill & Seal Machines
  - Blow Mold Fill & Seal Machines
  - Carton Form, Fill & Seal Machines
  - Deep Draw Form, Fill & Seal, Cold
  - Deep Draw Form, Fill & Seal, Thermoform
  - Deep Draw Form, Fill & Seal Machines
  - Group Package Form, Fill & Seal Machines
  - Horizontal Edge Sealing
  - Horizontal Flowrapping
  - Horizontal Form, Fill & Seal Machines
  - Horizontal Lower Reel Flowrapping
  - Horizontal Sachet Form, Fill & Seal
  - Mandrel Flexible Package Form, Fill & Seal Machines
  - Tubular Bag Form, Fill & Seal Machines
  - Vertical Form, Fill & Seal Machine For Cartons
  - Vertical Form, Fill & Seal Machines
  - Vertical Form, Fill & Seal Machines for Pillow Bags or Stand-up
  - Vertical Sachet Form, Fill & Seal
  - Vertical Strip Packing

- Group Packaging Form, Fill & Seal Machines
  - Bottom Load Casepacking
  - End Load Casepacking
  - Plastic Ringing

Appendix A

- Top Load Casepacking
- Wraparound Casepacking
- Wraparound Sleeving
- Wraparound Traypacking
- Inspecting Machines
  - Aerosol Testing
  - Cap Inspecting
  - Case Inspectors
  - Checkweigher
  - Color Matching
  - Empty Bottle Inspection
  - Fill Height Inspection
  - Foreign Body Detecting
  - Label Inspecting
  - Leak Detectors
  - Open Flap Detector
  - Seal Checking
  - Torque/Tightness Testing
- Labeling Machines
  - Heat Seal Labeling
  - Hot Melt Glue Labeling
  - Pre-Gummed Label Applicator
  - Pressure Sensitive Labeling
  - Print & Apply Labeling
  - Tamper Evident Seals & Bands
  - Weigh Price Labeling
  - Wet Glue Labeling
- Marking Machines
  - Bar Code
  - Coders, Daters & Imprinters
  - Emboss Coder
  - Hot Foil Coder
  - Ink Jet Coder
  - Laser Coder
  - Solid Ink Coder
  - Wet Ink Coder
- Packaging Support/Specialty Equipment
  - Adhesive Equipment
  - Aerosol Equipment
  - Ampule
  - Aseptic Packaging Systems
  - Assembling

- Bag Compression
- Blister Equipment
- Bottle Uncasers
- Card Packaging
- Cartridge Machinery
- Count/Cotton/Stack
- Foam In Place
- Hooding
- Identifying
- Lab/Pilot Plant
- Label Stripping
- Lid Banders
- Lidding - Non-Capping
- Material Handling
- Measuring
- Netting Machinery
- Protective Packaging
- Rework/Reverse Operation
- Spraying
- Stacking
- Tablet & Capsule Packaging
- Tea Bag
- Tying
- Vacuum Packaging
- Vial Packaging
- Weigh/Count/Print
- Weighing Systems

⊟ Pallet Forming, Dismantling & Securing Machines
  - Bag Palletizers
  - Can Stacking
  - High Level Palletizer
  - Layer Pad Dispenser
  - Layer Pad Removing
  - Lift-Off Depalletizer
  - Lift-Off Palletizer
  - Low Level Palletizer
  - Multi-Position Palletizer
  - Pallet Dispenser
  - Pallet Inverter
  - Palletize/Depalletize
  - Palletize/Unitize
  - Pallet Shrink-Wrapping

Appendix A

- Pallet Stacker
- Pallet Strapping
- Pallet Stretch Wrapper
- Robot Depalletizer
- Robot Palletizer
- Sweep-Off Depalletizer
- Top Sheet Dispenser
- Pasteurizing Machines
  - Batch Pasteurizer
  - Continuous Pasteurizer
- Rebuilt/Used Equipment
- Shrinking Equipment
  - Shrink Frame
  - Shrink Oven
  - Shrink Tunnel
- Sterilizing Machines
  - Batch Sterilizer
  - Container Sterilizing
  - Continuous Sterilizer
  - Irradiation
  - Retort
- Strapping Machines
- Wrapping Machines
- Banding
- Edge Sealing
- Flowrapping
- Foil & Band Wrapping
- Fold Wrapping
- L-Sealing
- Overwrapping
- Pleat Wrapping
- Roll Wrapping
- Skin Packing Machine
- Sleeve Wrapping
- Spiral Wrapping
- Stretch Banding
- Stretch Film Wrapping
- Twist Wrapping
- Wrapping Machines For Loading Units

Appendix B: Projects

Listed in this section of the text are some potential projects for individual students or teams of students to develop. The level of development will be a function of the number of people participating, the length of time available (10 week trimester, 15 week semester, etc.). The topics were compiled from many years of conducting these projects at RPI, and were selected due to the ease of purchasing parts.

If time permits, it has been found to be very useful for the student(s) to build a mockup of one or more facets of the prototype automation. This mockup can be made from materials such as wood rather than metal, and might be human powered rather than actuator driven. The key to choosing what to mock up lies in the needs to investigate a potential automation solution that has a fairly high risk factor, or is just an unknown. Experience shows that all of the CAD modeled designs in the world may not address these questions. After completion of the mockup, the student(s) and faculty would have a better sense that the proposed method would work.

Students are also encouraged to find their own project to develop. Faculty may be of great assistance also. All one has to do is to talk to local companies and offer a "what if" look at some automation, and the possible projects will seemingly fall out of the sky.

Whatever project is selected, it is highly recommended that there be sample products with which to work. Professional automation designers always insist on a few samples, both as loose parts and as a final product, before they will brainstorm an answer for developing a proposal. So student teams should be encouraged to purchase a few parts where it is logical, if they are not provided free by local companies looking for assistance.

PROJECT NUMBER AND NAME: 1, ASSEMBLE DOOR HINGES AND PACKAGE

- *Task* — Assemble standard door hinges from the two halves, and insert the pin. Care needs to be taken to keep pin in place while holding in place under vacuum head that seals warm plastic to printed cardboard backing material.
- *Recommended Production Rate* — 12 hinges per minute.
- *Recommended Items to Purchase* — Several door hinges on vacuum-sealed package. Pin must be able to be removed.
- *Possible Mockup* — Model hinge alignment and pin insertion.

PROJECT NUMBER AND NAME: 2, HANDLE, COUNT, AND PACKAGE FLEXIBLE DRINKING STRAWS

- *Task* — To accumulate, count, and place into a box 100 plastic drinking straws. Straws need to be handled in a relatively clean environment since they are used directly out of the box by consumers. Straws may be straight, or they may have a section capable of bending. System should work for both.
- *Recommended Production Rate* — Box or 100 straws should be filled every two seconds.
- *Recommended Items to Purchase* — Several boxes of straws. May need boxes from other products.
- *Possible Mockup* — Hand-powered singulation and counting station.

PROJECT NUMBER AND NAME: 3, DISPENSE PALLETIZING SLIP SHEET

- *Task* — A slip sheet is a piece of standard paper placed between layers of boxes on a pallet to create better structural rigidity. Paper can be as simple as newsprint. Paper can either come from a continuous roll or a stack of sheets. Student team is to select which. Automation needs to be very quick since this is a time where no boxes get placed on pallet. Rest of motion can be slower while a new layer is placed.
- *Recommended Production Rate* — One sheet placed in one second.
- *Recommended Items to Purchase* — Large drawing pad of artists paper.

Appendix B: Projects

- *Possible Mockup* — Hand-powered gripper to grab paper from roll or stack.

PROJECT NUMBER AND NAME: 4, PALLET DISPENSER

- *Task* — Pallets need to be placed at base of palletizer. Pallets can be wood (rough surface) or plastic (relatively smooth). Empty pallets will be stacked up to 4 ft high. Pallet to be loaded can be grabbed from top of the stack or the bottom.
- *Recommended Production Rate* — Pallet needs to be dispensed into working area in two seconds.
- *Recommended Items to Purchase* — Several wooden pallets. Can usually be obtained for free at larger supermarkets and department stores. Wooden pallets often burned rather than recycled.
- *Possible Mockup* — Hand-powered station to move one pallet from stack. Can create a scaled version.

PROJECT NUMBER AND NAME: 5, HANDLE BOTTLES OF SPICES AND RELABEL

- *Task* — Custom label needs to be placed on generically labeled bottle of spice. New label must cover the original smaller label. New label is self-adhesive type. Bottles of spices are packed 24 to a cardboard case. Bottles need to be removed from case, new labels attached, and repackaged into a similar cardboard case.
- *Recommended Production Rate* — One case per two minutes.
- *Recommended Items to Purchase* — Several bottles of spice. The type found in dollar stores is best. Find preprinted labels or stickers.
- *Possible Mockup* — Create a station to apply labels and attempt to achieve a smooth and aligned application.

PROJECT NUMBER AND NAME: 6, PIZZA STACKING SYSTEM

- *Task* — Stack six frozen pizzas, each in vacuum-type plastic wrap, into a box. Stacking process can be done pizza right side up or upside down, but if upside down then the box needs to be filled from the bottom. Pizzas need to be transported right side up. Box flaps need to be closed and taped.
- *Recommended Production Rate* — One box of six pizzas per ten seconds.

- *Recommended Items to Purchase* — Several frozen pizzas that are vacuum plastic wrapped.
- *Possible Mockup* — Material handing method to move pizzas with their irregular surface shape.

PROJECT NUMBER AND NAME: 7, ELECTRICAL WIRE STRIPPER AND BENDER MACHINE

- *Task* — 110 V and 220 V circuits are pre-wired into a circuit breaker panel. The wire is fairly stiff and the lengths are calculated by electricians. Wire from a spool needs to straightened, bent at 90°, and stripped at both ends.
- *Recommended Production Rate* — One stripped wire with two 90° bends every 5 seconds.
- *Recommended Items to Purchase* — Roll of #12 single conductor wire and several terminal strips to emulate circuit breakers.
- *Possible Mockup* — Hand-operated station to bend wire and strip one end.

PROJECT NUMBER AND NAME: 8, BOOK PICKING

- *Task* — E commerce companies like Amazon.com have thousands of pallets filled with books. Automation needs to be able to find the top book on a pallet and grab it without damaging the loose book jacket. The book needs to keep some identity while it gets transported to order processing location. Book location on the pallet is in interlocking rows.
- *Recommended Production Rate* — Grab one book from pallet and place onto transport system in five seconds.
- *Recommended Items to Purchase* — Several cheap hardcover books with jacket.
- *Possible Mockup* — Book grabbing device.

PROJECT NUMBER AND NAME: 9, BOOK PACKAGING

- *Task* — This Project takes the output of Project 8 and by using a database, places the order in the smallest box possible. Project should consider three different sizes of boxes, and a limit of six books. Boxes would need to be filled with packing materials (peanuts, etc.) to limit damage, and then closed and sealed.

Appendix B: Projects

- *Recommended Production Rate* — Fill a box of six books and close in 30 seconds.
- *Recommended Items to Purchase* — Several cheap hardcover books with jacket.
- *Possible Mockup* — Book transfer station so as not to drop in books and damage them.

PROJECT NUMBER AND NAME: 10, UNLOAD AND FLIP COFFEE CAKES AND STACK PANS

- *Task* — Coffee cakes are baked in a ring or Bundt pan, and cooked in a continuous oven. Cakes then go through a cooling station and must be unloaded by flipping them upside down. They must be placed on a stiff cardboard sheet within 1/2 in. of center. The empty pans are then to be stacked, with a limit of stack height of 3 ft.
- *Recommended Production Rate* — 16 per minute.
- *Recommended Items to Purchase* — Several ring or Bundt pans and several cake mixes.
- *Possible Mockup* — Hand-powered cake-flipping station.

PROJECT NUMBER AND NAME: 11, LOW-COST REMOTE PALLET STRETCH WRAPPER

- *Task* — To wrap full pallets of mixed goods with stretch wrap while the pallet is on a human-steered forklift truck. Current stretch wrappers require full pallet to return to home base, but mixed goods pallet might come loose getting there. So pallet needs to be wrapped as soon as it is filled by last goods on top. Starting the wrapping is a major concern.
- *Recommended Production Rate* — Wrap 5 ft high pallet with two or more wrap layers in 45 seconds.
- *Recommended Items to Purchase* — Several boxes of Saran wrap or generic equivalent.
- *Possible Mockup* — Scale model using Saran wrap and small pallet of mixed goods. Hand powered possibly.

PROJECT NUMBER AND NAME: 12, SCOOP ICE CREAM CONE FROM FREEZER

- *Task* — To make ice cream cones at dairy store without human intervention. Should work with larger freezer with sliding doors to open

the top. Ice cream is in square three gallon cardboard containers, stored in three rows of eight flavors each. Cones are square bottom type. Single scoop sized cones only. Gripper needs to be rinsed after each made cone.
- *Recommended Production Rate* — One cone every 20 seconds.
- *Recommended Items to Purchase* — Several different types of ice cream scoops and a few half gallons of ice cream.
- *Possible Mockup* — Hand-powered but designed for automation scoop.

PROJECT NUMBER AND NAME: 13, PACKAGE HEARING AID BATTERIES INTO VARIOUS PATTERNS

- *Task* — Take a single stream of hearing aid (calculator type) batteries and place them into several different package layouts. Placement is to be into the clear molded plastic part of a package, and can either be in a row or a circular pattern. Batteries cannot be mishandled significantly or they may be damaged or shorted out.
- *Recommended Production Rate* — Average eight batteries per four second cycle.
- *Recommended Items to Purchase* — Several different styles of packaged batteries.
- *Possible Mockup* — Create gripper device to transfer batteries into different patterns.

PROJECT NUMBER AND NAME: 14, TRANSFER PETRI DISHES FROM STORAGE TO AN INSPECTION SYSTEM

- *Task* — A stack of 3 in. diameter Petri dishes with attached covers need to be unstacked and manipulated so as to pass under a vision camera for inspection. The stack can be 25 dishes high, and each Petri dish is 0.75 in. in height. The stack of inspected dishes needs to be in the same order as the starting stack.
- *Recommended Production Rate* — One dish per five seconds through inspection station.
- *Recommended Items to Purchase* — Several Petri dishes from websites.
- *Possible Mockup* — Method to remove dish from stack as it gets shorter, or method to remove bottom dish from stack without stack falling over.

Appendix B: Projects

PROJECT NUMBER AND NAME: 15, REMOVE WRAP FROM BOTH SIDES OF PLASTIC SHEET

- *Task* — Polycarbonate sheet plastic is often used instead of glass in windows. Pieces over 1 in. thick are bullet proof. Sheets come with thin plastic film on each side to protect from scratches, and this film must be removed before use. The film is held on by what appears to be significant static attraction, although this is only a guess. Film cannot be harshly removed by screwdriver or scratches appear.
- *Recommended Production Rate* — Remove film from both sides of 1 ft × 1 ft sheet in five seconds. Finding film edge is most difficult part.
- *Recommended Items to Purchase* — Some samples of Polycarbonate (Lexan) from hardware store.
- *Possible Mockup* — Film starting device for one side.

PROJECT NUMBER AND NAME: 16, LOAD SIX DONUTS FLAT INTO BOX

- *Task* — Six donuts are placed into a box, with a second row of three placed on top of the first row a of three. Donuts are round in shape, and are not wrapped individually in plastic. Boxes are pre-formed and the lid must be closed after filling.
- *Recommended Production Rate* — One box of six donuts every five seconds.
- *Recommended Items to Purchase* — Several boxes of donuts.
- *Possible Mockup* — Hand-powered donut placement station.

PROJECT NUMBER AND NAME: 17, STACK AND BOX FROZEN COOKIE DOUGH

- *Task* — Chocolate chip cookies are frozen in ready-to-bake form. The cookies are fairly large (2 in. diameter) and flat. Baking does not make them significantly larger in diameter as opposed to home baking results. Cookies are stacked in three rows by four columns, six cookies high in each box. Stacks of cookies can tilt significantly due to chip location variation.
- *Recommended Production Rate* — One box full of cookies every five seconds. Assume cookies are produced in rows of four.
- *Recommended Items to Purchase* — Several dozen already baked cookies and several boxes.
- *Possible Mockup* — Cookie stacking station and box loading station.

PROJECT NUMBER AND NAME: 18, BREAD BAGGING

- *Task* — Standard 1 lb loaves of Italian bread have significant shape variation. Loaves are to be bagged in individual paper open end bags, and the ends twisted closed. Loaves are coming from a cooling conveyor in random position and orientation.
- *Recommended Production Rate* — One loaf bagged every two seconds.
- *Recommended Items to Purchase* — Eight to ten loaves of Italian bread in their paper bags.
- *Possible Mockup* — Model bread loaves and determining their position and orientation.

Index

Acceptance tests, 383, 384
Accidents, 160
Accumulators, 178–179, 209
Actuators, 194–241
   amount of work to be done, 199
   amplifiers, 219–221
   application issues, 199–204
   case studies, 221–240
   conveyors, 203–204
   dispensing frictional materials, 201, 202
   drivers, 219–221
   electric motors, 211–219
   frictional losses, 199–201
   gearing, 201–203
   hydraulics, 209–211
   industrial robots, 107–110
   inertia, 201–203, 216–217
   intermittent motion, 204
   pneumatics, 204–209
   problems, 240
   project assignments, 240–241
   servomotor sizing case study, 225–240
   sizing, 231–234
   stepper motor sizing case study, 221–225
   tuning, 219–221
   types, 194–198
Adept cycle, 348
Adept flex feeder, 338, 339, 340
Adept MV controllers, 291
Adeptnet, 297
Adept Technologies, 41
Adhesive dispensing, 112
Advanced pneumatic devices, 207
Aggressive companies, 39–40
Agile automation control systems case study, 289–300
Agile designs, 69–72
Air bladders, 147–148
Air conveyors, 174
Air cushions, 147, 148, 264–273
Air line diameter, 206
Air supply issues, 266
Air *see* Pneumatics
Alignment of parts, 100–101
Aluminium extrusions, 129
American Dixie Group, 15, 323
American National Standards Institute (ANSI) safety standards, 157

American Society of Mechanical Engineers (ASME) safety standards, 157
Amplifiers, 219–221, 234
Analogue signals, 219–220, 244
ANSI *see* American National Standards Institute
Applesauce specification example, 377
Arc welding, 110–111
Area determination, 345
Arm selection, 101–103
ASME *see* American Society of Mechanical Engineers (ASME)
Assembly
 industrial robots, 112–113
 machines, 52
 modules, 289, 290
Assembly Flex system, 289–300
Automated screwdriver workstation design case study, 152–155
Automation definition, 8, 9
Automation design process, 10–18

Backlash, 235
Bags, 399–400
Baking cup de-nester device, 15–16, 17
Ball bearings, 134
Banks of switches, 280–281
Base geometries, 104
Batteries, 62, 63
Bearing devices, 130–136
Bearing washers, 131, 132
Bellows design air bladders, 147
Belts, 136–138
Bending process, 195–197
Bid Award Package, bulk mail container unloader, 382–385
Bidding process, 381–385
Bid price, 382
Black box approach, 8
Bladders, air, 147–148
Blister packs, 397, 399
BMC *see* Bulk mail container
Boiler plating, 382
Bomb bay doors, 188
Bookshelves, 34
Bots, 221–225, 349–369

Bottles
 caps detection system, 250
 liquid filling machines, 386, 388–389, 390
 rotary continuous configurations, 65–66
Bowl feeders
 modified vibratory, 342–347
 vibratory, 154, 168–170, 337–338, 339
Bow tying, 13–14, 68
Boxes, 251, 389, 391–392
Bradley, Allen, 288–289
Brainstorming, 50–51
Brake bands, 139
Brakes, 139, 140
Bringmann, Bernhard, 264
Bulging cylindrical air bladders, 147
Bulk mail container (BMC) unloaders
 Bid Award Package, 382–385
 specifications, 378–381
 TBBL case study, 89–93, 94
Bull nose conveyors, 181–183
Burdened hourly rate, 30–31
Business plans, 371

Cable chains, 107–108
Cable drives, 229
Calculator batteries, 62, 63
Cambridge Valley Machining Inc., 335–336, 369
Cameras, 254
Cams, 139
 dual cam track mechanisms, 189–192
 timer-motor driven, 278, 280
 trackplate (Stackbot), 324–328, 330
Cancellation clauses, 377
Capacitance sensors, 244, 252–253
Capacitors, 18–21
Cap chutes, 389, 390
Capping machines, 389, 390, 402–403
Carbon fibers, 196
Cardboard box packaging machines, 389, 391–392
Cardboard trays, USPS mail tray palletizing case study, 317–318
Carlisle, Brian, 41
Cartesian robot bases, 104

Index

Cartons and cartoning machines, 113, 114, 389, 391–392, 401
Case opening TBBL case study, 81–85
Cases and case erecting machines, 393–394, 395, 401–402
Case studies
  actuators, 221–240
  agile automation control systems 289–300
  automated screw driver workstation design, 152–155
  control, 289–304
  donut loader machine, 187–192
  dropping cookies, 180–183
  feeders and conveyors, 180–192
  feeding TBBL cases, 184–187
  industrial robots, 117–122
  machine loading/unloading, 117–119
  OMAC automation control, 300–304
  palletizing, 317–336
  pallet leveling sensor system, 263–273, 274
  pants pressing robot, 120–122
  pouch singulation, 336–370
  precision automation, 306–317
  sensors, 257–273, 274
  servomotor sizing, 225–240
  stepper motor sizing, 221–225
  TBBL case opening, 81–85
  TBBL crossed four-bar BMC unloader, 89–93, 94
  TBBL label insertion and printing, 85–89
  TBBL workstation design, 141–152
  user input motion device, 257–263
Cassette cases, 76, 77
CAT *see* New York State Center for Automation Technologies
Centripetal feeders, 170
CEOs *see* Chief executive officers
Chain drives, 137, 138, 229
  Trackbot, 349, 354–355
Chain-style belts, 59, 60
Check-weighers, 53
Chief executive officers (CEOs), 1
Chips, 280

CLAMP *see* Closed loop assembly micro positioner
Clamshell pants press, 120
Classification
  machines by function, 51–55
  machines by transfer method, 55–72
Closed loop assembly micro positioner (CLAMP), 308–311, 312
Closed loop control, 214
Cloth marking case study, 225–240
Clutch brake systems, 139, 140
CNC *see* Computer numerical control
Color detection, 252
Color of machines, 11
Combined loading of rotating shafts, 130, 131
Commercially available workstations, 126–127
Commission for sales representatives, 36
Common drive shafts, 58
Company chief executive officers (CEOs), 1
Company workers automation impact, 2
Comparator motor controllers, 212–113
Competition, 375
Components systematic design, 10
Compressed air *see* Pneumatics
Compression springs, 309
Computer numerical control (CNC) machines, 51
Computer programs, 75
Computer vision systems *see* Vision systems
Conceptual design, 5–6
Conductive rails, 350–351, 353
Confidentiality, 383
Configuration editor, 293–295
Configuration trade-offs, 72–74
Conical holes, 308–309
Construction speed, 71
Consumers, automation impact, 2
Contact sensors, 244–245
Container handling machines, 403–404
Contingency funds approach, 37
Continuity (normally closed) notation, 282
Continuous belt conveyors, 56–57, 172

Continuous products without position
  registration, 59, 60
Continuous sensors, 243–244
Contracts, 377, 382
Control and controllers, 276–305
  agile automation control systems,
    289–300
  Assembly flex project case study,
    289–300
  case studies, 289–304
  chips, 280
  control panels, 220–221
  derivative, 236–237
  flexible pouch feeder, 346–347
  integral, 236–237
  ladder logic, 282–287
  OMAC automation control, 300–304
  pallet leveling sensor system case study,
    266–267
  problems, 304
  programmable logic controllers,
    280–287
  project assignments, 305
  proportional, 236 237
  proportional integral derivative,
    236–237
  robots, 103
  sensor output, 244
  timing diagrams, 278–280
  Trackbot, 363–366
Control panels, 220–221
Conveyors, 163–193, 404–405
  actuators, 203–204
  air, 174
  bull nose, 181–183
  case studies, 180–192
  continuous belt, 56–57, 172
  flexible parts feeder, 338, 339
  flighted (flited), 56–57, 172–173, 174
  linear asynchronous, 61–62
  linear continuous, 58–61
  linear indexing configurations, 56–57
  palletizing mail letter trays case study,
    322–323
  powered roller, 175–176
  problems, 192–193
  project assignments, 193
  segmented, 172–174
  serpentine, 65
Cookies, 59–61, 180–183
Cooper, Clay, 13, 15–16
Copyrights, 44
Corrupted feedback information, 238–239
Costs
  costing components, 36–37
  maximum profit cost estimation, 39–40
  overruns, 32–36
  patents, 43
  robots, 103
  Stackbot, 371
  structure for an automation builder,
    36–38
  system specification pressures, 375, 376
  Trackbot, 371
  traditional costing estimating, 32–38
  traditional project cost justification for a
    purchase, 29–32
  win–win purchasing philosophy, 38–39
  workers, 29–32
Couplings, shafts, 139–141
Creativity, 25
Crossed four-bar BMC unloader case
    study, 89–93, 94
Curved lines, 238–240
Cushions
  air bladders, 147, 148
  self-leveling pallets sensor case study,
    264–273
Custom design, 68–69
Customers' system specification
    expectations, 375–376
Cycle times
  palletizing mail letter trays case study,
    321, 322
  robots, 102–103
Cylinder diameter versus piston speed, 207
Cylindrical bearings, 133, 134
Cylindrical robot bases, 104

DC motors, 211
Deadbands, 236
Deburring, 114

Index

Decision criteria ranking, 77, 78
Dedicated custom design, 68–69
Defense Logistics Agency (DLA), 120
Definitions
   automation, 8, 9
   robot, 98
   workstations, 127–128
Delivery, safety responsibilities after delivery, 159
De-nester devices, 15–16, 17
Derivative controllers, 236–237
Design process, 10–18, 48–95
   brainstorming, 50–51
   machine classification by function, 51–55
   machine classification by transfer method, 55–72
   machine configuration trade-offs, 72–74
   mechanisms toolbox, 74–75
   system specifications, 49–50
   TBBL automation project, 75–93, 94
Design specifications for bulk mail container unloader, 379–381
Development
   groups involved, 37–38
   Stackbot time issues, 370
   timing, 37–38
DeviceNet communications system, 291, 296, 297
Dial machines, 62–65, 100–101, 204
Diffuse reflection, 248
Digital signals, 219–220, 244
Dimensional measurements, 52
Direct-drive coupling, 231
Directors, automation impact, 1
Discrete sensors, 243–244
Dispensing
   adhesives, 112
   frictional materials, 201, 202
Distance measurement, 252
Distance verification, 252
Distributed control, 363–366
Distributed palletizing, 329–334, 335
Dither effect, 307, 308
DLA *see* Defense Logistics Agency
Dockbot, 313–315, 316–317
Docking process
   CLAMP device, 308–311, 312
   Dockbot, 313–314, 315
Donut loader machine, 187–192
Door locking devices, 158
Double robotic grippers, 115–117
Down time reduction, 71
Drifting, 252–253
Drive belt and pulley system, 136, 137
Drive shafts, 58
Drive systems
   actuators, 219–221
   chips, 220
   workstations, 136–141
   X/Y axis, 229–230
Dropping cookies, 180–183
Drug discovery, 114
Dual arm tray handling robot, 298, 299
Dual axis inclinometers, 269–270, 271–272
Dual cam track mechanisms, 189–192
Dual clutching drive system, 356, 357
Duty cycles, 198, 216

ECO *see* Engineering Change Order
Electrical slip rings, 64, 65
Electric eye beams, 142
Electric motors, 211–219
   advances, 219
   DC motors, 211
   motion profiles, 218–219
   selection, 216–218
   servomotors, 212–213, 219, 225–240
   sizing, 215, 221–240
   stepper motors, 213–215, 216, 219, 221–225
   strengths and weaknesses, 219
   torque–speed curves, 216
   weight and inertia, 216–217
Electric relays, 281
Electric screwdrivers, 152–153
Employees' intellectual property rights, 41–42
Encoder counts, 239
Encoder discs, 250
End effectors, 99–100, 308–311, 312

Engineering Change Orders (ECO), 38, 377
Environment issues
 automation impact, 2
 sensors, 253
Error handling, 99
Escapement feeders, 166–167
ESTOP, 302, 303
E-stop button, 283, 284
Ethernet, 296, 297
Event viewers, 293
Existing knowledge/technology, 12–13, 25
Expectations, system specifications, 375–377
Experimentation, 14–16
Extrusion process, 23–24

Fast food workers, 30–31
FBD *see* Function block diagrams
FEA *see* Finite element analysis
Feature creep, 375–376
Feedback information, corrupted, 238–239
Feeders, 163–193
 Adept flex feeder, 338, 339, 340
 case studies, 180–192
 centripetal, 170
 escapement, 166–167
 flexible parts, 171–172, 337–338, 339
 flexible pouch, 342–347
 modified vibratory bowl, 342–347
 pick and place, 179–180
 problems, 192–193
 programmable reconfigurable parts, 338, 339, 341
 project assignments, 193
 robot and bowl, 338, 339, 340
 slide escapement, 167
 vibratory bowl, 154, 168–170, 337–338, 339
Feeding pouches, 338, 340–341
FFS *see* Form, fill and seal
Fiber optics alignment, 316
Fiducials, 168, 170
Filling machines, 405–406
Fill and seal machines, 405
Finger grippers, 99, 100

Finite element analysis (FEA), 130
Fin seals, 397, 398
Flanged bearings, 131, 132
Flaws in cloth, 226
Flexibility, modular automation goals, 70
Flexible automation
 Hansford Assembly Flex Project, 289–300
 justification, 40–41
Flexible designs, 69–72
Flexible parts feeders, 171–172, 337–338, 339
Flexible pouch feeder, 342–347
Flexible segmented belts, 173–174
Flighted (flited) conveyors, 56–57, 172–173, 174
Floor covering costs, 33
Floor space *see* Space issues
Focal point, optical sensors, 247, 248–249
Food processing, 113
Force, 199, 252
Fork lever, limit switches, 246
Form, fill and seal (FFS) machine, 392, 406–407
Four-bar linkage, 150–151
Four-bar mechanisms, 74–75
Four hand gripper, 121, 122
Four-pole stators, 212
Frames, 129
Friction, 130, 199–201
Frictional materials dispensing, 201, 202
Functional specifications, 384
Function block diagrams (FBD), 288

Gains, 220, 237
Gantry robot system, 66, 67
Gantry style lifting devices, 258
Gantry units, palletizing mail letter trays case study, 320
Gears
 gear ratio, 203
 gear reduction, 233–234
 inertia, 201–203
 slippage, 239
Generalized automatic machine, 10, 11
Generic robot types, 103–106

Index

Generic six-step process, 10, 12
Geneva mechanism, 204
Geometries, robots, 104
Goals of modular automation, 69
GRASP Inc., 311–313
Grippers, 99–100
  double robotic, 115–117
  finger, 99, 100
  four hand, 121, 122
  two finger, 99, 100
Guarding, 156–158
Guide rails, 58–59
Guides (plow), 186

Handling of parts, industrial robots, 98–101
Hansford Assembly Flex Project, 289–300
'Hard' automation, 40
Hard stops, 141, 142
Harmonic drive, 108, 109
Heat shrink bottle capping, 389, 390
Heat shrink labels, 393
Height determination, 251
Helical cut shaft couplings, 140, 141
High-speed donut loader, 190–192
HMI *see* Human machine interface
Hoses, 206
Host computers, Trackbot, 363–364, 367
Human machine interfaces (HMI), 292
Human performance observation, 13–14
Human presence detection, 252
Hydraulics
  actuators, 209–211
  overload detection, 252
  power supplies, 198
  strengths and weaknesses, 219

IDLE, 302, 303
Idler pulleys, 137, 138
Image processors, 254
Impacts of automation, 1–2
Inclinometers, 269–270, 271–272
Indexing
  linear, 55–58
  rotary, 62–65, 73
Industrial robots *see* Robots

Inertia, 201–203, 216–217
Inflatable air cushions, 264–273
Infrared (IR) detection, 252, 345
Insertion of labels, 85–89
Inspection machines, 52–53, 407
Intangibles, workers, 30, 31
Integral controllers, 236–237
Intellectual property, 41–45
Intelligence, robots, 103
Interface standardization, 70
Intermittent duty, 216
Intermittent motion, 204
Inverse kinematics, 105–106
IR *see* Infrared

Joining shafts, 139–141
Jointed spherical robots, 104, 110
Joint variables, kinematics, 105–106
Justification for automation, 28–47

Kinematic mount, 308–309
Kinematics, robots, 105–106
Kinetic friction, 200
Kitting, 356
Knowledge of process, 18–21

Labeling, 85–89, 393, 407
Laboratory robots, 114
Ladder Diagrams (LD), 283–287
Ladder Logic, 282–287
Lamp notations, 281
Lap seals, 397, 398
Laser interferometer, 310, 312
Laser lines, 256–257
Laser range finding, 252
Latch opening case study, 81–85
Lawsuits, 156, 160
LD *see* Ladder Diagrams
Lead screw drives, 229–230
Leash leading method, 260–263
Licensing, 371, 383
Life cycle *see* Product life cycle
Life expectancy, automation machine, 197–198
Lift device, sensor case study, 257–263
Light beams, 141, 142

Lighting
  fixture costs, 34
  importance, 255
Limbo bars, 186
Limit switches, 142, 245–246, 258–259, 282
LINCAGES, 75
Linear asynchronous conveyors, 61–62
Linear bearings, 134, 137
Linear continuous conveyors, 58–61
Linear indexing, 55–58
Linear potentiometers, 259
Linear robotic modules, 108–110
Linear slides, 144
Linear track systems, 321–323
Linear viscous friction, 200–201
Line scan vision systems, 255
Line of sight sensors, 249–251
Linkage design tools, 91–92
Liquid filling machines, 386, 388–389, 390
Load bearing rods, 143–144, 145
Load cells, 65
Loading
  cases, 394, 395
  industrial robots, 115–119
  rotating shafts, 130, 131
Load relays, 283, 284
Locking process, Dockbot, 313–314, 315
Lockout/tagout arrangements, 158
Logic controllers, 280–288

Machine modes, OMAC, 302–303
Machines
  classification by function, 51–55
  classification by transfer method, 55–72
  configuration trade-offs, 72–74
  design and workstations safety, 155–160
  industrial robots loading/unloading, 115–117
  Pack Expo categories, 387–388
Machining centers case study, 117–119
Maglev see Magnetic levitation
Magnetic actuation, 148–150
Magnetic levitation (maglev), 149–150
Magnetism, rodless air cylinder, 207

Mail carts, USPS mail tray palletizing case study, 317–336
Mailing labels, TBBL case study, 85–89
Manufacturing directors, automation impact, 1
Markets
  CLAMP device, 311–313
  Dockbot, 316–317
  fickle, 371
  pouch singulation case study, 369–370
  Stackbot, 334–336
  which to approach, 32
Marking machines, 225–240, 407
Material cost overruns, 32–36
Material excess detector, 251
Material handling, 112–113
Matlab, 238
Maximum profit cost estimation, 39–40
MDI see Multiple document interface
Mechanical actuators, 107–110
Mechanical Power Transmission Apparatus (MPTA) safety standards, 157
Mechanical slippage, 239
Mechanisms toolbox, 74–75
Mechatronics system design, 16–18
Members, 129–130
Metal
  bending plates, 195–197
  detection, 53, 252
  position sensors, 252
Micro positioners, 308–311, 315–316
Microprocessors, 220
Mini robots, 307
Mirrors, 256–257
Model trains, 284–287
Modular automation goals, 69
Modular designs, 69–72
Modular Stackbot, 328–334
Module palette, 293, 295
Module size, 70
Momentary contact, 245
Moment of inertia, 233–234
Monopolies, 40
Morrison Berkshire Inc., 324–325

Index

Motion
  analysis, 360–363
  Bots, 360–363
  controllers, 19
  detectors, 252
  intermittent, 204
  profiles, 218–219, 225
  standard robot, 348, 361
  Trackbot, 356–363
Motion control system sizing, 231–234
Motors *see* Electric motors
Moveable stops, 58, 59, 60, 175
Moving clamp bars, 231
Moving members, 129–130
MPTA *see* Mechanical Power Transmission Apparatus
Multiple document interface (MDI), 292
Multiple implementation of sensors, 253
Multiple Stackbot, 328–334

NC *see* Normally closed
Needle bearings, 133, 134
Net weight filling, 388, 390, 399, 400
Newspapers, 98
New York State Center for Automation Technologies (CAT), 76
New York Talking Book and Braille Library (NYSL/TBBL), 76, 77
The nip, 156
NO *see* Normally open
No continuity (normally open) notation, 282
Noise pollution, 198
Noncontact sensors, 245
Normally closed (NC)
  continuity notation, 282
  limit switch notation, 282
  push button notation, 282
  relay contact notation, 281
  toggle switch notation, 282
Normally open (NO)
  limit switch notation, 282
  no continuity notation, 282
  push button notation, 282
  relay contact notation, 281
  toggle switch notation, 282

Novel approaches, 14–16
Number detection, 345
NYSL/TBBL *see* New York Talking Book and Braille Library

Off-the shelf parts, 71
Open loop control, 214–215
Open Modular Architecture Controls (OMAC)
  automation control case study, 300–304
  Packaging Working Groups, 300, 301–304
Operational task lists, 78, 79–81
Optical sensors, 252
Optical switches, 246–251
Optimal performance, 220
Orientation detection, 100–101, 346
Oscillating press, 156
Outsourcing, 4
Oval tracks, 349, 362
Overhead gantry units, 320

Package counting, 249
Package tracking, 252
Packaging machines, 55, 386–409
  industrial robots, 113, 114
  OMAC automation control case study, 300–304
Packaging and Packaging Related Converting Machinery, ANSI standards, 157
Packaging support machines, 407–408
Pack Expo, 386
Pack Expo packaging machine categories, 387–388, 401–409
Packing patterns, 395, 396
Paddle wheel transfer, 189
Painting
  costs, 34
  spray, 111
Pallets
  distributed palletizing, 329–334, 335
  forming, dismantling and securing machines, 407–408
  leveling sensor system case study, 263–273, 274

[Pallets]
  linear asynchronous systems, 61–62
  mail letter trays case study, 317–336
    commercial robotic prototypes, 319–320
    linear track system, 321–323
    multiple Stackbot solution, 328–334
    Stackbot design, 323–328, 329, 330
    Stackbot production and market acceptance, 334–336
    USPS mail tray project, 317–319
  packing patterns, 394, 395, 396
  palletizing, 394–397
  trade offs, 73
  types, 395, 396
  wrapping, 395
Panic stops, 293
Pantograph linkage, 325
Pants pressing robot case study, 120–122
Paper bags, 399
Paper tape, 279–280
Parts
  flexible parts feeders, 171–172, 337–338, 339
  off-the shelf, 71
  orientation, 100–101
Patents, 42–43, 371
Payloads, 101–102
Payments
  groups involved, 37–38
  timing, 37–38
Peanut butter chocolate kiss cookies process example, 21–25
Pedestal robots, 319–320
Pendulum effect, 259
Performance check machines, 54–55
Permanent magnet stepper motors, 213–214
Pianos, 279–280
Pick and place feeders, 179–180
Pickup efficiency, 347
PID *see* Proportional integral derivative
Piezoelectric actuators, 316
Pillow blocks, 133, 136
Pinch points, 156–158
Piston speed versus cylinder diameter, 207

Pixels, 254, 255
P-labels, 78
PLC *see* Programmable logic controllers
Plows, 186
Plunger rods, 245
Pneumatics
  actuators, 204–209
  advanced devices, 207
  air cylinders, 143–145, 205–207
  capacity issues, 198
  line checking devices, 252
  rotary joints, 64
  screw feeding, 154
  self-leveling pallets sensor case study, 271–272, 273
  strengths and weaknesses, 219
  vacuum generators, 207–209
Polishing, 114–115
POPM *see* Proof of Principle Models
Postal workers, 31–32
Pouches
  formation, 397, 398
  singulation case study, 336–370
    design considerations, 353–356
    distributed control, 363–366
    feeding pouches, 338, 340–341
    initial test model, 350–353
    lessons learned, 366–368
    modified bowl, 342–347
    motion analysis, 360–363
    second control architecture, 368–369
    standard parts feeding, 337–338, 339
    Trackbot invention, 347–350
    Trackbot motion strategies, 356–360
    Trackbot production and market acceptance, 369–370
    vacuum pickup, 341–342
Power
  actuator work to be done, 199
  source availability, 198
Powered lift device, sensor case study, 257–263
Powered roller conveyors, 175–176
Precision automation case study, 306–317
  CLAMP device details, 308–311

CLAMP production and market acceptance, 311–313
Dockbot, 313–315
market need, 306–307
micro positioners, 315–316
Pressure sensors, 252
Pretzels, 59
Printing labels, 85–89
Problems
  actuators, 240
  control, 304
  design process, 94–95
  feeders and conveyors, 192–193
  industrial robots, 123–124
  justifying automation, 45–46
  sensors, 274–275
  steps to automation, 25–26
  workstations, 160–161
Processing unit, Trackbot, 363–366
Process knowledge, 18–21
Process modules, Assembly Flex system, 289, 291
Product life cycle, 69, 71
Profit, 39–40
Programmable logic controllers (PLC), 280–287
Programmable reconfigurable parts feeder, 338, 339, 341
Programming
  function block diagrams, 288
  programmable logic controllers, 280–287
  sequential function charts, 288
  structured text, 288
Project assignments
  actuators, 240–241
  control, 305
  design process, 95
  feeders and conveyors, 193
  industrial robots, 124
  justifying automation, 46–47
  sensors, 275
  steps to automation, 26–27
  workstations, 161–162
Projects for students, 410–417
Proof of Principle Models (POPM), 68–69

Proportional controllers, 236–237
Proportional integral derivative (PID) controllers, 236–237
Proximity sensors, 252
Puck systems, 61–62, 63
Pulleys, 136, 137
Push button notation, 282
Pushers, 174–175
Push roller, limit switches, 246

Rack and pinion drive system, 230
Radial loading of rotating shafts, 130, 131
Radio frequency (RF) tags, 252
Rail and chain system, Trackbot, 350–353
Ramping curves, 225
Reasons to automate, 3–4
Reflection, diffuse, 248
Relays, 281
Remote computer, Trackbot, 364–366, 367, 368
Repeatability, 310, 312
Repetitive motion injury, 32
Representatives (reps), 2, 36, 376
Reprogramability, 321–322
Request for Proposal (RFP), 49, 381–385
Request for Quote (RFQ), 49, 381–385
Reservoirs, hydraulic, 209
Resonance modes, 129
Responsibilities of safety after delivery, 159
Return motion, 348
RF *see* Radio frequency
RFP *see* Request for Proposal
RFQ *see* Request for Quote
Rigid shaft couplings, 139–140
Risk assessments, 158–159
Robot and bowl feeder, 338, 339, 340
Robot code, 99
Robots
  1970s and 1980s, 9–10
  applications, 110–117
  arm selection, 101–103
  Assembly Flex system tray handling robot, 296–299
  case studies, 117–122
  code, 99

[Robots]
  generic types, 103–106
  handling of parts, 98–101
  industrial, 97–124
  linear robotic modules, 108–110
  machine loading/unloading case study, 117–119
  mechanical actuators, 107–110
  mini, 307
  motion of standard robots, 348, 361
  pants pressing case study, 120–122
  precision (repeatability and accuracy), 307
  problems, 123–124
  project assignments, 124
  robot and bowl feeder, 338, 339, 340
  robot centered configurations, 66–68, 73
  RS/1, 210–211
  Selective Compliant Articulated Robot Arms, 112–113, 312
  types, 103–106
  workspace analysis, 106–107
Rodless air cylinder, 207
Roller bearings, 131, 133
Rollers, 56, 57
  limit switches, 245–246
  the nip, 156
  powered roller conveyors, 175–176
Rolling process, 22–23
Room refinishing example, 32–35
Rotary continuous configurations, 65–66
Rotary indexing, 62–65, 73
Rotary joints, 64–65
Rotary motion issues, 64–65
Rotary output power, 199
Rotary position devices, 260–261
Rotary tables, 204
Rotary to linear drive application, 203
Rotary viscous friction, 200–201
Rotating shafts, 130, 131
Rotating stator fields, 212
RS/1 robot, 210–211
RSLogix 500, 288

Safety, 72, 155–160

Safety covers for swimming pools case study, 225–240
Safety standards, 157, 160
Sales forces, internal, 36
Sales representatives, 2, 36, 376
Scan Time, 282–283
SCARA *see* Selective Compliant Articulated Robot Arm
Scissor jack lifting, 146–147
Screwdriver workstation design case study, 152–155
Screw drives, 144, 145
Screw feeding, 154, 155
Segmented conveyors, 56, 172–174
Selective Compliant Articulated Robot Arm (SCARA), 66, 67, 112–113, 290, 312
Self-leveling pallets sensor case study, 263–273, 274
Self-sticking labels, 393
Self-tuning amplifiers, 220
Sensors, 242–275
  case studies, 257–273, 274
  failure, 242–243
  limit switches, 245–246
  optical switches, 246–251
  pallet leveling sensor system, 263–273, 274
  problems, 274–275
  project assignments, 275
  robots, 99, 100
  types, 243–245
  user input motion device, 257–263
  vision systems, 253–257
Sequential control, 280
Sequential function charts (SFC), 288
Serpentine accumulators, 178
Serpentine conveyor belts, 65
Servo control, air cylinders, 207
Servomotors, 212–213, 219, 225–240
SFC *see* Sequential function charts
Shafts, joining, 139–141
Shielding, 156–158
Shipping labels, 85–89
Shrink fit labels, 393
Shrink wrapping, 395

Index

Side pushers, 174–175
Side rotary, limit switches, 246
Signals, 219–220
Silicon dispensing process, 24
Silicon wafer detection, 250
Simon, Matthew, 75
Single arm tray handling robot, 297
Single rung starter motor circuit, 283–284
Singular processes, 70
Singulation case study, 336–370
Six-jointed robots, 66–67
Six-step process, 10, 12
Size of modules, 70
Sleeve bearings, 131, 132, 133, 135–136
Slide escapement feeders, 167
Slider crank mechanism, 150–151
Sliding, 143–146
Slipping, 215, 239
Slip rings, 64, 65
SoftLogix 5 controller, 288–289
Software tuning, 237
Solenoids, 148–149
Sortation machines, 32
Space issues
  floor, 70–71
  palletizing mail letter trays case study, 319–320
  workspace analysis, 106–107
Specifications *see* System specifications
Speed
  construction, 71
  robots, 102–103
  torque speed curves, 215
Speed reduction coupling, 231–232
Spherical bearings, 133, 135
Spherical robot bases, 104
SphinxPC *see* SYNTHETICA
Spiral accumulators, 178–179
Split sleeve bearings, 133–134, 136
Spot welding, 110–111
Spray head test machines, 54–55
Spray painting, 111
Springback, metal plates, 196–197
Spring constant, 238
Spring-loaded rollers, 314, 315
ST *see* Structured text

Stackbot
  design, 323–328, 329, 330
  development time issues, 370
  module concept, 328–334
  production and market acceptance 334–336
  system specification expectations, 376
Stainless steel frames, 129
Stall torque, 217
Standard robots
  motion analysis, 361
  motion cycle, 348
Standard vision systems, 254–255
Stand up pouches (SUPs), 397
Start buttons, 283, 284
Star wheels, 177
Static friction, 200
Stators, 211, 212
Steel belts, 326
Steel frames, 129
Stepper motors, 213–215, 216, 219, 221–225
Step process, 10, 12
Steps to automation, 7–27
Step-up transformers, 224
Stereo vision systems, 256–257
Stockholders, 2
Stops, 175
Strain gages, 252
Stretch wrap, 395
Structural members, 128–130
Structured text (ST), 288
Suction cups, 15–16, 87, 99, 100, 341–342, 344–345
SUPs *see* Stand up pouches
Swimming pool safety covers case study, 225–240
Switches, 280–282
SYNTHETICA, 75, 91–92
Systematic design of components, 10
System executive, Assembly Flex control suite, 292–293
System information box, Assembly Flex control suite, 293
System specifications, 374–385
  bulk mail container unloader

[System specifications]
    Bid Award Package, 382–385
    specifications, 378–381
    cancellation clauses, 377
    competition, 375
    contract agreements, 377
    cost pressures, 375, 376
    design process, 49–50
    Engineering Change Orders, 377
    examples, 378–381, 382–385
    expectations, 375–377
    feature creep, 375–376
    feature list, 376–377
    lessons to be learned, 385
    Request for Quote, 381–385
System tuning, 235–238

Table speed, 218
Tagout arrangements, 158
Talking Book and Braille Library (TBBL)
    automation project, 75–93, 94
    case feeding case study, 184–187
    workstation design case study, 141–152
Tapered cylinders, 221–222, 224
Tapered roller bearings, 133, 135
TBBL *see* Talking Book and Braille Library
Teams, 3–4
Technical simplicity, 71
Telescoping cylinder, 221–224
Temperature sensors, 252
Term misuses, 98
Test machines, 54–55
Thermal detection, 252
Threaded shafts, 221–224
Three-track encoder disc, 250
Through beam sensors, 249–251
Thrust bearings, 131, 132
Thrust loading of rotating shafts, 130, 131
Tilt sensors, 266, 269–270
Time issues
    development and payments, 37–38
    limitations, 11
Timer motors, 278, 280
Timing belts, 137–138, 326, 327
Timing diagrams, 278–280
Timing screws, 176

Toggle switch notation, 282
Top rotary, limit switches, 247
Torque
    calculation, 233
    stall torque, 217
    torque–speed curves, 215, 216
Total systems approach, 17
Trackbot, 221–225
    costs, 371
    design, 348–350, 353–356
    distributed control, 363–366
    first control architecture, 363–366
    initial test model, 350–353
    invention, 347–350
    motion analysis, 360–363
    motion strategies, 356–360
    production and market acceptance, 369–370
    second control architecture, 368–369
Tracks, dual cam track mechanisms, 189–192
Trademarks, 43
Trade-offs, design process, 72–74
Trade secrets, 44
Traditional costing estimating, 32–38
Traditional project costs, 29–32
Traffic lights, 244
Transfer devices, 176
Transformers, 224
Translation, 143–146
Travel costs, 37
Tray handling robot, 296–299
Trays
    erecting/handling machines, 401–402
    USPS mail tray palletizing case study, 317–336
Tripods, 308
Tuning, 219–221, 235–238
Two finger grippers, 99, 100
Two-rung software ladder, 284

Ultrasonic detectors, 252
United States Patent and Trademark Office (USPTO), 42–43
United States Postal Service (USPS), 31–32, 317–319
Universal joints, 140

Index

Unlatching mechanism case study, 81–85
Unloading
   bulk mail containers TBBL case study, 89–93, 94
   industrial robots, 115–119
Unspooling schemes, 230–231
User input motion device case study, 257–263
User selectable modes, 302
USPS *see* United States Postal Service
USPTO *see* United States Patent and Trademark Office

Vacuum
   cups, 15–16, 87, 99, 100, 341–342, 344–345
   generators, 207–209
   pouch pickup, 341–342, 353
   sensors, 252, 253
Valves, 206, 209
V belts, 137
Velocity, 199
Velocity–time profile, 357, 358
Venturi vacuum generators, 208
Vibrations, 129, 307
Vibratory bowl feeders, 154, 168–170, 337–338, 339
Video acquisition and processing unit, 366
Viscous friction, 200–201
Vision systems, 253–257
   flexible feeders, 171
   guided product processing, 59–61
   inspection machines, 52
   line scan, 255
   packet areas determination, 345
   standard, 254–255
   stereo, 256–257
   Trackbot, 366
Voice coil technology, 316
Volumetric filling, 388, 389

Walking beams, 57–58
Wall anchors, 35
Warning labels, 158, 159
Warrantee costs, 37
Washers, 131, 132

Washing machines, 278–279
Waterfall sorting, 185
Web handling excess material detection, 251
Weight issues
   electric motors, 216–217
   palletizing mail letter trays case study, 323
Weight sensors, 252
Welding, 110–111
Wheels, 354, 355–356
Windshields, 112
Win–win purchasing philosophy, 38–39
Wobble sticks, 245, 247
Wooden pallets, 395
Workcell design box, 294, 295
Workcells, 289
Workers
   automation impact, 2
   cost, 29–32
Working volume, 66, 67
Work motion, 348
Workspace *see* Space issues
Workstations, 125–162
   automated screwdriver workstation design, 152–155
   bearing devices, 130–136
   building blocks, 128
   case studies, 141–155
   commercially available, 126–127
   definition, 127–128
   drive mechanisms, 136–141
   frames, 129
   machine design and safety, 155–160
   moving members, 129–130
   problems, 160–161
   project assignments, 161–162
   structural members, 128–130
   TBBL workstation design, 141–152
Wound capacitors, 18–21
Wrapping machines, 409
Wrapping pallets of cases, 395
Wrist joints, 99

X/Y axis drive system, 229–230